Springer Series in Materials Science

Volume 237

The Springer Series in Materials Science covers the complete spectrum of materials physics, including fundamental principles, physical properties, materials theory and design. Recognizing the increasing importance of materials science in future device technologies, the book titles in this series reflect the state-of-the-art in understanding and controlling the structure and properties of all important classes of materials.

More information about this series at http://www.springer.com/series/856

M.G. Krukovich · B.A. Prusakov
I.G. Sizov

Plasticity of Boronized Layers

Springer

M.G. Krukovich
Moscow State University of Railway Engineering
Moscow
Russia

I.G. Sizov
East Siberia State University of Technology and Management
Ulan-Ude
Russia

B.A. Prusakov
Bauman Moscow State Technical University
Moscow
Russia

ISSN 0933-033X ISSN 2196-2812 (electronic)
Springer Series in Materials Science
ISBN 978-3-319-40011-2 ISBN 978-3-319-40012-9 (eBook)
DOI 10.1007/978-3-319-40012-9

Library of Congress Control Number: 2016943853

Printed on acid-free paper

This Springer imprint is published by Springer Nature
The registered company is Springer International Publishing AG Switzerland

Preface to the English Edition

The process of boriding is widely used for the purpose of improving the wear resistance of details and tools produced from a range of steels and smelts. The present edition is the translation of the original book—"The Plasticity of Boronized Layers"—which was first published in 2010 by Fizmatlit Publishing House in Russia and has achieved certain recognition and demand. The book deals with the production of wear-resistant layers as a process aimed at finding ways for improving the only disadvantage of borides—their brittleness.

The book contains the systematic analysis of various data on the theory, technology, and usage of boriding and boronized layers. It has finally allowed to specify the usage of different morphological structures of boronized layers for certain loading types of the details and tools under hardening. Moreover, the book has given the impulse for the further theoretical development and has opened new directions in the technology of applying boronized layers. In particular:

- the improvement of statistical and thermodynamic calculation methods of eutectic temperatures and concentrations in two- and multicomponent systems which assist in determining the temperature for boriding of technical construction materials and smelts;
- the invention of technological processes of simultaneous saturation by boron and nitrate, boron and phosphorous, etc.
- the expansion of the studies aimed at obtaining boronized layers with different morphological structures involving concentrated energy sources (electrophysical, laser, electron beam impact).

The English edition contains several corrections to the previous edition; thus it now involves the analysis of the studies which either were overlooked in the original book or were published after 2010. The main focus is done on the improvement of the technology and expansion of boriding application. Finally, the authors offer a new original chapter which covers the analysis of the scientific and practical results achieved over the last years.

Moscow, Russia M.G. Krukovich
Moscow, Russia B.A. Prusakov
Ulan-Ude, Russia I.G. Sizov

Contents

Chapter 1
Introduction

This book was written for several reasons. First, boriding has always been the object of interest due to the fact that it allows to obtain the unique combination of mechanical, corrosion and tribology properties of surface layers in ferrous alloys. Boriding as a method of surface hardening attracts specialists from different industry areas. The only restriction which prevents its wide spread is the insufficient plasticity of boronized layers. It impedes boriding from being used for hardening details that work under conditions of alternating loads and risks of mechanical and thermal shocks.

The critical attitude towards the plasticity of boronized metals and alloys began to appear in the earliest works in this area. The attempts to enhance the plasticity by specific alloying and using thermal methods have been made many times. Quite frequently they gave good results.

This book is dedicated to the study of boriding in the context of enhancing the plasticity of boronized layers. For the last dozen of years specialists have been given a chance to obtain layers with enhanced plasticity by using concentrated energy sources. For instance, the pivotal change in the layer structure and hence the changes in its plasticity can be achieved by the radiation-thermal impact of electron streams under boriding or by treating the preliminary obtained traditional boronized layers.

Boriding combined with concentrated energy sources changes the structure of layers, the morphology of obtained phases, the character and spread of bonds formed in the near-surface area. Moreover it turns out that the intensity of these changes can be effectively controlled by introducing different modes, for instance, the mode of radiation-thermal impact. It led to a fundamental possibility of creating a composite boronized layer with a heterogenic dispersed morphology and a Chinese-script eutectic structure of boride crystals instead of a columnar (needle or

M.G. Krukovich et al., *Plasticity of Boronized Layers*,
Springer Series in Materials Science 237,
DOI 10.1007/978-3-319-40012-9_1

tooth-like) morphology. The constructing of different composition levels which combine "solid" and "soft" phases in different ways allowed to obtain boronized layers with different ductility on steel surfaces.

The most recent Russian-language book on boriding was published quite a long time ago—about 25 years ago. It was L.G. Voroshnin and L.S. Lyakhovich's study which can be translated as "Steel Boriding" [1]. It successfully combined scientific and technological data on boriding processes which is still relevant.

The authors of the book presented to you don't mean to repeat already recognized facts. This study is an attempt to add some new information. To give the reader more thorough knowledge on the development of boriding, the book contains the analysis of double and triple B-Fe and B-C-Fe systems, components' properties and basic phases: solid boron solution in iron and iron borides. Moreover it shows the evolution of binary and triple phase equilibrium diagrams in time, beginning with the first built diagram; two complete double B-Fe diagrams created in different periods and the authors' variant of the full diagram are given. The work also deals with the method of building equilibrium diagrams with four and more components and the use of such diagrams in the process of boriding.

The authors hope that this book would be useful not only for specialists working in the fields of chemical and thermal treatment but for graduate- and postgraduate students, engineers and scientists in machine building, metallurgy, mining, textile and other industries.

When the book was almost finished, one of the authors, a prominent scientist and a brilliant teacher, a famous metal industry worker—Boris Alekseevich Prusakov—passed away. Boris Alekseevich initiated the work on this book and wanted it to be useful for everyone who was interested in the problem of metals and alloys hardening. We hope that he was (as always) right. Unfortunately, for the last time…

Reference

1. *Voroshnin L.G., Lyakhovich L.S.* Borirovanie stali. Moscow, 1978. 239 p. [Voroshin L.G., Lyakhnovich L.S. Steel Boriding]

Chapter 2
General Classification of Boriding Processes

Abstract The chapter presents a classification scheme which covers the theory and practice of boriding processes and their interconnection and allows to determine the areas of practical use for obtained boronized layers. The mechanism of atom formation includes the initial state of boron, the state of saturating medium, the formation conditions of saturating atoms and the means for their formation and transportation to the processed surface. The technological peculiarities involve the methods and ways, types of heating and processing specifics. The thermal interval of boriding processes is divided into high-temperature, medium-temperature and low-temperature boriding; it also reflects the possibilities of combining boriding with other processes of thermal treatment and processing of different details and tools. The analysis of phase composition covers the type of borderline and hardness of obtained layers of different morphology. All the technical terms are properly defined.

Theory and practice of boriding processes have been developed by a significant number of scientists and engineers over the years. There are several famous scientific schools working on the issue: German school is lead by H. Kunst, O. Schaaber, A. Graf von Matuschka, H. Vetters, P. Mayr; in the UK—T. Bell; in France—J.L. Andrieux, M. Gantois, M. Gæuriot; in Japan - Kawamaki Mamoru, Muta Akinori, Takamoto Nabujiro; in the USA—N.C. Cook et al. The leading companies in the fields of chemical and thermal treatment of metals and alloys have been dealing with the process implement: Degussa, Leybold Durferrit, Sandvik AB, Stahlwerke Röchlong-Burbach, HEF and others.

The Soviet science gave start to several schools of thought based in the range of scientific centers: Moscow (A.N. Minkevich, Yu. M. Lakhtin, N.S. Gorbunov), Belorussian (L.S. Lyakhovich, L.G. Voroshnin, E.I. Belski, M.G. Krukovich), Kiev (G.V. Samsonov, A.P. Epic), the Urals (N.G. Ilyuschenko), Voronezh (V.D. Korotkov, V.G. Permyakov, V.F. Loskutov) etc. For many years they have obtained a large number of materials on different boriding processes and their peculiarities.

M.G. Krukovich et al., *Plasticity of Boronized Layers*,
Springer Series in Materials Science 237,
DOI 10.1007/978-3-319-40012-9_2

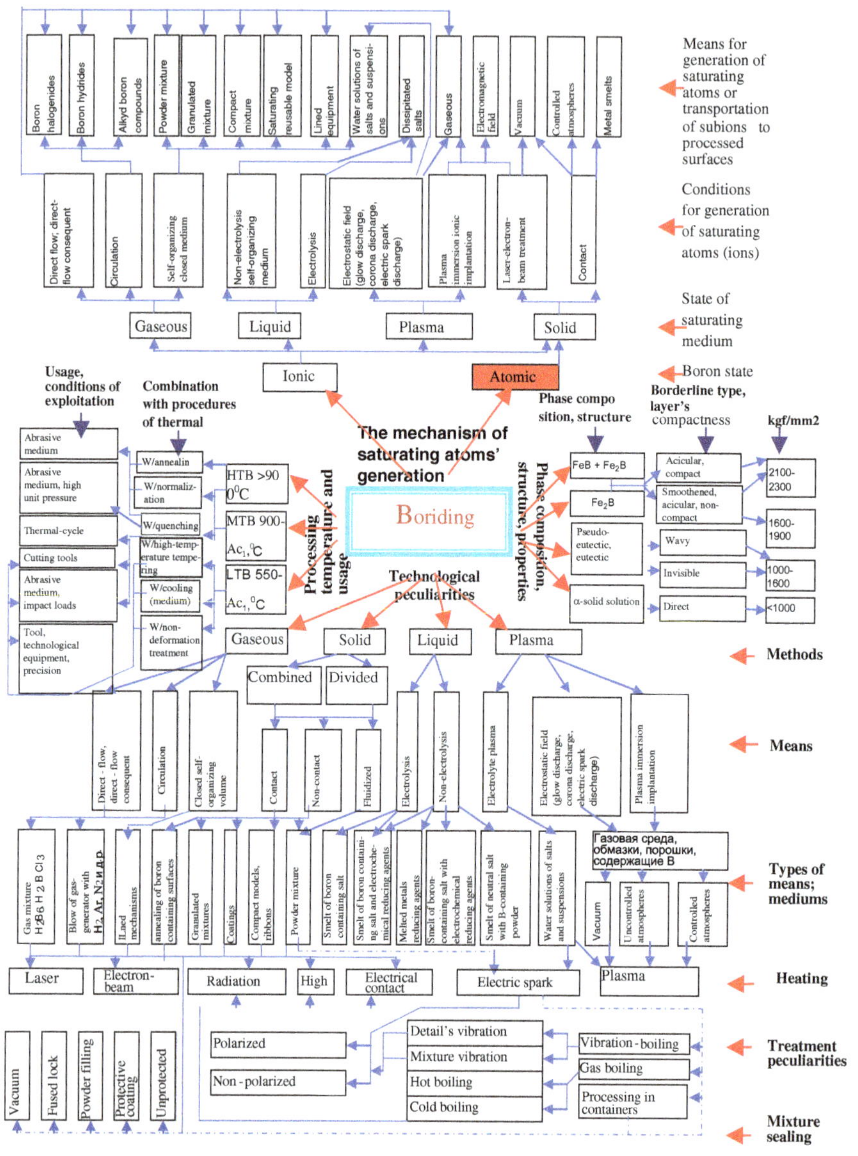

Fig. 2.1 The general classification scheme of boriding processes

In accordance with different time demands, processes have been classified by the structure of obtained layers and the process technology. This classification was intertwined with classifying saturating atoms' development mechanisms which prevented scientists from discovering all interconnections. These classifications helped to sum up the results, understand the process of treatment in more detail and find ways of its rational use.

The current state of boriding can be illustrated by the classification scheme given in the work [1]; the authors of the present book have developed it for this study (Fig. 2.1, colour illustrations set).

The proposed scheme introduces new interpretations, allows to see the process' theory and practice in all its connections, find ways for intensifying different technological variants and determine the areas of their application taking into account the structure of obtained layers.

The given classification studies the process by 4 criteria

- The mechanism of saturating boron atoms development;
- The technological characteristics, including all known techniques;
- The phase composition and structure;
- The temperature of the process and its purpose.

The analysis of the initial saturating mediums revealed the presence of boron in two states: ionic and atomic. The medium itself can exist in all four states of matter (gaseous, liquid, solid and plasma).

Sub-compounds[1] of boron are generated in the interaction of medium components during its circulation, self-organizing processes, under electrolysis and without it, under electro-physical impact etc.

The classification of sub-ions' formation and their transportation to the processed surface is the final stage of classifying saturating atoms by the mechanism of their development. The ionic state of boron in saturating mediums indicates that intensification of boriding and its stability can be obtained by using electrolysis or by spontaneous impact of short-circuited micro-galvanic elements.

The classification of the processes by their technological characteristics means that the saturating medium's state of matter demands certain techniques and determines the number of boriding methods. Each method is characterized by certain ways of the process procedure and certain types of saturating mediums. Some mediums (gaseous, powder, coating) are used differently. Developing this interconnection, it might be possible to broaden the technological scope of boriding procedures. For instance, electrolysis may be used in gas and solid methods; combined gas and liquid processing with the use of gas etc. The final stage of technological principles classification is the interaction between a saturating medium and a kind of containment and a type of heating for processing.

The structural classification divides boronized layers into four types: two-phase ($FeB + Fe_2B$), one-phase (Fe_2B), multi-phase (eutectic and pseudo-eutectic) and α-solid Fe(B) solution based layers; it deals with the relations of phase composition and layer's structure with its compactness, a type of border and micro-solidness.

---

[1]Sub-compounds—compounds occurring in non-equilibrium thermo-dynamical conditions. They are formed by low valent ions of boron (sub-ions). For oxohalide systems it can be BO, BCl, BCl_2, BF, BF_2 which form ad-atoms and/or boron ad-ions on the processed surface due to the reactions of disproportioning.

This classification partly coincides with the classification given by A.N. Minkevich [2] and complements it. α-solid Fe(B) solution based layers are generated in weak saturating mediums (2–2.2% B). For example, in a preliminarily electrolyzed borax smelt; in saline (salt) mixtures, containing small quantity of reducing agents or boron-containing substances (0.5–1.5%); in special powder mixtures. These layers are generated while processing, among all, nickel and cobalt based high-chrome steel (XI2 M, for example), precipitation-hardening, high-temperature, stainless steels and alloys etc. The layers have a strictly special function, for example, of prolonging stainless and high-temperature steels' durability.

Boronized boride layers can be compact and non-compact. In the latter, boride needles are divided by α-solid Fe(B) solution. The maximum phase disintegration allows to obtain eutectic and pseudo-eutectic structures. They are less solid than compact layers; the hardness is determined by the specific number of phases. Yet they are characterized by the enhanced plasticity and thickness that makes them more suitable for mechanical treatment; they can be used for hardening details that work under different loads.

The borderline type determines the strength of a layer and a substrate's adherence. The maximum length of the borderline, which is determined by the acicular structure of a boride layer, influences the intensity of adherence though the hardness values of borides and substrate are of different kind. Thus a layer can't be disintegrated along its borderline during mechanical tests or exploitation. This effect can also be observed with blurred or invisible borders.

The given structural classification concerns only high-quality boronized layers obtained in stable conditions. Other classifications [3, 4] deal with a wider spectre of structures which can be defective structures of boriding processes, or related to a multicomponent uncontrolled saturation, or obtained as a result of additional treatment. These structures are formed under general or local changes in conditions of a layer formation: temperature, surface boron concentration, alloying elements concentration in a surface or a sub-layer.

The technologies of obtaining boronized layers various in properties and purpose can be divided into high-temperature (HTB), medium-temperature (MTB) and low-temperature (LTB) saturation. Each interval demands the use of certain compositions and technological methods. The temperature classification further determines the opportunities of hardening and heating treatment integration for different types of thermal processing or even the use of saturating mediums (for example, melted) as cooling ones. In the end this leads to obtaining a certain complex of details' properties and, together with the layer's phase composition, provides recommendations on the layer's thickness and conditions of the use of obtained boronized layers.

The explanation of some technical terms used in the classification scheme of boriding process are given below.

2.1 Plain: The Mechanism of Saturating Atoms Generation

- The characteristics of saturating atoms generation process:

A direct flow process occurs under saturation during continuous blowing of a saturating medium by a carrier gas or a boron-containing gas;

A direct-flow consequent process occurs under continuous blowing of a saturating medium in a certain consequence by different gases or in different modes of blowing;

A circulation process occurs under a continuous circulation of a gaseous medium with a consequent zone passing; the continuous circulation forms boron ions and saturates in the closed cycle mode.

A self-organizing closed medium is a closed gaseous medium, characterized by the cyclic character of ion formation and atoms saturation which present themselves as self-organizing processes.

A non-electrolysis self-organizing medium is a liquid medium, where a self-organizing atoms saturation process goes according to the electro-chemical laws excluding electrical circuit conduction.

An electrolysis process is a formation of saturating atoms in a liquid medium according to the laws of electrolysis; the process excludes conducting electrical currents.

An electrostatic field (*glow discharge, corona discharge, electric spark discharge*) is a plasma medium, where saturating atoms are generated on a hardening surface due to the charges occurring in a gas medium at the border between a surface and a saturating medium (gaseous, liquid, solid or their combination).

Plasma-immersion ionic implantation is a plasma medium which accelerates the bombardment of a hardening surface with boron ions in the conditions of a high voltage glow discharge between a surface and an anode in a gaseous medium. The pressure is lower than atmospheric.

Laser electron-beam treatment is a heating of a surface covered with a boronizing solid medium (coating) or heating of a surface which is preliminary boron-saturated in several mediums in order to obtain a boronized layer or change its structure by a stimulating laser or electron-beam effect;

A contact process is a formation of a boronized layer by a direct contact between a working surface and boron-containing components of a saturating medium (solid and liquid) in the presence of a boron concentration gradient.

- Means for generation of saturating atoms or transportation of subions to a processed surface:

Boron halogenides—compounds of boron and halogens (BCl_3, BF_3, BJ_3, BBr_3);
Boron hybrids—compounds of boron and hydrogen (B_2H_2, B_2H_6);
Alkyd boron compounds—organic boron compounds ($(CH_3)_3B$, $(C_2H_5)_3B$ etc.);
A powder mixture is a solid multi-component mixture in the form of powder;

A granulated mixture is a solid multi-component mixture in the form of granules of a particular size (usually 0.5–5.0 mm);

A compact mixture is a composite material obtained from coatings, slips and suspensions after applying on a working surface and further drying;

A saturating reusable model is a detachable reusable technological device which replicates the relief of a zone or a whole surface; it is coated with a special saturating mixture or a porous material soaking.

Lined mechanisms are containers, retorts or muffles; their inner walls are covered with a saturating boride mixture (fettling).

Water solutions of salts and suspensions are water mediums which contain boron salts in a solution state and/or boron compounds or boron in a colloid or suspended state.

An electromagnetic field is an external electromagnetic field applied in order to activate the process of saturating boron atoms generation and increase their diffusive mobility in a metal.

Metal smelts are smelts which contain soluted boron and ensure the formation of a layer on a surface only by a gradient of concentrations.

2.2 Plain: Technological Specifics

- Methods and means of boriding and their types:

A solid combined method is saturation in metal-thermic mixtures as a result of combining processes of boron reduction from its oxides and details' boriding.

A solid divided method is saturation in mixtures as a result of separating metal-thermic boron reduction from its oxides and the later use of these powder mixtures for details' boriding.

A contact method is when boriding of details is carried out in a container after their covering and following capsulation/sealing.

A non-contact method is when boriding is carried out in a container where a saturating mixture is separated from details or is placed into a special gas-forming container (gas-generator).

A fluidized method is saturation in a powder mixture in a pseudo-liquefied state.

A gaseous saturation method in a closed volume as a self-regulating process is saturation in a closed container, muffle or converter where inner walls are covered with a saturating mixture (saturation in lined mechanisms).

An electrolysis method in boron-containing salts smelts is saturation of metals with boron in smelts during the process of electrolysis under electrocuting.

An electrolysis method in boron-containing salts smelts with the addition of powder electrochemical reducing agents is saturation of metals with boron in smelts with the addition of powder electrochemical reducing agents for the additional generation of boron sub-ions which reduce easier on a working surface during electrolysis.

A non-electrolysis method in boron-containing salts smelts with the addition of powder electrochemical reducing agents is saturation of metals in smelts, where boron atoms generation is carried out in a self-organizing mode under a work of short-circuit galvanic elements.

A non-electrolysis method in metal smelts is liquid saturation when boron in metal smelt is in a dissolved or a suspended state.

A non-electrolysis method in boron-containing salt smelts with the addition of powder boron-containing electrochemical reducing agents is liquid saturation in salts smelts which ensure a maximum speed of boron sub-ions generation in a self-organizing mode under a work of short-circuit galvanic elements.

A non-electrolysis method in neutral salts smelts with the addition of powder boron-containing electrochemical reducing agents is saturation in liquid mediums where boron atoms generation is self-organized and is carried out by electrochemical mass carry during the generation of boron sub-ions.

An electrolysis-plasma method is boriding in water solutions of boron-containing salts and/or salts containing boron in a colloid and/or an atom state and/or in compounds, which takes place due to the generation of spark plasma in a steam coating on the border of a metal and a solution under electrocuting.

A vibration fluidized method is saturation in a powder mixture in a fluidized state which was acquired after vibrating a detail or vibrating a container (stove, converter, container) with details and a mixture.

Hot boiling is creation of a fluidized layer by adding a hot gas medium into a converter below a powder mix.

Cold boiling is creation of a fluidized layer by adding a cool gas medium into a converter below a powder mix; heating of details can be done with any known method (radiation, alternating current, electrocontact, electric spark).

Electric spark polarized heating is heating of details with electric spark charges on the border of a powder mixture and a working surface under a constant current.

Electric spark non-polarized heating is heating of details with electric spark charges on the border of a powder mixture and a working surface under an alternating current.

- Ways of sealing a saturating medium:

A fused lock is a smelted medium in a container above a saturating mixture used to prevent air from coming through into a saturating medium with details.

A powder filling is an additional powder medium above a saturating mixture used to prevent air from coming through into a saturating medium with details due to its big resemblance to oxygen and other air components;

Unprotected saturation is saturation in an open container with a partial protection from oxidation by a skin formed as a result of fusing, oxidation and smelting of the most of saturating mixture's surface.

2.3 Plain: Phase Composition, Structure and Properties

- The structure of layers and borderline types:

A compact structure with acicular borders is a common structure in iron-based boronized layers and FeB + Fe2B or one phase Fe2B based steels.

A non-compact structure with acicular borders is a layer's structure divided into zones or separate needles; spaces between them are filled with α-solid Fe(B, C) solution.

A smooth acicular border is an acicular structure of a boride layer which has a smooth borderline between borides and a metal substrate due to the deceleration influence of alloy elements and carbon on the growth of borides.

A pseudo-eutectic structure is a disperse structure of iron borides and alloy elements in α-solid solutions without a clear borderline with a substrate layer due to the high resemblance of alloy elements to boron. This structure is formed as a result of the diffusion crystallizing mechanism of boronized layer generation.

A eutectic structure is common for eutectics, the structure of iron borides and alloy elements in α-solid solutions with a low level of dispersibility which have a wavy borderline.

α-solid solution structure is a structure and a borderline common for α-solid solutions; may appear when there is more α-solid solution than borides, for example, when there is a eutectic structure of boronized layer.

2.4 Plain: Processing Temperature and Usage

High-temperature boriding (HTB) is high temperature boriding which is carried out at temperatures >900 °C.

Medium-temperature boriding (MTB) is medium temperature boriding which is carried out in the interval Ac_1—900 °C (Ac_1—a referential critical point for a particular steel).

Low-temperature boriding (LTB) is low temperature boriding which is carried out in the interval 550—Ac_1, °C;

Cooling (a medium) is the use of baths during low temperature boriding as tempering cooling mediums. For example, for tempering high-speed steels.

Non-deformation treatment is strengthening which is carried out by boriding at temperatures lower than phase transformations in a processed steel.

References

1. *Krukovich N.G.* Razrabotka teoreticheskih i prikladnyh aspektov upravleniya strukturoi i svoistvami borirovannih sloyov i ih ispolzovanie pri proizvodstve transportnoi tehniki // Diss. dokt. techn. nauk. Moscow, 1995. 416 p. [Krukovich N.G. The Development of Theoretical and Practical Aspects of Managing the Structure and Properties of Boronized Layers and Their Usage in Transport Machinery]
2. *Minkevich A.N.* Khimiko-termicheskaya obrabotka metallov i splavov. Moscow, Mashinostroenie, 1965. 491 p. [Minkevich A.N. Chemical-Thermal Processing of Metals and Alloys]
3. *Voroshnin L.G., Lyakhovich L.S.* Borirovanie stali. Moscow, 1978. 239 p. [Voroshin L.G., Lyakhnovich L.S. Steel Boriding]
4. *Kunst H. und Schaaber O.* Harterei-Technische Mitteilungen. 1967. Bd. 22. H. 3. H. 4. S. 275–292.

Chapter 3
The Components and Phases of Systems 'Boron-Iron' and 'Boron-Carbon-Iron'

Abstract Basic chemical and physical properties of the components and phases in B-Fe and B-C-Fe systems are analyzed and the conditions of their existence and formation during the process of boriding of iron and its smelts are studied. Different types of iron borides (Fe_3B, Fe_2B, FeB, FeB_2), their crystall lattices, concentration and temperature formation conditions are thoroughly investigated. The chapter also provides the values of hardness and chemical properties which form the exploitation characteristics of boronized layers.

The plasticity of boronized layers depends on the type, content, quantity, size and mutual position of phases in the layers. One more additional factor that influences the properties is a substrate layer's material. This makes the description of existing phases in B-Fe and B-C-Fe systems especially important for the properties' evaluation.

The specifics of initial studies on iron and boron alloys was that scientists were not certain if the investigated iron-boron alloys actually contained boron. The reason was that the analysis methods of alloys' chemical contents were far from precise and were not appropriate for alloys that contained boron in small quantities. This is applied to the earliest studies of F. Osmond who was investigating allotropic metamorphosis in iron alloys generally and iron-boron alloys in particular in 1890–1895 [1]; J. O. Arnold's study on the physical influence of different chemical elements on iron (including boron) [2]; H. Moissan and G. Charpy's works on steels with different boron contents [3].

One of the earliest works based on the chemical analysis which confirmed the presence of residual boron was the paper of Guillet [4], published in 1907. The paper was devoted to changes in the structures of steels containing 0.2 and 0.5% C in dependence of the quantity of injected boron in the range of 0.2–1.5%. The result was the discovery of a specific structural component; its quantity was increasing with the rising quantity of carbon and boron. It was defined as an iron boron carbide.

The analysis of equilibrium diagrams for B-Fe and B-C-Fe systems, built on the base of various results provided by different scientists, reveals that the difference between them are defined by the diagrams' type but mostly by the positions of

M.G. Krukovich et al., *Plasticity of Boronized Layers*,
Springer Series in Materials Science 237,
DOI 10.1007/978-3-319-40012-9_3

critical points and the presence of certain phases. This is connected to the fact that in each case the result was influenced by the chosen study methods and their precision.

The further decades have made it possible to develop methods of investigation and deepen the knowledge of physical and chemical states of system components (boron, carbon and iron) and the phase composition and structure for double, triple and multiple iron and boron alloys.

3.1 Boron (B)

Boron is a chemical element №5 in Mendeleev periodic table (group III, period 2); its atomic weight is 10,811; boron atom diameter—1.78 kX [5]. There are 6 isotopes with mass numbers 8, 10–14; typical oxidation states—+III and −III; clear crystal boron has a greyish-black colour; the melting point = 2352 K; it is dissolvable in a concentrated HNO_3 bath; it oxidates into B_2O_3 at 973 K; it is concentrated on Earth in boric acid and various borates; it was isolated by Joseph Louis Gay-Lussac and Louis Jacques Thénard (France) in 1808.

Boron is used in metallurgy as an alloy element which increases the wear resistance and high-temperature strength of steels. Boron is combined with hydrogen to form hydrides (boranes); hydrides in solid objects (e.g., metal objects) are effectively used for hardening metal details' coatings. Boron interacts with metals to form solid solutions and chemical compounds—borides [6].

The special physical property of boron is its ability to absorb neutrons. The cross section area of heat neutron absorption for a natural isotopes mixture is 750; for isotope B^{10} – 3990 bn.

3.2 Carbon (C)

Carbon is a chemical element №6 in Mendeleev periodic table (group IV, period 2); its atomic weight is 12,01115; carbon atom diameter—1.54 kX [7]. There are 8 isotopes with mass numbers 9-16, typical oxidation states −IV, −II, 0, +II, +IV; the melting point = 4003 K (graphite). Carbon-12 is of particular importance as it is used as the standard from which atomic masses of all nuclides are measured: its mass number is 12 by definition.

Carbon is found in different allotropic modifications: graphite, diamond, carbyne, fullerene. The chemistry of carbon compounds is singled out into a separate field of science—organic chemistry.

Carbon is known since ancient times and has a wide-spread use: it can be a part of many alloys (steel, cast iron), it is used for producing electrodes, crucibles, black powder, metal smelting from ore (charred coal) and as an adsorbent (activated coal) in chemical industry; isotope ^{14}C is used as a radioactive indicator [6].

3.3 Iron (Fe)

Iron (Fe) is a chemical element №26 in Mendeleev periodic table (group VIII, period 4); its atomic weight is 55,847; iron atom diameter is 2.54 kX [5]. There are 12 isotopes with mass numbers 49, 52–62, typical oxidation states are +II, +III, +VI; it is a silver-white ductile metal; the melting point = 1808 K (1539 °C); easily adaptable for forging and rolling; oxidates and covers in rust in dump air; interacts with most of chemical elements; easily dissolves in dilute acids and becomes passivated in concentrated acids; one of the most widespread elements; found in the form of ore; known from ancient times (the beginning of its use dates back to VIII-VI centuries B.C.—Iron Age); used as a base for most construction materials—cast iron, steel—as a component of special alloys, as a catalyst [6]. Iron is characterized by its polymorphism. The biggest significance for thermal and chemical-thermal processing is applied to α- and ϒ-Fe, which exist at the temperatures corresponding to these technological processes.

3.4 Boron Solid Solution in Iron

In 1917 N. P. Chizevsky discovered that iron, heated in boron or ferroboron powder, acquires a solid boron enriched surface layer [8]. Later boron's 'cementing' effect was confirmed and it was revealed that boron diffusion into a steel is similar to carbon diffusion in many ways [9, 10]. The diffusion of boron into iron and steel was first systematically studied in work [11]. It introduced the coefficient of boron diffusion in austenite:

$$D_{\Upsilon} = 2 \times 10^{-3} e^{-21000/RT}.$$

It was also established that D_{Υ} doesn't depend on the carbon content below 0.43% and boron content lower than 0.003%. Later the coefficient of boron diffusion into Fe_{α} was determined as [12]

$$D_{\alpha} = 10^{6} e^{-62000/RT}.$$

The comparison of obtained diffusion coefficients allowed to state that boron forms a solid interstitial solution in the lattice of Fe_{Υ} and a solid substitutional solution in the lattice Fe_{α}. This conclusion was previously made by other authors [13] but based only on theoretical arguments. The implication that boron forms a solid interstitial solution in austenite could be also found in paper [14], where the researchers studied the speeds of deboronizing and decarbonizing of steel. It was established that the boron content in a decarbonized steel layer decreases with the fall in carbon concentration.

The boron solubility in iron is insignificantly low and many times less than the carbon solubility: at temperature 710 °C it is 0.0002%, at 906 °C—0.0082%, i.e. it largely depends on the temperature.

Till the 1960–1970s it was believed that boron, while dissolving in iron, can form solid interstitial solutions in Fe_{Υ} and substitutional solutions in Fe_{α}. However, at that time, with the introduction of the internal friction method, it was shown that boron in areas of low content and iron can form only solid interstitial solutions [15–17]. The comparison of boron and iron electron shells' states leads to the same conclusion as boron in its electron structure differs from metals in such a way that it can't possibly take the metal's place in its lattice's point without changing the nature of mid-atomic bonds (i.e. without turning a metal in a metal-like compound —boride) [18–20]. Nevertheless, Japanese scientists, while studying the internal friction of alloys in Fe-B and Fe-Ni-B systems in order to recognize the nature of the solid solution in Fe_{α}, discovered that boron atoms in Fe_{α} grains are found in interstitial and substitutional positions [21].

Despite the low solubility in iron, boron is quite effectively used for alloying construction steels. Microsolidness of a boron-saturated α-solid iron solution is ~233–236 HV [9].

Boron in its influence on iron can be compared to carbon, but it is several times more active. Relatively small boron addition (0.02%) makes iron more adaptable for quench hardening and improves the milling of its crystal structure [22]. Boron in a solid solution significantly increases the steel's hardenability. This increase demands the boron content of 0.0005–0.0012% in a solid solution. Boron additives of more than 0.007% lead to hot shortness and complications during hot rolling and forging of steel [5].

The significantly important data for explaining increased hardenability of thermally processed steels was obtained in the study of boron distribution in Fe_{Υ} grains; especially—the discovered inhomogeneity of such distribution. It was shown [23] that boron is adsorbed in Fe_{Υ}: boron atoms try to segregate on borders of ϒ-grains; the quantity of segregating atoms increases with the rise of temperature. E.g., the number of boron on the borders of ϒ-grains in steel AISI86B40 (~40XГH) increases in double with the rise of saturation temperature from 1800 °F (982 °C) to 2200 °F (1204 °C). It is well known that boron in steels under hardening slows the beginning of austenite decomposition. The increased concentration of boron doesn't effect it. Only saturation of grain borders up to the critical concentration leads to its appearance.

The introduction of boron into steel aimed at increasing its hardenability allows to save a significant number of rare and expensive alloying elements. The incision of 0.002% B (20 grams of boron for 1 ton of steel) allows to save 1,2–2.4% Ni or 0.15–0.35% Mo. Due to the large size of boron's atom diameter (1.78 kX) in comparison with carbon (1.54 kX), nitrate (1.40 kX) or oxygen (1.32 kX), its solubility in iron (atom diameter −2.54 kX) is lower that other mentioned elements'. However, the influence of boron atoms on the iron lattice, due to its big atom diameter, is stronger than solid interstitial solutions of carbon, nitrate or oxygen with iron have. Moreover, a low carbon content in steel makes boron more effective in increasing hardenability. Normally boron is added to low- or medium-carbonated steels which contain less than 0.60% C.

Boron doesn't increase the hardenability of eutectic and over-eutectic steels, thus it is not added to these kinds of steels. The addition of boron is most effective in steels which contain no more than 0.30% C [5].

3.5 Iron Borides

Iron borides are chemical compounds of boron and iron. The main peculiarity of iron borides is their high density and low ductility. Iron borides, when heated in ammonia currents at 400–600 °C, decompose into iron nitrides and boron; when heated at 700–768 °C, iron Fe_{α} and boron nitride BN are acquired [22, 24].

Borides found in the structure reduce the hardenability of steels as they are the center of austenite transformation, e.g. into perlite [5, 25, 26]. If borides are assembled in a row according to the increasing level of boron content, four borides will be found there—Fe_3B, Fe_2B, FeB and FeB_2—all of them are studied in connection with iron and steel boriding.

Iron Boride Fe_3B The study [27] states that borides like Fe_3B exist only in the presence of carbon. This way they are the product of the substitution of carbon by boron in the cementite lattice [28, 29] and are defined by the formula $Fe_3(B,C)$. The data in the work [28] shows that boron can substitute more than 40% of carbon atoms in $Fe_3(B,C)$ phase without transforming its crystal structure. This fact allowed the authors in the paper [17] to doubt the legitimacy of attributing this compound to borides.

However other works [30–32] inform that Fe-B system contains a compound Fe_3B. It can be obtained by hardening of alloy $Fe_{76}B_{24}$ from liquid state at cooling speeds which suppress the equilibrium Fe_{α} + Fe_2B and allows to get a crystal structure. The obtained Fe_3B phase exists in the thermodynamic equilibrium at high temperatures. The temperature range of existence is established between 1150 and 1250 °C. Fe_3B Boride has two stable modifications.

Iron Boride Fe_2B Iron boride Fe_2B contains 8.83% (8.34% [33]) of boron and has a body-centered tetragonal structure type—$CuAl_2$ with 12 atoms in a cell and with periods of lattice: a = 0.5109 nm, c = 0.4249 nm, c/a = 0.831 [18, 27, 34–36] or a = 0.5078 nm, c = 0.4249 nm, c/a = 0.836 [37]. Microsolidness is 1800–2000 HV [37], or 1320–1680 HV [9], or 1500–1800 HV [38].

The phase equilibrium diagram of B-Fe doesn't clearly show [39, 40] if boride Fe_2B is a steady compound. It has the elastic modulus of (285–295) × 10^3 MPa, the thermal-expansion coefficient—7.65 mln^{-1}/°C and 9.2 mln^{-1}/°C in the temperature range from 200 to 600 °C and from 100 to 800 °C correspondingly [37, 41, 42], or 14 × 10^{-6} $grad^{-1}$ [38], density—7.43 t/m^3. The general opinion is that boriding should better ensure the generation of a one-phase monolayer Fe_2B with columnar saw-shape morphology and not the generation of a two-phase layer with FeB phase

[33, 37, 42–48]. The formation of columnar structure is witnessed due to the anisotropy of the diffusion coefficient in borides.

A monophase layer Fe_2B may be obtained from a two-phase FeB-Fe_2B layer by special vacuum processing or processing in a salt bath for several hours at the temperature ~800 °C which can be followed by hardening in oil in order to improve the plasticity properties of a substrate [37, 49].

Iron Boride FeB Iron boride FeB (monoboride) contains 16.23% of boron and has an orthorhombic lattice with 8 atoms in a cell (4 boron atoms and 4 iron atoms) and with lattice periods: $a = 0.4061$ nm, $b = 0.5506$ nm, $c = 0.2952$ nm or $a = 0.4053$ nm, $b = 0.5495$ nm, $c = 0.2946$ nm [1, 6, 51]. It is characterized by polymorphism and exists in two variants: low-temperature and high-temperature (lower and higher 1135 °C correspondingly). However this fact requires additional proof as it was the existence of only one variant which was experimentally proved —the modification of FeB which has a B27 structure type.

The microsolidness is 1900–2100 HV [33, 37, 46] and 1850–1900 HV [33] or 1800–2200 HV [38]. The melting point is 1390 °C [33, 37]. The elastic modulus is 590×10^3 MPa, the density—6.75 t/m^3 [37], the thermal-expansion coefficient—23 mln^{-1}/°C between 200 and 600 °C [37, 41, 42] or $\sim 9 \times 10^{-6}$ grad^{-1} [37]. Boride FeB is the most chemically stable phase of B-Fe system. It is stable and doesn't decompose under hot pressing in the mix with carbide B_4C up to 1450 °C and in this environment there is no interaction between FeB and B_4C.

Boride FeB is easily formed under hot pressing (~1300 °C) of powder mixtures of iron and its carbide (Fe_3C) with boron or boron carbide (B_4C) [7, 16]. Iron boride FeB dissolutes in nitric acid [22].

The hard exploitation, especially under alternating loading, makes the presence of a boron-saturated whole outer layer of a FeB phase in a boronized layer almost unacceptable [33, 37, 42, 46, 50]. The presence of the sharp variation of residual stresses on the border of these phases leads to formation of cracks in the layers with FeB and Fe_2B on or near the surface of transition FeB-Fe_2B [13, 33, 37, 42, 51–53] (Fig. 3.1). The conclusion is that it is necessary to avoid or significantly limit the quantity of FeB phase in boronized layers. Good exploitation properties are characteristic for boronized layers where the correlation of phases FeB:Fe_2B is no more than 1:3 [38].

The presence of carbon in steel doesn't affect its hardness; alloying elements Ti, Mo, W increase the hardness of boride FeB while the presence of Al or Cr decreases it [33, 54]. Under heating boride FeB is stable up to 800 °C.

Iron Boride FeB_2 Iron boride FeB_2 (diboride) contains 27.9% of boron and can be attributed to hexagonal systems, structural type AlB_2. The parameters of its elemental cell: $a = 3.045$ nm; $c = 3.035$ nm; $c/a = 0.997$. The melting point of FeB_2 is 2070 ± 50 °C [33, 55].

The number of atoms in a rhombohedra lattice is 1 (3 cells—2 boron atoms and 1 iron atom). The atom coordinates are Fe (000), B1 (1/3 2/3 1/2), B2 (2/3 1/3 1/2). Boron atoms form graphite-shape lattices, which dimensions are

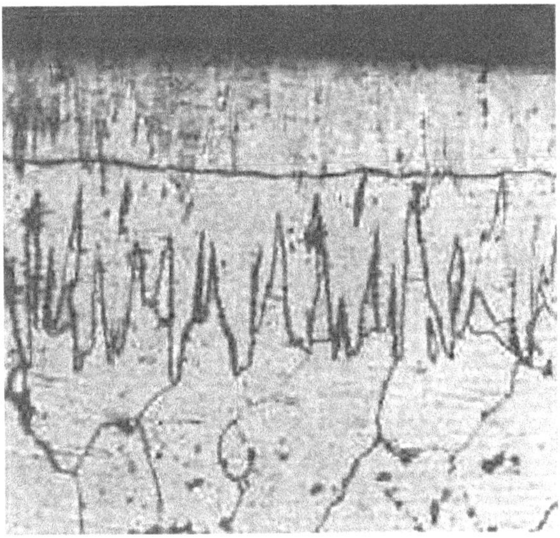

Fig. 3.1 The microstructure (x200) of a two-phase boride layer in a low-carbon steel (boriding at 900 °C for 4 h)

perpendicular to axis Z. Boron atoms' layers are alternated with iron atoms' layers which contain two times less boron atoms [17, 56].

The microsolidness is 2740–2800 HV [17] or 1650 HV_{100} [54]. The elastic modulus is 25.3×10^3 kg/mm^2, the melting point is 2070 ± 50 °C. The presence of carbon or alloying elements Ti, Mo, W, Cr doesn't affect the hardness of FeB_2 boride [33, 54]. FeB_2 boride is stable under heating up to 1000 °C.

References

1. *Osmond F.* On the Critical Points of Iron and Steel // J. Iron Steel Inst. 1890. V. 37. I. P. 38–80.
2. *Arnold J.O.* The Physical Influence of Elements on Iron // J. Iron Steel Inst. 1894. V. 45. I. P. 107–155.
3. *Moissan H. et Charpy G.* Sur l'acier au bore // Compt. Rend. Acad. Sci. Paris. 1895. V. 120. P. 130–132.
4. *Guillet L.* Boron Steels // J. Iron Steel Inst. 1907. V. 74. II. P. 207– 218.
5. *Bolkhovitinov N.F.* Metallovedeniye i termicheskaya obrabotka / Uchebnik dlya vuzov. Moscow, 1958. 431 p. [Physical Metallurgy and Thermal Processing]
6. *Kershenbaum V.Ya., Prusakov B.A.*, ed. Mnogoyaychnyi tolkovyi slovar' 'Metally'. Moscow, 1999. 710 p. [Multi-language explanatory dictionary 'Metals']
7. *Glaser F.W.* // Journ. Metals. 1952. V. 4, № 4. P. 391.
8. *Tschischewsky N.* The Case-Hardening of Iron by boron // J. Iron Steel Inst. 1917. V. 95. P. 185–187.
9. *Thomas W.R. and Leak G.M.* Condition of Boron in Alpha Iron // Nature. 1955. July 2. V. 176, № 4470. P. 29–31.
10. *Campbell T.P., and Fay H.* The Case-Hardening of Steel by Boron and Nitrogen. //Ind. Eng. Chem. 1924. V. 6. P. 719–723.

11. *Busby P.E., Warga M.E., and C. Wells.* Diffusion and Solubility of Boron in Iron and Steel // J. Met. 1953. November. P. 1463–1468.
12. *Busby P.E., and C. Wells.* Diffusion of Boron in Alpha Iron // Trans. Am. Inst. Mining Met. Engrs. 1954. V. 200. P. 972
13. *Digges T.G., Irish C.R., and Carwill N.L.* Effect of Boron on the Hardenability of High-Purity Alloys and Commercial Steels // Bureau of Standards / J. Research. 1948. V. 41. P. 545–574.
14. *McBride C.C., Spretnak J.E., and Speiser R.* A study of the Fe-Fe2B System // Trans. Am. Soc. Metals. 1954. V. 46. P. 499–524.
15. *Krishtal M.A.* et al. Vnutrennee trenie v metallah i splavah. Moscow, 1964. 245 p. [Internal Friction in Metals and Alloys].
16. *Krishtal M.A., Turkeltaub G.M.* Mikrostruktura splavov zhelezo-bor // Metallovedeniye i termicheskaya obrabotka. 1967. №8. Pp. 63-64 [Mictrostructure of iron-boron smelts]
17. *Voroshnin L.G., Lyakhovich L.S., Panich G.G., Protasevich G.F.* Struktura splavov sistemy Fe-B // Metallovedeniye i termicheskaya obrabotka metallov. 1970. №9. Pp. 14-17 [The Structure of Fe-B system's smelts]
18. *Samsonov G.V.* Tugoplavkiye soedineniya. Moscow, 1963. 397 p. [High-melting point compounds]
19. *Portnoy K.I. Boridy* // Bolshaya Sovetskaya Entcyklopediya. Moscow, 1970. Vol. 3. P. 567. [Borides].
20. *Samsonov G.V., Vinitckiy I.M.* Tugoplaviye soedineniya. Moscow, 1976. 559 p. [High-melting point compounds].
21. *Hayashi Y. and Sugeno T.* Nature of Boron in α-Iron // Acta Metallurgica. 1970. June. V. 18. P. 693–697.
22. *Vol A.E.* Bor-zhelezo (B-Fe) // Stroeniye i svoistva dvoinyh metallicheskix system. Moscow, 1959. Vol. 1. Pp. 679-686. [Boron-Iron (B-Fe)].
23. *Goldhoff R.M. and Spretnak J.W.* Distribution of Boron in Gamma Iron Grains // J. Met. 1957. October. P. 1278–1283.
24. *Kiessling R., Liu Y.H.* // Journ. Metals. 1951. V. 3. № 8. P. 639.
25. *Samsonov G.V., Serebryakova T.I., Neronov V.A. Boridy.* Moscow, 1975. 375 p. [Borides].
26. *Gulyaev A.P.* Metallovedeniye. Moscow, 1986. 544 p [Physical Metallurgy].
27. *Matyushenko N.N.* Kristallicheskiye struktury dvoinyh soedineniy. Moscow, 1969. 302 p. [Crystal structures of binary compounds]
28. *Fichtl W.J.G.* // Heat Treat. Met. 1983. P. 79–80.
29. *Koifman I.S. et al.* Rentgenograficheskii analiz borotsementita. // Metallovedeniye i termicheskaya obrabotka. 1969. №2. P. 59-60. [X-ray analysis of boron-cementite].
30. *Lyakishev N.P.*, ed. Diagramy sostoyaniya dvoinyh metallicheskih system. Moscow, 1996. Vol. 1. Pp. 442-444 [Equilibrium diagrams of binary metallic systems].
31. *Khan Y., Kneller E., Sostarich M.* // Z. Metallkunde. 1981. Bd. 72, № 8. S. 553–557.
32. *Khan Y., Kneller E., Sostarich M.* // Z. Metallkunde. 1982. Bd. 73, № 10. S. 624–626.
33. *Voroshnin L.G., Lyakhovich L.S.* Borirovaniye stali. Moscow, 1978. 239 p. [The Boriding of Steel].
34. *Kiessling R.* // Acta chem. scand. 1950. 4. № 2. P. 209–227. 26
35. *Hendricks S.B., und Kosting P.R.* // Zeitshr. Kristallogr. 1930. 74. S. 511–545.
36. *Aronson B., Lundstrom T., Rundqvist S.* Borides, Silicides and Phosphides. — London, « Methuen » . 1965. P. 230.
37. *Anvil Kumar Sinha.* Boriding (Boronizing) // ASM Handbook. V. 4. Heat Treating. 1998. P. 437–447.
38. *Dubinin G.N., Kogan Ya. D.*, ed. Progressivnye metody khimiko-termicheskoi obrabotki. Moscow, 1979. 184 p. [Progressive methods of chemical-thermal processing].
39. *Mildleham T. and others.* // Iron and Steel. 1957. V. 30. P. 12.
40. *Khansen M., Anderko K.* Struktury dvoinyh splavov. Moscow, 1962. Vol. 1. 608 p. [The structures of binary smelts]
41. *Serebryakova T.I., Neronov V.A., Peshev P.D.* Vysokotemperaturnye boridy. Moscow, Chelyabinsk, 1991. 368 p. [High-termperature borides]

42. *Kunst H. und Schaaber O.* // Harterei-Technische Mitteilungen, 1967. Bd. 22. H. 3. H. 4. S. 275–292.
43. *Lyakhovich L.S.*, ed. Khimiko-termicheskaya obrabotka metallov i splavov. Spravochnik. Moscow, 1981. 424 p [Chemical-thermal processing of metals and smelts. Reference book.]
44. *Krukovich M.G.* Issledovaniye zhidkostnyh bezelektroliznyh protcessov khimiko-termicheskoi obrabotki. Minsk, 1974. 298 p. [The study of liquid non-electrolysis processes of chemical-thermal treatment].
45. *Lyakhovich L.S., Krukovich M.G., Turov Yu. V., Kosachevskii L.N.* Sovremennoe napravlenie primeneniya zhidkostnogo borirovaniya dlya povysheniya stoikosti izdelii. Minsk, 1975. 20 p. [Modern trends of applying liquid boriding for increasing wear-resistance of details].
46. *Krukovich M.G.* Razrabotka teoreticheskih i prikladnyh aspektov upravleniya strukturoi i svoistvami borirovannyh sloev i ih ispolzovanie pri proizvodstve transportnoi tekhniki. Moscow, 1995. 416 p. [The development of theretical and practical aspects of controlling the structure and properties of boronized layers and their usage in producting transportation vehicles]
47. *Tsipas D.N., Rus J., and Noguerra H.* // The Metals Society. 1988. P. 203–210
48. *Bernstein M.L., Rachstadt A.G.*, ed. Metallovedeniye i termicheskaya obrabotka. Spravochnik. Moscow, 1962. Vol. 2. Pp. 1041–1043 [Metallurgy and thermal processing. Reference book]
49. *Dearnley P.A., Farrell T., and Bell T.* // J. Mater. Energy Sys. 1986. V. 8, № 2. P. 128–131
50. *Krukovich M.G.* Odnofaznoe borirovaniye stalei // Poverkhnostnye metody uprochneniya metallov i splavov v mashinostroenii. Moscow, 1983. P. 81–84. [One-phase steel boriding]
51. *Minkevich A.N.* Khimiko-termicheskaya obrabotka metallov i splavov. Moscow, 1965. 491 p. [Chemical-thermal processing of metals and alloys]
52. *Glukhov V.P.* Boridnye pokrytiya na zheleze i stalyah. Kiev, 1970. 208 p. [Boride coatings on iron and steels]
53. *Fichtl W.J.G.* // Hart.-Tech. Mitt. 1974. Bd. 29. № 2, S. 113–119
54. *Lakhtin Yu. M., Kozlovskiy I.S.* Borirovaniye // Termicheskaya obrabotka v mashinostroenii. Moscow, 1967. Pp. 344–350. [Boriding]
55. *Bokiy G.B.* Vvedeniye v kristallokhimiyu. Moscow, 1971. 425 p. [Introduction to the chemistry of crystals]
56. *Mirkin L.I.* Spravochnik po rentgenostrukturnomu analizy polikristalov. Moscow, 1961. 863 p. [Reference book on the x-ray diffraction analysis of polycrystals]

Chapter 4
The Equilibrium Diagram of 'Boron-Iron' Binary System

Abstract The chapter deals with the evolution of studies and construction of the binary equilibrium diagram of B-Fe in the areas of low (<0.25% B) and high (>0.25% B) concentration of boron according to the scientific results from different countries. The properties of structural components in B-Fe diagram are presented. Three diagrams are given: Marder's diagram, the diagram created by Tylkina and Bochvar and the complete diagram proposed by the authors along with the description of reactions in B-Fe system. According to the equilibrium diagram built under the boron accumulation on the surface and saturation of α or γ-solid iron solutions (one tenths of a procent), atoms of iron and boron form the initial areas of boride layers (Fe_2B) by fluctuation. The permanent activity of saturating medium and boron accumulation provides the concentration sufficient for the formation of a FeB boride in the place of fluctuation. Generally, the initial generation of a new phase is determined by the further rebuilding of the lattice in the volume of crystal nucleus. The process is developed in zones of increased concentration of diffusing elements, i.e. along the borders of initial structure (grains, subgrains, blocks, nucleus). This regularity in the growth of boride layers in iron and steels is accompanied by the anisotropy of boron diffusion coefficients in borides which leads to the formation of acicular structure of the layer which is only partially connected to the structure and crystal composition of the substrate.

An equilibrium diagram determines the conditions of temperature and concentration of boriding processes and the laws of structure-formation during saturation. The complete picture of this correlation allows to control the plasticity of obtained boronized layers with the maximum effectiveness.

M.G. Krukovich et al., *Plasticity of Boronized Layers*,
Springer Series in Materials Science 237,
DOI 10.1007/978-3-319-40012-9_4

4.1 The Evolution of the Equilibrium Diagram at High Concentrations of Boron (>0.25%)

One of the first most complete equilibrium diagrams [1] (18% limit of boron content) based on the studies of Hannesen, Chizevsky and Gerdt [3, 4] and Wever and Muller [5] is shown in Fig. 4.1. The experimental data, presented in works [2, 3], is the result of thermal and microscopic analysis. They are similar to each other in many ways but strongly differ in the values of boron solubility in solid iron. Moreover, Hannesen discovered that B-Fe system has a stable boride which melting point is 1350 °C; the formula Fe_5B_2 (7, 19%) was applied to it.

Chizhevsky and Gerdt introduced the formula Fe_2B and determined its melting point as 1325 °C.

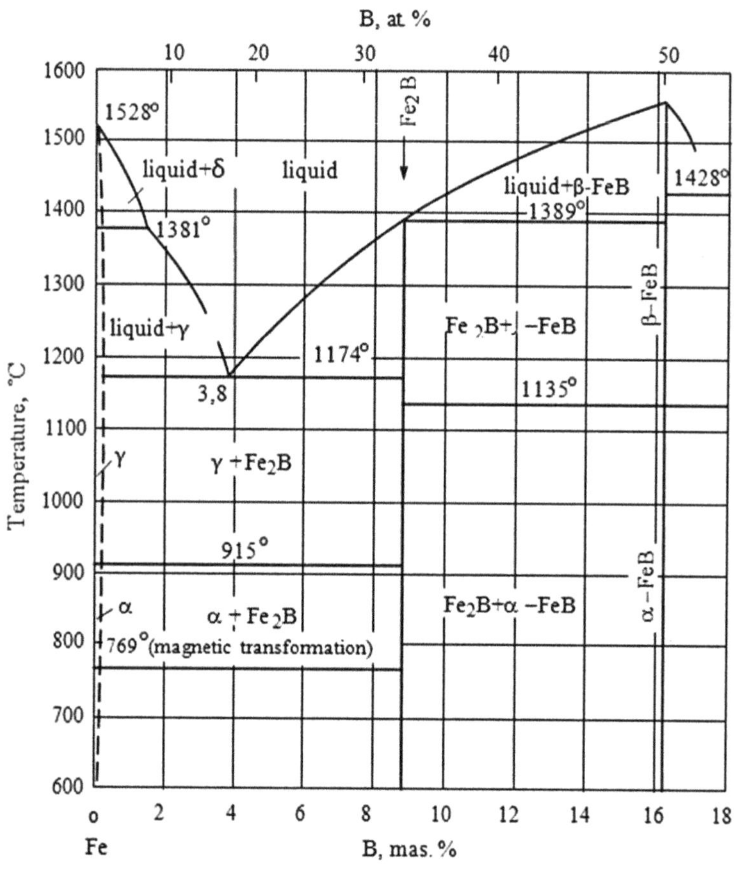

Fig. 4.1 The equilibrium diagram of B-Fe system at high concentrations of boron (>0.25) introduced by Wever and Muller (Hansen) [1, 6]

Boron solubility in Fe_γ was determined by Hanessen: ~0.8% B at 713 °C. Chizhevsky and Gerdt's data showed its solubility as 3.6% at 760 °C; the solubility of boron in Fe_α was 0.08% at the eutectoid temperature.

In contrast with these three authors, Wever and Muller [6] expressed the opinion that boron presence only slightly effects the polymorphic transformations of iron. Later they showed [7] that the diagrams, built by their predecessors, contained mistakes as the studied alloys of "iron and boron" were contaminated by carbon, aluminum and silicon at such proportion that the boron influence was annihilated in the areas of low concentration. Wever and Muller studied B-Fe system by using the methods of thermal, microscopic and X-ray analysis applied to alloys with low concentration of carbon (still with rather high aluminum and silicone content). To eliminate their influence, they studied separately the influence of aluminum and silicone on iron transformation points' positions.

Weber and Muller discovered the presence of chemical compound FeB (16.23% B) in alloys with boron content >8.83%. Thus, the equilibrium diagram of B-Fe acquired two borides—Fe_2B and FeB. The existence of these two borides was proved by the X-ray studies, carried out by Bjurström and Arnfeldt [8]. The obtained results were taken into account by Wever and Muller while correcting the results of Hanessen, Chizhevsky and Gerdt. Eventually they constructed an "ideal" equilibrium diagram for boron and iron alloys free of admixtures (Fig. 4.1).

The analysis of the general diagram leads to the conclusion that in the range of boron content up to ~9%, it is very similar to Fe-Fe_3C system diagram. This part of the diagram has a self-contained austenite area (up to ~915 °C temperature) where a Fe_2B compound (similar to Fe_3C) is formed as a result of peritectic transformation at 1389 °C. The eutectic transformation takes place when the boron content is 3.8% B and the temperature is 1174 °C. This part of the diagram can also be characterized by the temperature of magnetic transformation of two phase (α + Fe_2B) alloys equal to 769 °C. When the boron content is higher than 9%, the system acquires the chemical compound FeB, that congruently melts at the point of 1540 °C and exists in two polymorphic variants with the transformation temperature of 1135 °C: α-FeB up to 1135 °C and β-FeB over 1135 °C.

B-Fe system has always been characterized by the principal question if a row of iron borides consists only of Fe_2B and FeB phases. It was known that all the transition metals had no less than 3 borides and generated borides in contact with boron—phases similar to MeB_2 [9–12]. Such phases are generated by the elements in the groups of palladium and platinum which can be seen as iron counterparts in the sense of their electron structure. These facts allowed Lyakhovich and his colleagues [13] to assume that FeB system is capable of generating diborides. The study showed that alloys, which contained more than 16% B, begin to generate eutectics (FeB + unknown phase) with the temperature rise and later—the primal precipitation of this phase. The unknown phase differed from other known borides by several properties: it was resistant to pickling, had a greyish-blue colour before and after pickling in spirit solution HNO_3 and high microsolidness (Table 4.1). The X-ray diffraction analysis of the alloys containing 15–38.7% B revealed the presence of lines not attributable to the known borides.

Table 4.1 Phase composition, structure and hardness of structural composites of Fe-B alloys [13]

No.	Boron content, mass (%)		Phase composition of the alloy	Structural composites	Hardness of structural composites, HV				
	Calculated	Chemical analysis			α-solid solution	Eutectics	Fe_2B	FeB	FeB_2
1	2.0	2.0	–	α-solid solution + eutectics	233	440	–	–	–
2	4.0	3.6	–	Eutectics + α-solid solution	236	465	–	–	–
3	7.0	5.4	–	Eutectics + Fe_2B	–	580	1680	–	–
4	9.0	6.8	–	Fe_2B + eutectics	–	570	1320	–	–
5	13.0	9.7	Fe_2B + FeB	Fe_2B + FeB	–	–	1500	1850	–
6	16.5	–	Fe_2B + FeB	Fe_2B + FeB	–	–	1505	1850	–
7	20.0	13.9	FeB + Fe_2B	FeB + Fe_2B	–	–	–	1900	–
8	25.0	18.1	FeB + FeB_2	FeB + eutectics (FeB + FeB_2)	–	2350	–	1950	–
9	30.0	22.6	FeB + FeB_2	Eutectics + FeB_2	–	2160	–	–	2750
10	35.0	27.0	FeB + FeB_2	FeB_2 + eutectics	–	–	–	–	–
11	40.0	31.5	–	FeB_2 + eutectics	–	2200	–	–	2800
12	50.0	38.7	–	FeB_2 + eutectics	–	2100	–	–	2740

Note Alloys 1–4—eutectics is formed by phases Fe_2B and iron-based α-solid solution; alloys 8–10—phases FeB and FeB_2; the third eutectics (alloys 11–12) is possibly formed by FeB_2 and boron

The juxtaposition of the calculated and experimental results of X-ray studies (interplane spacing) led the authors to the conclusion: iron di-boride FeB_2 is generated at the concentration 27.9% B. The presence of a large amount of precipitation (up to 0.1 mm) in high-boron alloys made the scientists [14] believe that phases FeB and Fe_2B interact in a eutectic way, i.e. both compounds must melt congruently (without decomposition) in contrast to iron boride FeB2 which melts incongruently.

The obtained results led the authors to build FeB equilibrium diagram which considered the existence of the third iron boride FeB_2 (Fig. 4.2). Moreover, the authors made a proposition that the diagram might also contain some third eutectic transformation at high boron concentrations between FeB_2 and pure boron.

About 10 years later Khan, Kneller and Sostarich announced the existence of Fe_3B in the system B-Fe [14–16]. This compound was obtained as a result of $Fe_{76}B_{24}$ alloy hardening from the liquid state at a cooling speed which suppressed the equilibrium of Fe_α + Fe_2B but allowed to get a material with a crystal structure. The temperature range of existence of this phase was established in the range of 1150–1250 °C. Fe_3B compound has two stable modifications.

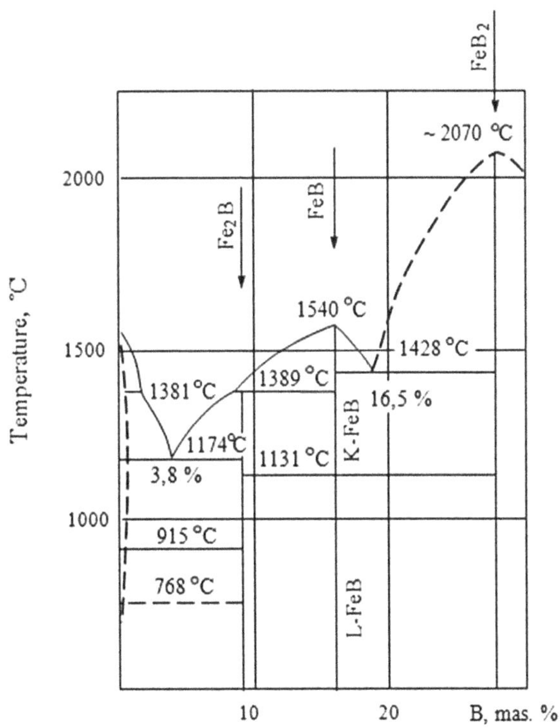

Fig. 4.2 B-Fe equilibrium diagram acknowledging the existense of FeB_2 iron boride (Voroshnin and Lyakhovich) [13]

4.2 The Evolution of the Equilibrium Diagram at Low Boron Concentrations (<0.25%)

The necessity of studying Fe-B system in the area of low boron concentrations arose when it became clear that the influence of boron on the structure and properties of iron is significant only if the amount of additives is small. It was first recognized by Walter in 1921 [17, 18]. He found out that the boron content >0.20% makes steel very brittle and it becomes impossible to analyze the effect of boron. The amount of 0.001–0.10% B refines the grain structure of the cast steel, but the boron amount of 0.007–0.01% makes carbon steel self-hardening.

The first results in studying Fe-B equilibrium state in the area of low boron concentrations were obtained by Wever and Muller. They published [5] some new data on the boron solubility in different iron modifications which can be seen in Table 4.2.

These results demonstrate that the limiting solubility of boron in Fe_α and Fe_Υ doesn't exceed 0.15%. However, as mentioned in work [18], several peculiarities of boron steels showed even then that these numbers were too high. As it was established later by Spretnak and Speiser, boron solubility in Fe_Υ is only 0.003% at 970 °C [6, 19].

Wever and Muller also revealed the fact of point A_3 increase (transformation temperature $\Upsilon \leftrightarrow \alpha$) at 0.15% B concentration up to 915 °C which corresponds to the peritectic reaction $\Upsilon + Fe_2B \leftrightarrow \alpha$. The results of x-ray studies led to the conclusion that a substitutional solid solution is generated by boron and Fe_α. Wever and Muller's results are reflected in Fig. 4.3 as a part of B-Fe diagram in the area of boron concentration up to 0.0025%.

Further studies have added some precision to B-Fe diagram in the area of low boron concentrations. Nicholson's data [20, 21] revealed that eutectoid transformation $\Upsilon + Fe_2B$ takes place at 1165 °C while peritectic reaction $\Upsilon + Fe_2B \leftrightarrow \alpha$— at 911 °C. A_3 temperature for an alloy which contains 0.015%B is higher than for pure iron. Maximum boron solubility in oxygen-free Fe_Υ is 0.021% at 1165 °C and decreases to 0.001% at 911 °C.

The maximum boron solubility in Fe_α is witnessed at 911 °C and is 0.002%; with the temperature fall it decreases less abruptly than the solution of boron in Fe_Υ. Nicholson studied boron solubility in Fe_Υ and Fe_α by annealing boronized iron rods and determining the amount of solute boron by the chemical analysis taking into

Table 4.2 Boron solubility in different iron modifications at different temperatures

No.	Iron modification	Iron solubility, mas. %, at temperature (°C)			
		1381	1174	915	880
1	$Fe_д$	0.15	–	–	–
2	$Fe_г$	0.10	0.15	0.10	–
3	$Fe_б$	–	–	0.15	0.10

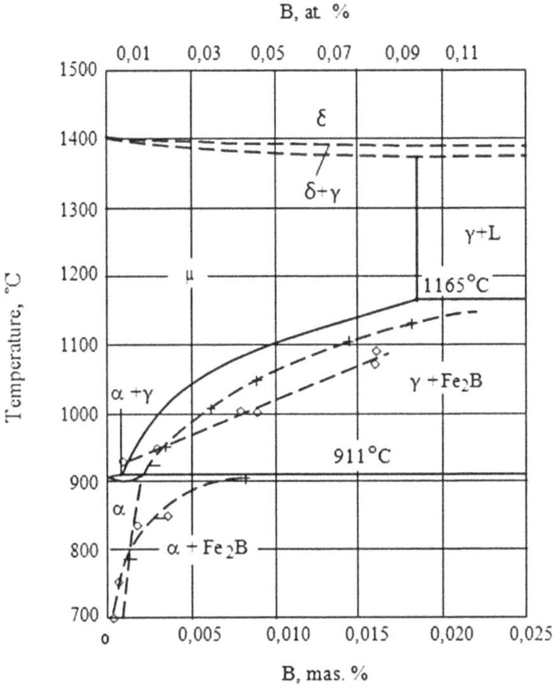

Fig. 4.3 A section of Fe-B equilibrium diagram at low concentrations of boron [5, 7]. Additional lines of boron solubility: - + - according to [21]; -o- according to [28, 29]

account oxygen [17, 20]. As a result, the diagram was built (Fig. 4.3) which also reflected the data of thermal and metallographic analysis of B-Fe alloys.

Around the same time another equilibrium diagram was created by McBride, Spretnak and Speiser [21]. They studied the area of low boron concentrations in B-Fe system using different methods; one of them was sampling of welded Fe_2B-Fe-Fe_2B sandwiches—they were soaked at different temperatures until reaching equilibrium. Metallographic and spectral methods of desorbing boron in iron core were used.

The authors discovered peritectic transformation at the minimum of 835 °C and 0.001% B. Boron solubility in Fe_{γ} at different temperatures in the range of eutectic (1149 °C) and peritectic (906 °C) reactions changed from 0.0182% at 1131 °C to 0.001% at 835 °C.

The Curie point for iron doesn't depend on boron concentration (773 °C), the transformation point for iron (A_{c3} and A_{r3}) decreased in boron presence by ~4 °C. The maximum boron solubility in Fe_{α} was 0.0082% at 906 °C; the temperature fall made it 0.0002% at 710 °C. This work also contained the hypothesis that boron forms a substitutional solid solution in Fe_{α} and an interstitial solid solution in Fe_{γ}. Solute lines of boron in iron are drawn as strokes in Fig. 4.3.

The analysis of works [22–27] leads to the conclusion that boron solubility in ferrite and austenite is significantly low (<0.008% at 900 °C). At the same time $\alpha/\gamma/Fe_2B$ peritectic reaction at ~912 °C is based on a bigger boron solubility in

Table 4.3 Boron solubility in Fe α in relation to temperature [6]

Temperature (°C)	850	835	751	700
Boron solubility, mas. (%)	0.0035	0.0018	0.0006	0.0003

ferrite (than in austenite) at the reaction temperature (as discovered by the authors). However further works revealed a stronger solubility of boron in austenite proving that the reaction is eutectoid by its nature [22, 25].

Higher levels of boron solubility in both modifications of iron (at the temperature below 1000 °C) were obtained in the works of Busby and his colleagues [6, 28, 29]. They determined the solubility by deboriding iron containing more than 0,003% B. The boron concentration was measured using a spectral method. It was found out that boron solubility in Fe_{Υ} at 982 °C is 0.0035%B, in Fe_{α}—the index changes depending on the temperature (as shown in Table 4.3.).

These data are also reflected in Fig. 4.3 in a stroke-dotted solubility curve. The mentioned works also proved the existence of peritectoid transformation which was discovered by McBride and colleagues [21].

Despite some contradictions between the results in the three analyzed works (which are certainly caused by imperfections of determining low quantities of boron), they revealed an extremely limited boron solubility in Fe_{α} and Fe_{Υ} at usual heat treatment temperatures.

4.3 The Complete Equilibrium Diagram of Fe-B Binary System

Figures 4.4, 4.5 and 4.6. Show the three variants of complete equilibrium diagrams of Fe-B system published in 1973, 1996 and in the present book accordingly.

The diagram in Fig. 4.4 was built by Marder in 1973 [30] based on the data of Hansen [31], Eliot [32], Shunk [33] and Pearson [34]. The scientific results of Russian specialists (Portnoi, Levinskaya, Romashov [35]) were also taken into account.

The diagram in Fig. 4.5 was presented by Tylkina and Bochvar in 1996; the basis was data published before 1985 in the book by Kubashevski [36] and in scientific papers published in Russian [13, 35] and foreign [15, 16] journals.

The diagram in Fig. 4.6 is a scheme of the complete equilibrium diagram of Fe-B system built by the authors of this book. It combines the results of thorough analysis presented in the Russian language works published before 1996. Positions of the critical points (temperature and concentration), taken from different works, are quite relative and presented as average arithmetical figures.

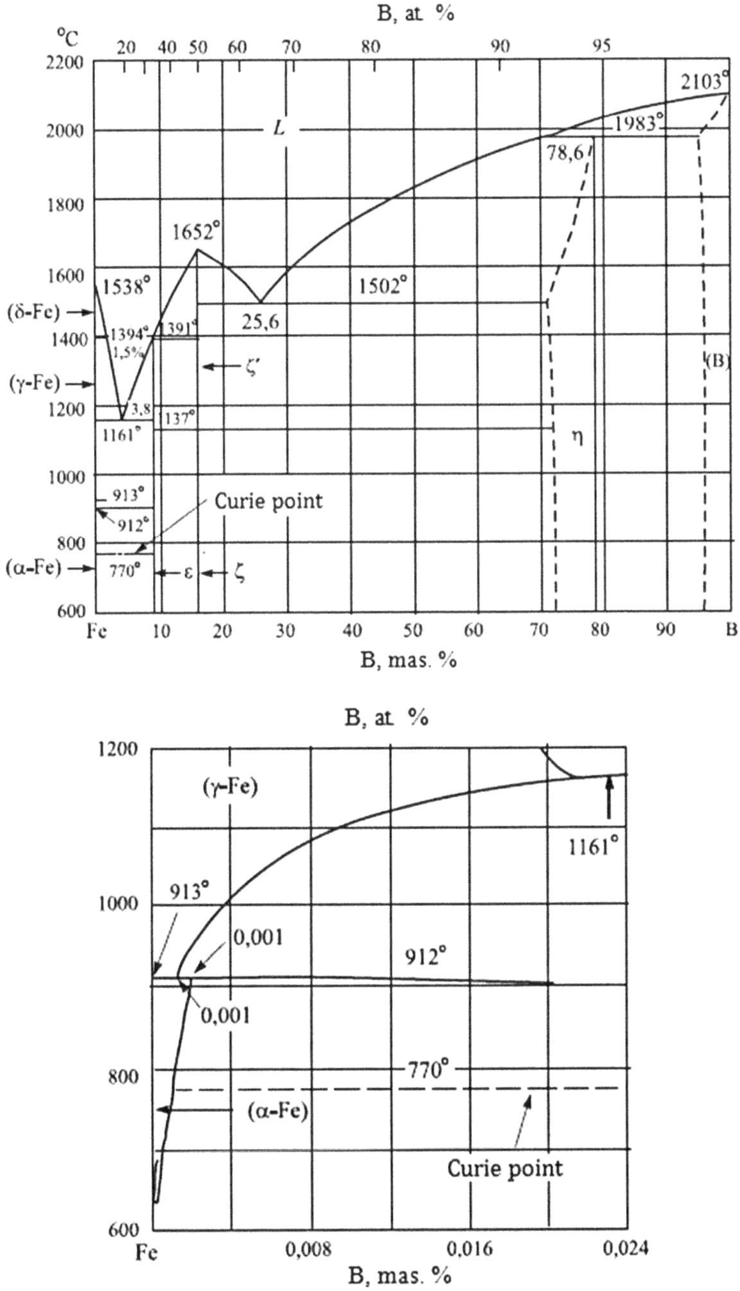

Fig. 4.4 The equilibrium diagram of Fe-B system by Marder

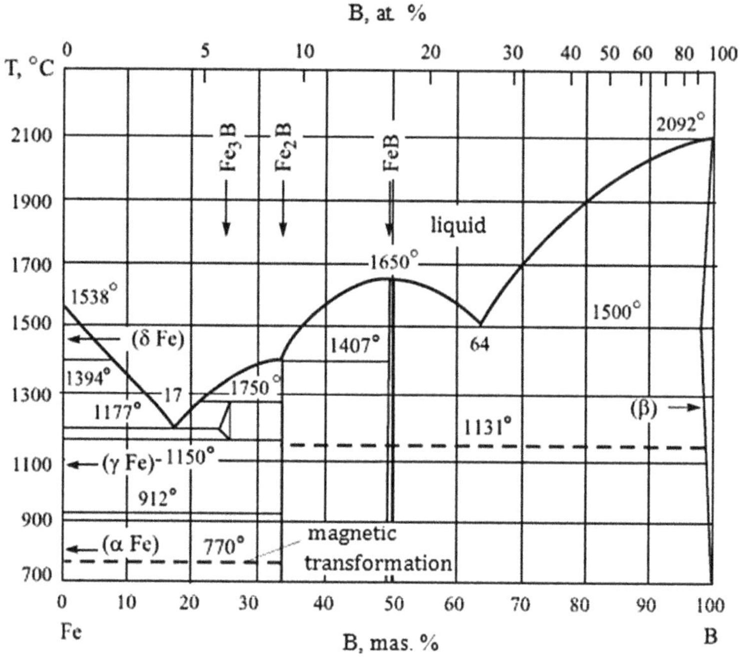

Fig. 4.5 The equilibrium diagram of Fe-B system which takes into account the existence of boride Fe_3B (Tylkina, Bochvar)

4.3.1 Marder's Diagram

The analysis of this diagram (Fig. 4.4) shows that only three phases are indicated in the area of high boron concentrations (>0,25%): two of them (ε and ζ) are with a fixed boron content and one (η)—with a variable one. The two first phases match two borides: ε-Fe_2B and ζ-FeB; their crystal lattices, symbols and prototypes are defined.

Phase ε (Fe_2B) has a body-centered tetragonal lattice with symbol C16 and prototype Al_2Cu. Phase ζ (FeB) has a crystal orthorhombic lattice with symbol B27 and prototype FeB. Phase (η), which is characterized by a variable boron content, is defined as FeB_{19}; its lattice type, symbols and prototypes are not defined. Phase ζ' is a high-temperature variant of phase ζ. It is presented as a result of a polymorphic transformation of phase ζ into phase ζ' at 1137 °C under heating. Unfortunately, the author of this diagram didn't take into consideration the results published by Russian scientists in 1970 [13]; these data are analyzed in detail in Part 2.2.1. If these results had been taken into account, the diagram of FeB in the area of high boron concentrations would have looked completely different—compound FeB_2 would have been introduced.

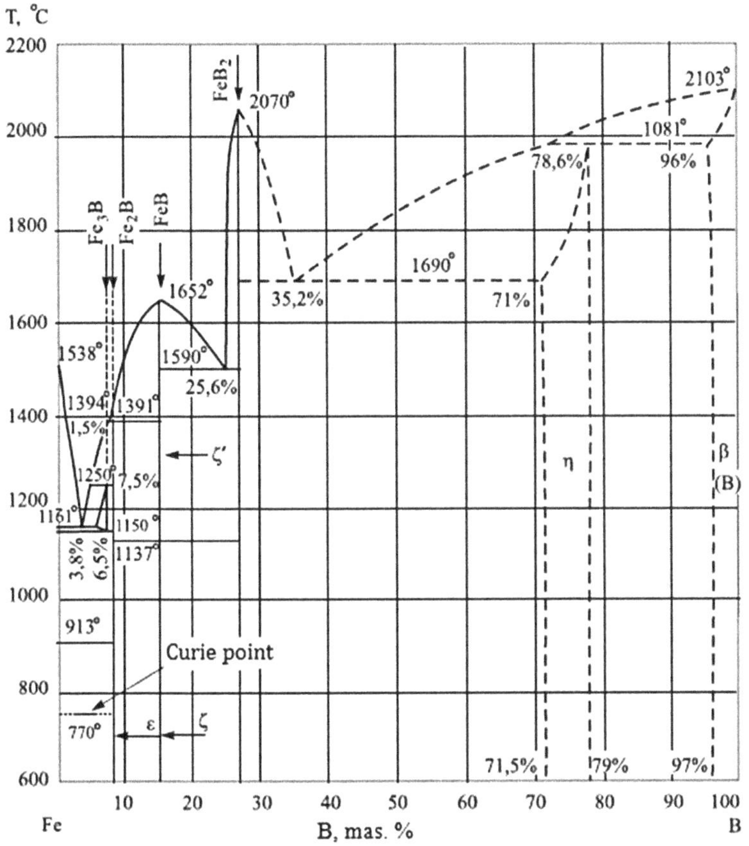

Fig. 4.6 The scheme of Fe-B equilibrium diagram which considers the existence of Fe_3B and FeB_2 borides (built by the authors)

Marder's diagram proves maximum boron solubility in Fe_γ at 1161 °C in the range between 0.021–0.022% and 0.002% in Fe_α at 913 °C in the area of low boron concentrations. The temperature fall decreases boron solubility in Fe_γ and Fe_α; the solubility of boron in Fe_γ at 913 °C is ~0.001% and in F_α at 600 °C it is practically absent.

4.3.2 Tylkina and Bochvar's Diagram

The diagram (Fig. 4.5) was introduced by Russian authors [14] and is supposed to reflect all the results of the studies published in the Russian language until 1985. However, despite the reference made in work [14] to the papers of Voroshnin, Lyakhovich et al. [13], the data on phase FeB_2 aren't reflected in the diagram.

Nevertheless, in comparison to Marder's diagram, it contains a new third phase (Fe_3B) which exists in the range of temperatures between 1150 and 1250 °C. It has two stable variants [15, 16]. Table 4.4 shows the temperatures of borides' formation and non-variant reactions in Fe-B system.

Boron solubility in Fe_α and Fe_γ is quite small. At temperatures 500 and 910 °C, boron solutes in Fe_α in 0.001 or 0.13% B accordingly; in Fe_γ—0.008% and 0.025% at temperatures 910 and 1150 °C. Solubility Fe in (B) is limited by a limiting content of FeB_{49} and is about 2% Fe [14, 37].

The Studies [14–16, 37] are focused on metastable phases in system Fe-B. At a high cooling speed several metastable phases are formed: $Fe_{23}B_6$—a phase with a cubic structure type $Cr_{23}C_6$ (Pearson's symbol cF116, dimension group Fm3 m), a = 1.069 nm; a phase with a rhombic structure (a = 0.6726 nm, b = 0.4311 nm, c = 0.5468 nm); a phase with a body-centered tetragonal structure (a = 0.8620 nm, c = 0,4720 nm) [37]. The data on generation of metastable phase Fe_3B with a structural type Fe_3P can be found in works [15, 16].

4.3.3 *The Scheme of the Complete Equilibrium Diagram (Including the Formation of Fe_3B and FeB_2 Borides)*

Figure 4.6 shows a scheme of the complete equilibrium diagram of system Fe-B that takes into account the existence of 4 experimentally defined iron borides Fe_3B, Fe_2B, FeB and FeB_2. This diagram is a complex of the three equilibrium diagrams represented in Figs. 4.2, 4.4 and 4.5. The basis for the complete diagram is Marder's diagram; two missing borides Fe_3B and FeB_2 were added to it according to the diagrams in Figs. 4.2 and 4.5. Moreover it also reflects the hypothesis made in [13] that the diagram should contain the third eutectic transformation between FeB_2 and pure boron—it should be between FeB_2 and η (FeB_{19}).

Several remarks on borides Fe_3B and η (FeB_{19}) should be made here; they don't completely coincide with the authors' conclusions in [14, 30]. It is known [37] that peritectic reactions are characterized by the formation of solid solutions and not chemical compounds. This makes us question the reliability of reaction 3 in Table 4.4. According to it, a peritectic reaction leads to the generation of a new boride Fe_3B. Moreover the temperature fall leads to a change in solubility by ~2% which is not acceptable while forming a chemical compound with a defined stoichiometric composition. Consequently, reaction 3 should lead to the formation of a solid solution of boron in iron (e.g. solid solution ε) where the temperature fall causes a decline in boron solubility (down to 1177 °C) and then a slight increase (up to 1150 °C). At temperature 1150 °C, ε-solid solution transforms into an eutectoid, which comprises of Fe_γ + Fe_2B (it is also necessary to make changes to reaction 4 in Table 4.4).

Table 4.4 Reactions in the system B-Fe [14]

No.	Reaction	Temperature (°C)	Reference
1	L ↔ FeB	1650	[35]
		1590	[36]
2	L + FeB ↔ Fe_2B	1410	[35]
		1407	[36]
		1389	[1, 6]
3	L + Fe_2B ↔ Fe_3B	1250	[16]
4	Fe_3B ↔ Fe_γ + Fe_2B	1150	[16]
5	L[a] ↔ Fe_γ + Fe_3B	1200	[16, 36]
		1177	[16, 36]
6	L[b] ↔ FeB + (B)	1500	[35]
		1497	[36]

[a] Concentration 17 at. % B
[b] Concentration 64 at. % B

Similar argumentations can be applied to η boride (FeB_{19}) [30]. It seems that in this case it should be a solid solution of iron in boron. It is formed as a result of a peritectic reaction at ~2000 °C and the temperature fall down to a room temperature doesn't effect iron solubility in boron. In this case a peritectic reaction can look like this

$$L + \beta \leftrightarrow \eta,$$

where L—liquid, β—solid solution of iron in boron.

As mentioned in [14], iron solubility in boron is approximately 2 at. %. In connection with the mentioned facts, reactions 3, 4, 5, 6 would be changed and transformed as follows:

3—L + Fe_2B ↔ ε;
4—ε ↔ $Fe_г$ + Fe_2B;
5—L* ↔ $Fe_г$ + ε;
6—L** ↔ FeB + FeB_2.

The complete diagram requires two more additional reactions 7 and 8:

7—L*** ↔ FeB_2 + η (~1780 °C);
8—L + β ↔ η (~1983 °C).

The third eutectics is formed when the boron content is ~34% and the temperature is ~1780 °C; the third peritectics—when the boron content is 78.6% and temperature is 1983 °C. The concentration of iron in η-solid solution can change from ~29 to 21.4% if the temperature falls. If one tries to compare an alternating solubility of boron in ε-solid solution with a limiting solution of boron in Fe_α and Fe_γ and an alternating solubility of iron in η-solid solution with a limiting solubility of iron in β-solid solution, they can see a significantly higher solubility in ε and η (three times higher). These results make scientists doubt the conclusions on the existence of borides Fe_3B and η (FeB_{19}) and require a physical-chemical analysis to explain the fact of an increased solubility.

Hence three out of four borides (Fe_2B, FeB and FeB_2) has precise stoichiometric compositions and nearly full absence of homogeneity areas (~1 at. %). Two of them (FeB and FeB_2) melt congruently, one (Fe_2B)—incongruently. Simultaneously ε-solid solution (or "boride" Fe_3B) exists when the boron content is 6–7% at the range of temperatures 1150–1250 °C and η-solid solution (or "boride" FeB_{19}) must exist at the boron content of ~71–79 mas. % (both at a high—1983 °C—and a normal room temperature). However areas of FeB_2 and FeB_{19}'s existence haven't been experimentally studied.

Following the sequence of phase formation in a diffusion layer when the boron content on the surface is below 20%, according to the diagram at boriding temperature 1050 °C, it can be described like this: under the temperature rise (up to 1050 °C), two processes happen. The first process is an α-iron saturation (below A_3), then in isothermal conditions—γ-iron saturation (over A_3). When boron concentration is ~9%, the saturation process at the isotherm of 1050 °C ends with the formation of tetragonal boride Fe2B. The further saturation leads to the generation of boride FeB (16% B) with an orthorhombic lattice and a boride FeB_2 (27.9%) with a hexagonal lattice.

The X-ray microanalysis shows the stability of all borides' compositions [38]. It allows to conclude that the generation of a diffusion layer, consisting of borides Fe_2B, FeB and FeB_2, is a reaction accompanied with obtaining a chemical compound with an almost precise stoichiometric composition.

The mechanism of a diffusion layer's formation in this case is the following: after saturating α and γ-solid solutions of iron (tenths of a percent), boron and iron atoms form first sections of a boride layer (Fe_2B) by means of fluctuation. This way fluctuations of concentrations are generated on the outer border of the solid solution layer. The more the layer grows, the slower the generation speed is. If a saturation medium and boron atoms accumulation are constantly active, then the concentration of boron items in the areas of the fluctuation turns out to be sufficient for a boride FeB formation. Boron and iron diffusion inside boride phases that is lower than in α- and γ-solid solutions of iron decelerated the generation of a boronized layer, despite the use of high temperatures.

In the general case of the formation of concentration fluctuations, a primary emerge of a new phase is determined by a further restructuring of the lattice in the body of the crystal nucleation center. In case of fluctuation, connected to the rebuilding of the crystal structure and in the body equal to the crystal nucleation center, the generation of a new phase is set by the saturation element's atoms' diffusion. In both cases the process goes in zones of increased concentration of diffusing elements, i.e. on the borders of the initial structure (grains, subgrains, blocks, nucleation centers). This pattern is supported by the anisotropy of boron diffusion coefficients in borides under the growth of boride phases in iron and steels which leads to the formation of acicular structure of a layer (only partly connected to the structure and crystal composition of a substrate).

The formation of new phase particles mandatorily requires an adherence to the principle of size and orientation compatibility (Dankov). The experiments showed that the initial phase orientation of an iron nitride particle under nitriding needs the minimum energy for distortion of the initial crystal lattice [38]. It must be assumed that the same happens under boriding. The limiting saturation of boron austenite by boron leads to the fluctuation transformation accompanied with the generation of a boride. After limiting saturation of Fe_2B boride by boron in separate sections, a FeB boride is generated according to the complete equilibrium diagram.

References

1. *Hansen M.* Struktury binarnyh splavov. Moscow, 1941. 267 p. [Structures of binary alloys].
2. *Hannesen G.* // Zeitschr. anorg. u. allg. Chemie, 1914. №88. p. 257.
3. *Chizhevsky N.P., Gerdt A.I.* // Zhurnal russkogo metallurgicheskogo obshestva. 1915. №1. p. 535.
4. *Chizhevsky N.P., Gerdt A.I., Mikhailovskii I.M.* Sistemy: zhelezo - bor i zhelezo-nikel-bor // Zhurnal russkogo metallurgicheskogo obshestva. 1915. №4. p. 533–559. [Systems: iron-boron and iron-nickel-boron].
5. *Wever F., Muller A.* // Z. anorg. Chemie, 1930. №192. p. 317.
6. *Vol A.E.* Bor-zhelezo (B-Fe) // Stroenie i svoistva dvoinyh metallicheskih sistem. Moscow, 1959. Vol. 1. pp. 679–686.
7. *Wever F. und Muller A.* Uber die Zweistoffsysteme Eisen-Bor und Eisen-Beryllium mit einem Beitrag zur Kenntnis des Zweistoffsys-tems Eisen-Aluminium // Mitt. Kaiser-Wilhelm Inst. Eisenforsch. 1929. Bd. 11. S. 193–223.
8. *Bjurstrom T., und Arnfelt H.* // Z. phys. Chemie. 1929. Bd. 4. S. 469–474.
9. *Samsonov G.V.* Tugoplavkiye soedineniya. Moscow, 1963. 397 p. [High-melting point compounds]
10. *Matyushenko N.N.* Kristallicheskie struktury dvoinyh soedinenii. Moscow, 1969. 302 p. [Crystal structures of binary compounds]
11. *Aronson B., Lundstrom T., Rundqvist S.* Borides, Silicides and Phosphides. L: Methuen, 1965, p. 230.
12. *Tumanov A.T., Portnoy K.I., ed.* Tugoplavkiye materialy v mashinostroenii. Reference book. Moscow, 1967. 392 p. [High-melting point materials in mechanical engineering].
13. *Voroshnin L.G., Lyakhovich L.S., Panich G.G., Protasevich G.F.* Struktura splavov sistemy Fe-B // Metallovedeniye i termicheskaya obrabotka metallov. 1970. №9. pp. 14–17. [The structure of Fe-B system smelts].
14. *Lyakishev N.P., ed.* Diagramy sostoyaniya dvoinyh metallicheskih sistem. Moscow, 1996. Vol. 1. pp. 442–444. [The equilibrium diagrams of binary metal systems]
15. *Khan Y., Kneller E., Sostarich M.* // Z. Metallkunde, 1981. Bd. 72,№8. S. 553–557.
16. *Khan Y., Kneller E., Sostarich M.* // Z. Metallkunde. 1982. Bd. 73,№10. S. 624–626.
17. *Mchedlishvili V.A., Khovrin V.V., transl. Vinarov S.M., ed.* Bor, kalcii, niobii i tsirkonii v chugune i stali. Moscow, 1961. 460 p. [Boron, calcium, niobium and zirconium in cast iron and steels].
18. *Walter R.* British patent №160792 (1921), USA patent №1519388, August 13 (1921).
19. *Spretnak J.W., Speiser R.* // Technical Report № 1, Contract № AF-18, (600), - 94. The Ohio State University Research Foundation, June 1952 // J. Metals. 1953. V. 5, №3. p. 445.
20. *Nicholson M.E.* // Journ. Metals. 1952. V. 4, №2. p. 148.

21. *McBride C.C., Spretnak J.E., and Speiser R.* A study of the Fe-Fe2B System // Trans. Am. Soc. Metals. 1954. V. 46. p. 499–524.
22. *Anvil Kumar Sinha.* Boriding (Boronizing) // ASM Handbook. 1998. V. 4. Heat Treating. p. 437–447.
23. Binary Phase Diagrams / Edited by Massalski T.B. et al. / American Society for Metals. Metals Park. OH, 1986. V. 1. — 356 p.
24. *Brown A. et al.* // Metall. Sci. 1974. V. 8. p. 317–324.
25. *Cameron T.B. and Morral J.E.* // Met. Trans. A. 1986. V. 17A. p. 1481–1483.
26. *Stadelmaier H.H. and Gregg R.A.* // Metall. 1963. V. 17. p. 412–414.
27. *Massalski T.B., Okamoto H., Subramonian P.R. and Kacprzak L.* Binary Alloy Phase Diagrams. Second Edition / ASM International, 1990. V. 1–3.
28. *Busby P.E., Warga M.E., Cyril Wells.* // Journ. Metals. 1953. V. 5, №11. p. 1463.
29. *Busby P.E., Cyril Wells.* // Journ. Metals. 1954. V. 6, №9. p. 972.
30. *Marder A.R.* B-Fe (Boron-Iron) // Metal Handbook. Metallography, Structures and Phase Diagrams / 8th Ed. V. 8. p. 270. 347.
31. *Hansen M.* Constitution of Binary Alloys / 2nd Ed. (prepared with the cooperation of K. Anderko), Mc.Graw-Hill, 1958. 460 p.
32. *Elliott R.P.* Struktura dvoinyh splavov. Moscow, 1970. Vol. 1., 470 p. Vol. 2, 490 p. [The structure of binary smelts]
33. *Shunk F.A.* Constitution of Binary Alloys / Second Supplement. McGraw-Hill, 1965. — 759 p.
34. *Pearson W.B.* Handbook of Lattice Spacing and Structures of Metals / Oxford: Pergamon Press, 1967. V. 2. p. 509.
35. *Portnoy K.I., Levinskaya M.Kh., Romashov V.M.* Diagramma sostoyaniya zhelezo-bor // Poroshkovaya metallurgiya. 1969. № 8. pp. 66–70. [The equilibrium diagram of iron-boron]
36. *Kubashevski O.* Diagramy sostoyaniya dvoinyh sistem na osnove zheleza. Moscow, 1985. 184 p. [The equilibrium diagrams of binary systems based on iron].
37. *Zakharov M.V., Rumyantsev M.V., Turkin V.D.* Diagramy sostoyaniya dvoinyh i troinyh metallicheskih sistem. Moscow, 1940. 232 p. [The equilibrium diagrams of binary and triple metallic systems].
38. *Blanter M.E.* Teoriya termicheskoi obrabotki. Moscow, 1984. 328 p. [The theory of thermal processing].

Chapter 5
Multicomponent Equilibrium Diagrams Used in Boriding Treatments of Steels and Alloys

Abstract The chapter studies the triple equilibrium diagram of B-C-Fe system and its isothermal sections at 700, 800, 900 and 1000 °C which are used for determining the thermal-concentration conditions for boriding and phase composition of estimated layers which are formed under the saturation of iron-carbon smelts. The isothermal section of three-component equilibrium diagrams of smelts like Me-B-Fe (where Me is Cr, Mn, Mo, V, W) which can be used for determining the phase composition of boronized layers in alloy steels and the properties of obtained protective layers. Temperature-concentration conditions of boriding of multicomponent smelts (n > 4) can be simplified by using the schemes of these multicomponent diagrams in traditional coordinates: temperature-concentration of components. The authors propose and analyze the original method for building these schemes of multicomponent diagrams of smelts by using the developed divergent coordination grid, the law of areas (for building the distribution stencil of alloy elements in the foundation of concentration polyhedron) and the rule of eutectic reaction. The use of developed methods resulted in the construction of four-component (Fe-Ni-Cr-B) and five-component (Cr-Mo-Fe-W-B) diagrams, liquidus and solidus surfaces and polythermal cuts of different sections. The obtained schemes of multicomponent equilibrium diagrams were used as a technological blueprint while determining the temperature of boriding and the phase composition of high-temperature smelts and high-speed steels.

Multicomponent equilibrium diagrams have a significant importance in defining temperature and concentration conditions for boriding (a saturating ability of a medium), evaluating phase compositions of boronized layers and choosing temperatures for final thermal processing of carbon and alloy steels.

First studies of triple equilibrium diagrams (Fig. 5.1) appeared in 1901 [1], only a few diagrams were built annually. The situation stayed unchanged till 1925–1927; later the interest for such studies significantly increased and by the end of the 1930s more than 100 diagrams were presented each year.

After a slight decrease in work intensity due to World War II, since 1950 the number of diagrams began to rise again and reached approximately 600 diagrams a

M.G. Krukovich et al., *Plasticity of Boronized Layers*,
Springer Series in Materials Science 237,
DOI 10.1007/978-3-319-40012-9_5

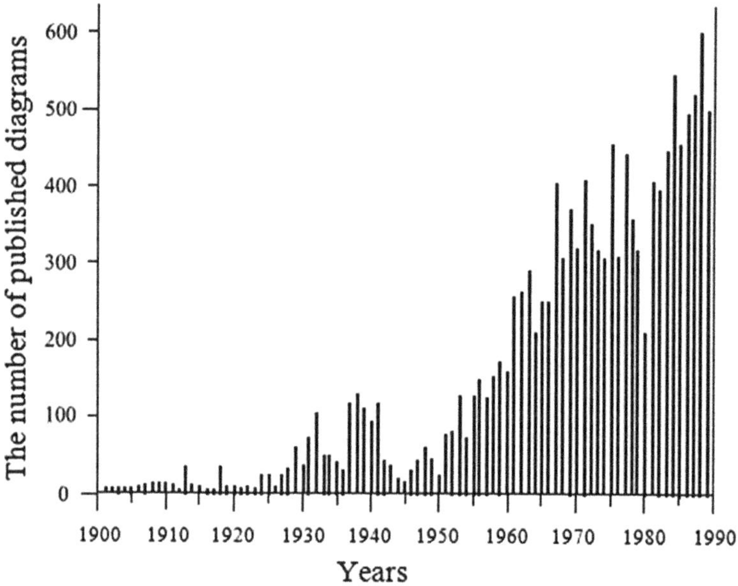

Fig. 5.1 The number of triple equilibrium diagrams, published over the years [1]

year by the end of the 1980s [2]. The encyclopedia on phase equilibrium diagrams of triple systems, published in 1994 [3], contained more than 15,000 diagrams of more than 7380 triple systems.

5.1 The Equilibrium Diagram of 'B-C-Fe' Triple System

B-C-Fe triple system is investigated rather poorly. One of the first relatively complete studies on B-C-Fe system was carried out by R. Vogel and G. Tamman in 1922 [4, 5]. They studied the triple diagram as a system consisting of two binary systems Fe-Fe_2B and Fe-Fe_3C and one quasi-binary system Fe_3C-Fe_2B (Fig. 5.2). The content figures of 2.9% B and 1.5% C allowed them to find triple eutectics with the melting point ~1100 °C. However, the result was obtained in smelts with uncontrolled purity with 0.2% B and higher.

Later, Busby, Warga and Wells [4, 6] while investigating boron diffusion in steels with 0.43% C, discovered that carbon doesn't influence boron solubility in Fe_α and Fe_γ. They also revealed that, if austenite is saturated with carbon, boron solubility may slightly decrease. These results had a big significance as they showed that much of the data for B-Fe system can be applied to the triple smelts of B-C-Fe system.

Another important step in this field was the discovery of iron boron-carbide in 1954 [4, 7]. This carbide has a composition which can be roughly reflected by the

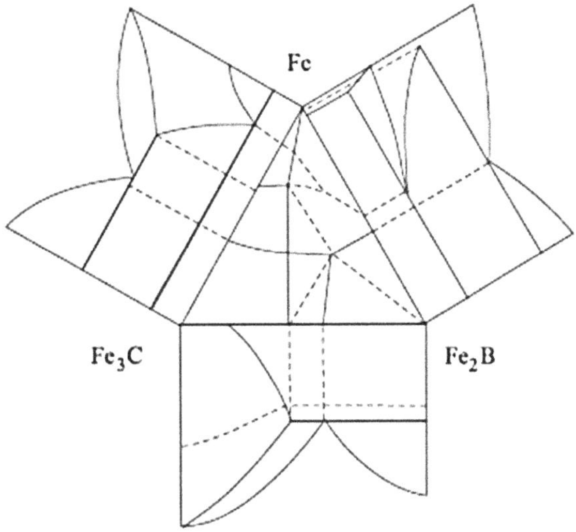

Fig. 5.2 The equilibrium diagram of B-C-Fe triple system [4, 5]

formula $Fe_{23}(C,B)_6$. Its difference from other known carbides is that boron substitutes carbon. In other complex carbides, that are found in steels, an alloying element normally substitutes iron. Three years later M. E. Nicholson [8] discovered that up to 80% of carbon in cementite at 100 °C can be substituted by boron; the corresponding formula for cementite would be $Fe_3C_{0,2}B_{0,8}$. The substitution of carbon by boron leads to the change in the period of the crystal lattice. As they showed, a lot of data for B-Fe binary system can also be applied to triple smelts of B-C-Fe system: one can witness the compression of the orthorhombic lattice by axis 'a' and'c' and its expansion by axis 'b'. Simultaneously with the increase of boron content, the magnetic moment of saturation and the value of Curie point increase.

Thus, boron-carbide is a boron-cementite where a part of carbon atoms in the cementite lattice are substituted by boron atoms [9–11].

Carroll and his colleagues found out that boron-carbide $Fe_{23}(C,B)_6$ is isomorphic with cubic chrome carbide $Cr_{23}C_6$ (D84 structure type). This carbide was studied in more detail by Stadelmaier and Gregg [3, 12] and also by Borlera and Pradelli [13, 14]. The aforementioned authors discovered that $Fe_{23}(C,B)_6$ melts congruently at the transition from a carbon saturated to a boron saturated compound after hardening at 800 °C and that the lattice period 'a' changes from 1.0594 to 1.0628 nm. The works of the latter authors [14] (Fig. 5.3a–d) contain isothermal sections of the diagram at different temperatures: 700, 800, 900, 1000 °C. The analysis of these sections leads to several conclusions.

The analysis of the section at 700 °C shows that $Fe_{23}(C,B)_6$ phase has a composition which is reflected by the following formula—$Fe_{23}(C_{0,73}B_{0,27})$; it is in the equilibrium with boron-cementite $Fe_3(C,B)$ and Fe_α. $Fe_{23}(C_{0,44}B_{0,56})$ phase is in the equilibrium with Fe_2B and $Fe_3(C,B)$.

The analysis of the section at 800 °C leads to the conclusion that phase $Fe_{23}(C,B)_6$ is present at a big range of carbon and boron content, between $Fe_{23}(C_{0,38}B_{0,62})$ and

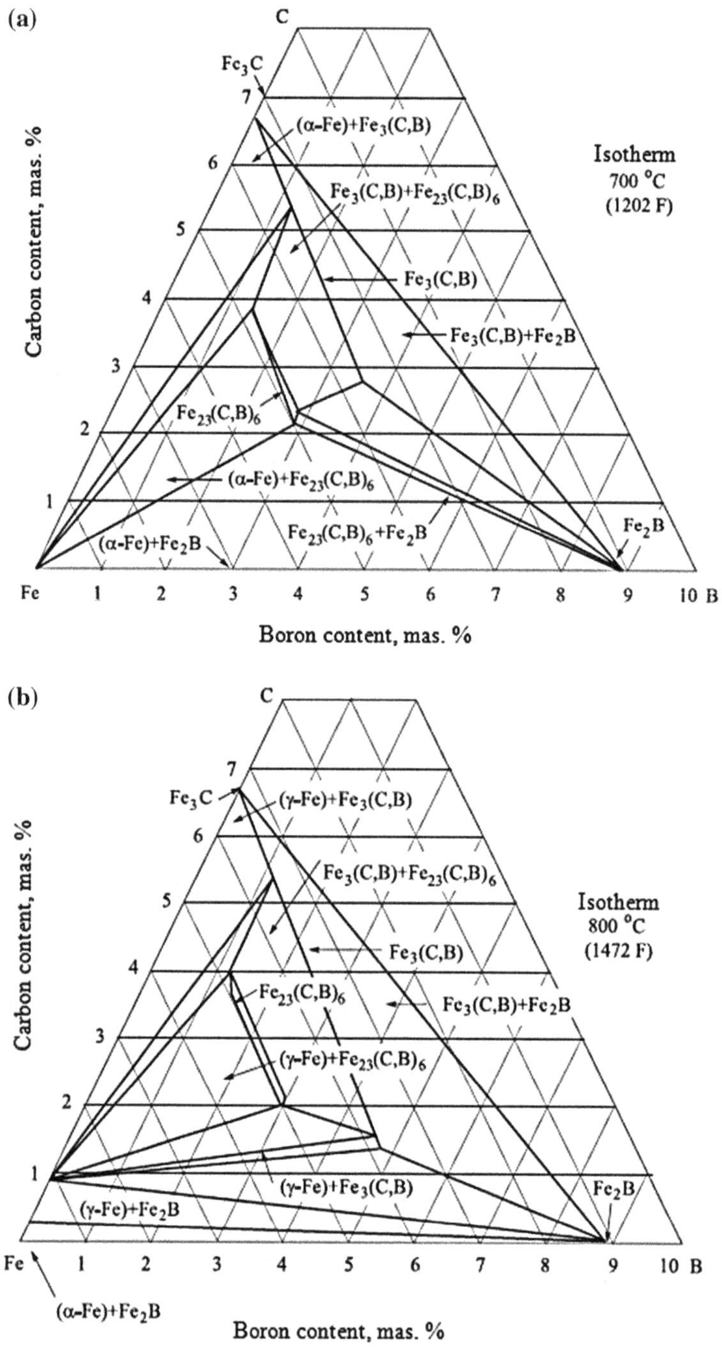

Fig. 5.3 **a** Isothermal sections of the triple equilibrium diagram of B-C-Fe at 700 °C [14], **b** Isothermal sections of the triple equilibrium diagram of B-C-Fe at 800 °C [14], **c** Isothermal sections of the triple equilibrium diagram of B-C-Fe at 900 °C [14], **d** Isothermal sections of the triple equilibrium diagram of B-C-Fe at 1000 °C [14]

(c)

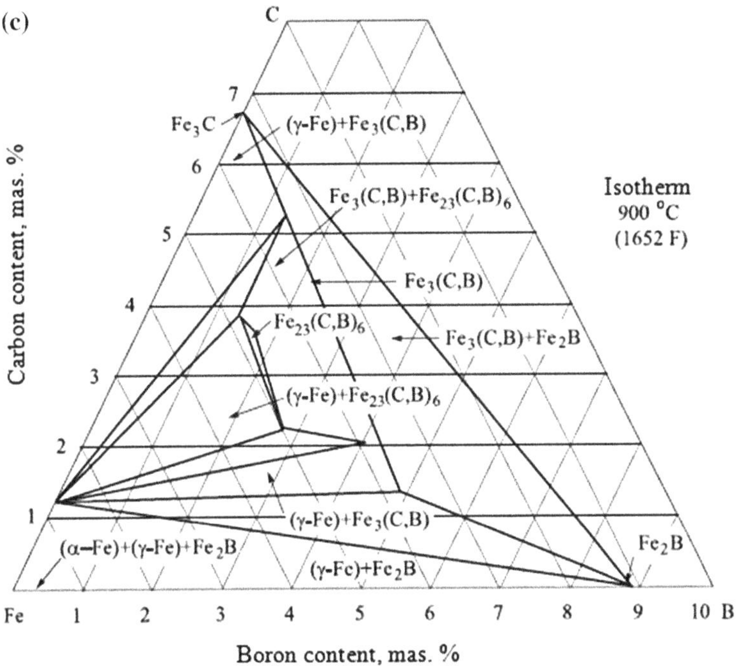

(d)

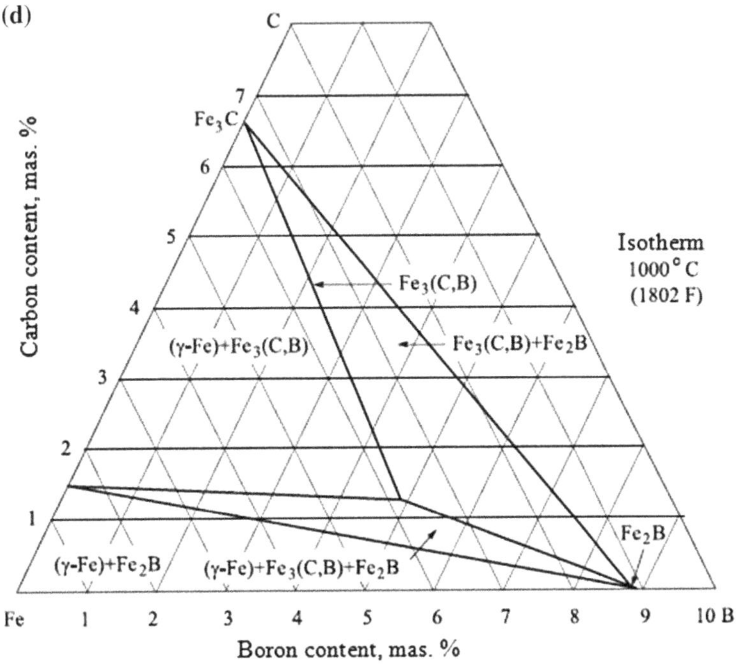

Fig. 5.3 (continued)

$Fe_{23}(C_{0,77}B_{0,23})_6$ and is in the equilibrium with phases Fe_γ and $Fe_3(C,B)$. Phase $Fe_{23}(C,B)_6$ is stable up to 965 + 5 °C.

5.2 Three-Component Diagrams of Boron-Containing Transition Metals

Boriding of alloy steels that contain chrome, nickel, wolfram, molybdenum and other elements requires two- or three-component diagrams with these elements to carry out a preliminary evaluation of phase compositions for boronized layers. The evaluation of mechanical properties of boronized layers should include an awareness of interconnection between metal and boron atoms and a type of the boride crystal lattice. In particular, the electron structure of boron and transition metals leads to the conclusion on the following bonds in borides: Me-Me form a mixed covalent-metal type; Me-B form a mostly metal type and in some instances—a covalent-metal type; B-B form a mostly covalent type [15, 16]. Thus, the solidness, ductility and electrical conductivity of boride phases are determined by the electron structure and types of formed bonds. Moreover the increase in number of B-B bonds makes the configuration of structural elements derived from boron atoms in crystal lattices of borides more complex: from simple isolated atoms or zigzag chains to two-dimension nets or three-dimension carcasses. As a rule this process is accompanied by the decrease in ductility and the increase in solidness. Consequently, the improvement of boronized layers' ductility requires such concentration conditions on the surface which facilitate the formation of low borides.

The establishment of conditions for alloy steels and smelts' boriding is preceded by analyzing either the equilibrium diagram which contains iron and the diagrams which contain alloy elements and boron. The prognosis on the boride formula and stages of their alloyness must be based on the knowledge of isomorphism of boride crystal lattices and mutual solubility of borides. The assumption must be made that the formation of one or another basic boride under smelts' boriding will depend on the level of affinity to boron and the amount of this basic element in the smelt.

Fe-Ni-B system is characterized by the unlimited mutual solubility of isomorphic borides Fe_2B-Ni_2B and the limited solubility of FeB-NiB [17, 18] which proves the possibility of boride formation on the basis of either Fe or Ni in conditions of boriding. However, the generation of more complex borides (like $Fe_3Ni_{20}B_6$, $Fe_{4,5}Ni1_{8,5}B_6$; $Cr_2Ni_3B_6$, Cr_3NiB_6 in the presence of chrome) is not excluded [19].

The following remark must be made: some borides which are described by one stoichiometric correlation can have different structure types. E.g., compound Ni_4B_3 in conditions of boron deficit may form a modification with an orthorhombic crystal lattice o-Ni_4B_3 (41.4 at. % B), in conditions of boron excess- a modification with a monoclinic lattice m-Ni_4B_3 (43.6 at. % B) [17].

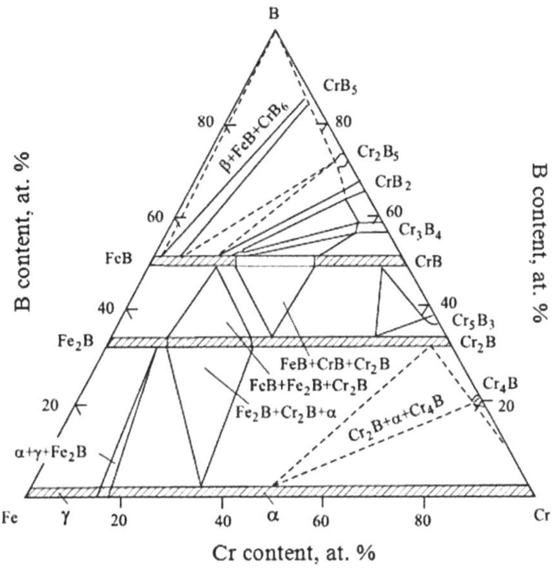

Fig. 5.4 The isothermal cross section of Fe–Cr-B system at 1100 °C

Fe–Cr-B system at 1000 °C in the equilibrium with γ-Fe contains α-Fe and Fe_2B phases. The equilibrium with α-phase is witnessed with γ-Fe, Fe_2B, Cr_2B and Cr_4B phases. Phases Fe_2B and especially Cr_2B form wide areas of solid solutions (Fig. 5.4) [25]. Triple intermediate phases in this system at the range of boriding temperatures between 700 and 1250 °C are not generated.

Cr–Mn-B system at 800 °C shows the existence of an unlimited range of solid solutions (Cr, Mn)B_2, (Cr, Mn)$_3B_4$ and between borides Cr_2B and Mn_4B (Fig. 5.5)

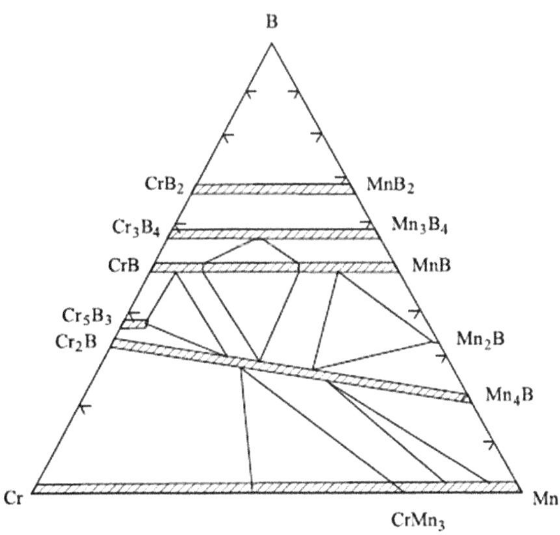

Fig. 5.5 The diagram of phase equilibriums of Cr–Mn-B system at 800 °C

[26]. Cr_5B_3 phase solutes 0.08 mass units of Mn_5B_3 boride; CrB—0.20 mass units of MnB. In its turn MnB solutes 0.4 mass units of CrB.

Another interpretation of Mn-B double equilibrium diagram's isothermal section at 1025 °C shows the existence of an unlimited range of solid solutions $(Cr, Mn)_2B$ instead of Cr_2B-Mn_4B. Moreover, solid solutions of chrome mono-borides and manganese were discovered; their border composition was: $Cr_{0,46}Mn_{0,54}B$ and $Mn_{0,60}Cr_{0,40}B$ [27].

A two-phase area between solid solutions on the basis of CrB and MnB exists in the compounds $Cr_xMn_{1\text{-}x}B$ ($0.35 < x < 0.80$) [28].

Cr-Mo-B system contains an unlimited range of solid solutions $(Cr,Mo)B_2$ [29]. The section CrB2-Mo has a compound Cr_2MoB_4 and two eutectics with melting points of 1960 and 2120 °C, which significantly exceed the temperature of boriding. A slight molybdenum solubility in CrB_3 and more noticeable CrB_2 solubility in molybdenum are also witnessed [30, 31]. The study of an isothermal section at 1000 °C allowed to discover another phase- $(Cr,Mo)_2B$—which is in the equilibrium with a continuous range of solid solutions (CrMo) [32]. Phase equilibriums at 1000 °C don't noticeably differ from the equilibrium at 1400 °C (Fig. 5.6) [33]. The discovered phase is a part of the solid solution on the basis of Cr_5B_3, a compound Cr_2MoB_4—of the solid solution on the basis of Cr_3B_4.

Cr-V-B system holds a continuous range of solid solutions of $(V,Cr)B_2$, (V,Cr)B and $(V, Cr)_3B_4$ (Fig. 5.7) [29, 33]. There is also a limited solubility between borides V_3B_2 and Cr_3B_2, Cr_5B_3 and V_5B_3. $V_{0.2\text{–}0.15}Cr_{1.80\text{–}1.85}B$ triple compound has the following structure type- $CuAl_2$.

Cr-W-B system has mutual solid solutions (Fig. 5.8). Chrome borides solute: Cr_5B_3—0.72 mass units of W_5B_3; CrB—0.74 mass units of W_5B_3; wolfram borides solute: W_2B—0.92 mass units of Cr_2B; WB—0.14 mass units of CrB and W_2B_5—0.17 mass units of CrB_2 [34, 35].

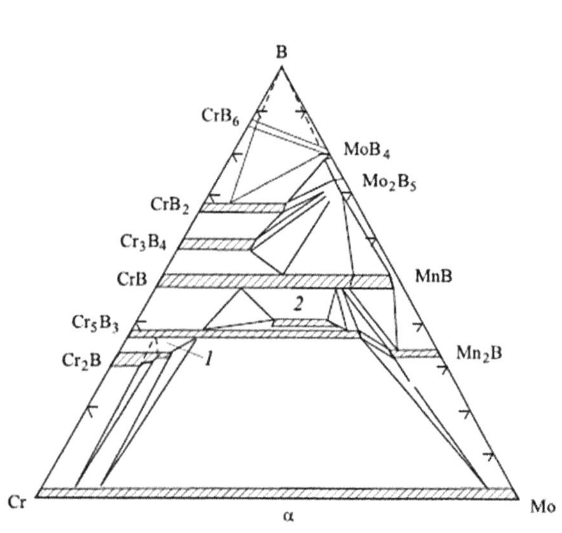

Fig. 5.6 Phases equilibrium diagram of Cr-Mo-B system at 400 °C

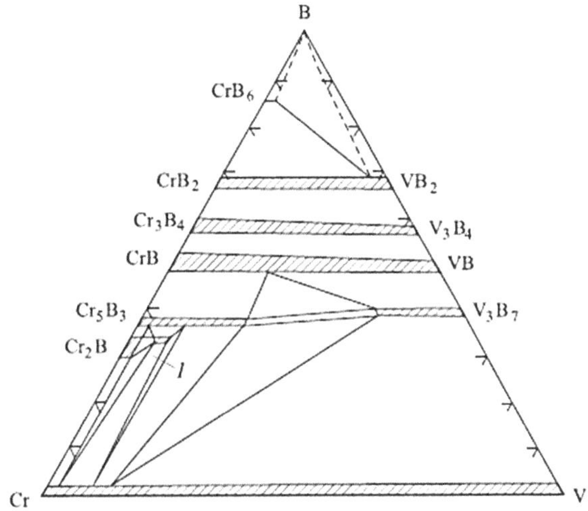

Fig. 5.7 Phases equilibrium diagram of Cr-V-B system at 1400 °C

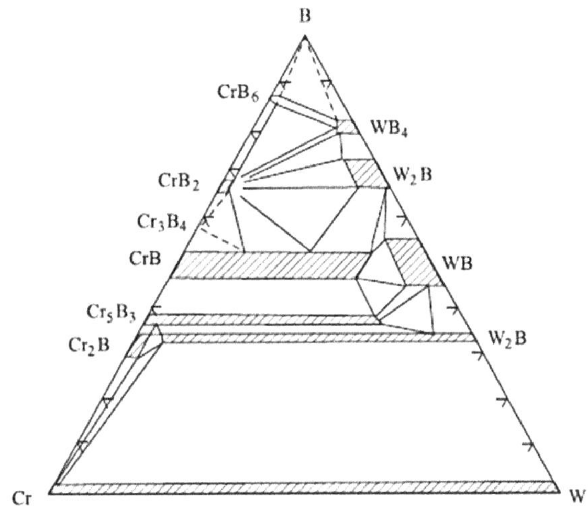

Fig. 5.8 Phases equilibrium diagram of Cr-W-B system at 1500 °C

In Fe-V-B system, V_3B_4 boride solutes up to 0.70 mass units of Fe_3B_4, V_3B_2—up to 0.08 mass units of Fe_3B_2, FeB—up to 0.10 mass units of VB. The isothermal section is built on the data of x-ray diffraction analysis and microstructural analysis (Fig. 5.9) [36].

Thus, the mutual solubility of most borides affects the analysis of phase compositions for boronized layers obtained in alloy steels and smelts under boriding; it must involve the alloying level of these layers and the character of changes happening to their mechanical properties, i.e. solidness, brittleness and ductility.

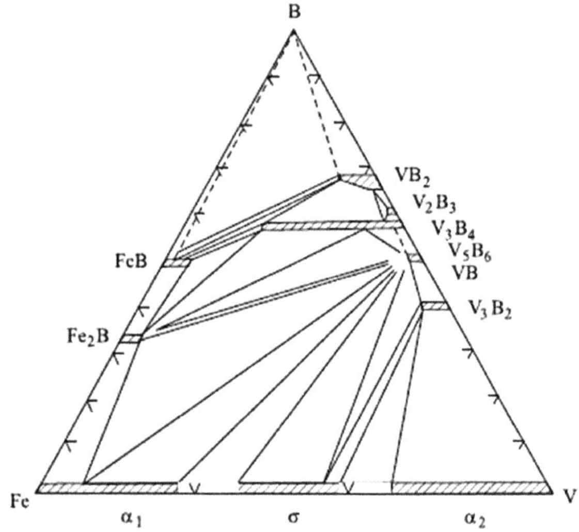

Fig. 5.9 Phases equilibrium diagram of Fe-V-B system at 800 °C

5.3 Four- and Five-Component Equilibrium Diagram Schemes for the Alloys Used in Boriding

Multi-component equilibrium diagrams serve as a foundation for discovering new alloys, improving existing ones and optimizing the technologies of their processing. Simultaneously there are certain obstacles in building multi-component diagrams that contain more than 3 elements; most of the metal and salt systems are not provided with these kinds of diagrams yet.

Multi-component equilibrium diagrams have a special significance for chemical-thermal processing of alloy steels and alloys as the alloys in surface layers obtain alternating chemical compositions and multi-phase structures under this treatment. A multi-component diagram allows professionals to predict a phase composition and layer's structure, find ways to control these properties by changing an amount, form and manner of the distribution for phases and structure components.

As a rule, multi-component diagrams are built in coordinates "concentration-temperature" with a fixed amount of components in all apexes of the concentration triangle; in some cases three-dimensional figures (tetrahedron etc.) are built for each temperature level. A complete representation of transformations in the system requires a large corpus of these diagrams. Moreover, such diagrams are difficult for perception and have very little use as a working technological plan.

5.3.1 *The Principles of Building Multi-Component Equilibrium Diagram Schemes in Traditional Coordinates*

One of the issues in representing multicomponent equilibrium diagrams of salt and metallic systems graphically is a creation of dimensional figures which are easy to perceive and can be used as a technological plan.

The search for this problem's solution hasn't led to any positive results. E.g., Jenike square [37] is used for building equilibrium diagrams of salt mutual triple systems where one of the exchange reactions goes between two salts without a common ion. However it is not effective for building multi-component equilibrium diagrams for metallic and most salt systems.

Bocke-Schout's method [38, 39] leads to obtaining a tetrad which represents the composition and state of a system in 2d, although it comes from a 4 dimensional figure.

The systems of Lodochnikov [40] and Bochvar [41] lead to a geometrical representation of multi-component systems' contents in 2d and are not useful in practice. The proposed ways of improving the general approach [42, 43] result in acquiring systems of flat projections instead of working diagrams themselves; the systems of flat projections obtained after a double planning are also difficult to use as working plans for developing technological processes.

The foundation for a new representation of equilibrium diagram schemes is a widely used method of putting them in a 3 dimensional space: two coordinates describe the change in the components' concentration. These axis are not perpendicular if the number of elements isn't equal to 4 (they form an obligue-angled coordinate system). The third coordinate is temperature; the axis is perpendicular to other axis. Thus, the basis for such schemes is regular flat geometric figures-squares, pentagons, hexagons etc. Polygon's apexes belong to particular components, and sides of a three-dimensional figure (prism) are double equilibrium diagrams. The general regularities in spreading certain components on the diagram (including the smooth decrease of its content with the increase in distance to the top) are witnessed; although it's necessary to point out several peculiarities which reflect the nature of components' distribution and allow to use one universal law for their determination.

In particular, the volume of component A (Fig. 5.10) changes from 100 to 0% at sides AB and AD. Simultaneously its volume on other sides of n-angle figure is 0. Consequently, component A is spread by the square of n-angle from two sides at $n = 4$ from AB to AD (Fig. 5.10), at $n = 5$ from AB to AE and at $n = 6$ from AB to AF (Fig. 5.11); in the volume of the whole diagram—from two corresponding sides of n-angle prism. Such interpretation allows to conclude that the regularity of system's components' spread is defined by some areas in the isothermal section of n-angle figure, which can be defined by the general area rule [44, 45].

This rule states that the total volume of all elements in any point of a concentration polygon is corresponding to the area of this polygon, and the concentration

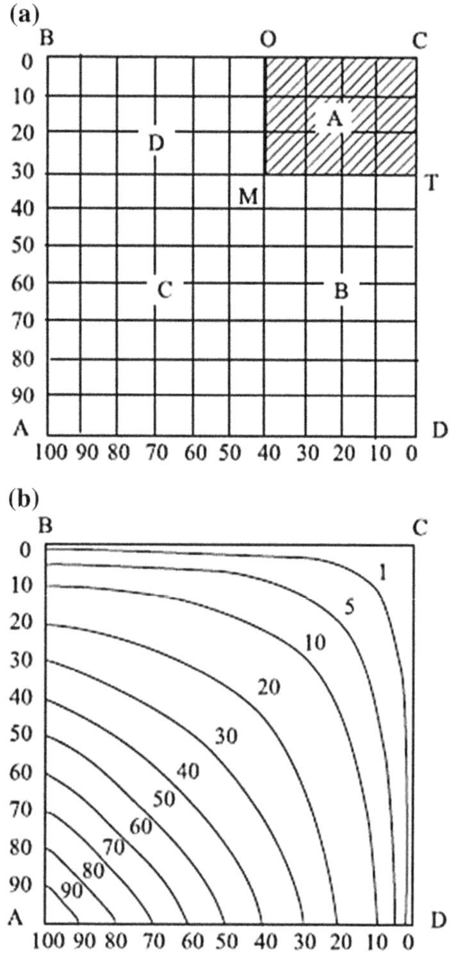

Fig. 5.10 The foundation for the 4-component equilibrium diagram scheme (**a**) and the concentration distribution of component A (**b**) (a distribution stencil for each element)

of a certain element in a figurative point is corresponding to a part of the total area defined in percent. This area subtends to the top of the initial element and lies between the lines of the grid chart that go through this figurative point and polygon sides; it is defined in percentage of the total polygon's area.

To sum up, we can establish that Krukovich's divergent concentration grid chart describes regularities of each element's distribution in the most precise way. This grid can be calculated and built geometrically [44, 45].

The calculation involves the determination of the divergence angle in grades (α) of each chart's line from the previous one. The formula is following:

$$\alpha = (1 - \frac{4}{n})\frac{180}{c} \quad (5.1)$$

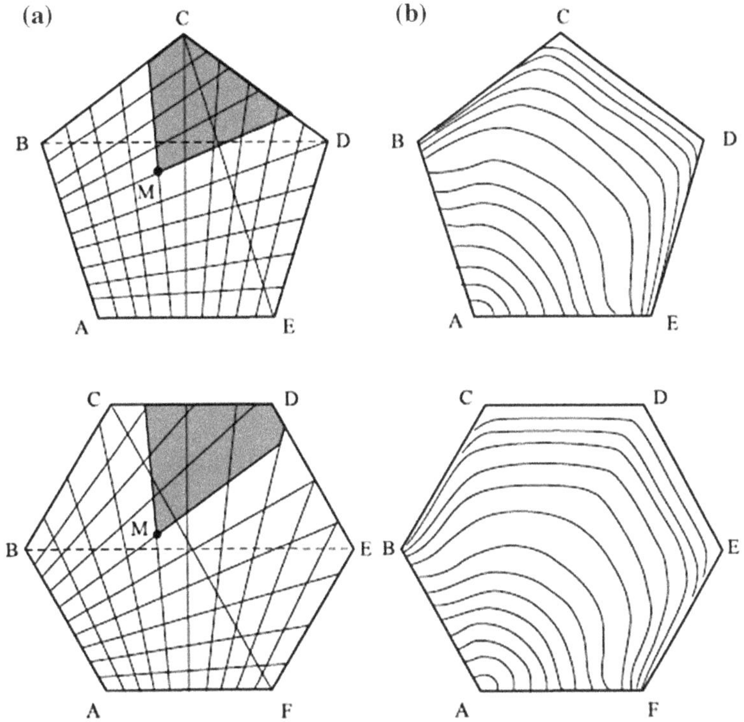

Fig. 5.11 The foundations for 5- and 6-compound equilibrium diagram schemes with 2-dimension divergent grid chart (**a**) and the regularity of concentration distribution of a component A (**b**) (distribution stencils for all elements)

where *n*- a number of components in the system (polygon sides); *c*- a number of points on the concentration scale.

Obviously, the divergence angle is 0 when $n = 4$, i.e. a 4-component system's concentration grid is rectangular and parallel to the square's sides. A divergent grid chart for each element is built from sides on n-angle figure which contain a particular element. E.g., element A—from sides AB and AE at n = 5; AB and AF at n = 6 (Fig. 5.11a).

The amount of element A in a figurative point M for a 4-component system is corresponding to the area of MOCT square (Fig. 5.10a) and is 12% (0.4 × 0.3 × 100%). The number of other elements in this point is: B = 28% (0.7 × 0.4 × 100%); C = 42% (0.6 × 0.7 × 100%); D = 18% (0.6 × 0.3 × 100%). In 5- and 6-component systems the amount of A in M would correspond to the area of the dashed section (Fig. 5.11a). This area is calculated geometrically by dividing a polygon into simple figures (e.g. triangles). The calculated distribution of component A by the areas of concentration polygons may be represented in iso-concentration lines: in the form of a stencil which is similar to all of the system's elements (Figs. 5.10b and 5.11b). By applying a stencil to each top we can determine the amount of elements in the figurative point.

The disadvantages of a given grid chart are the following:

- the irregular distribution of elements by the area of the concentration polygon;
- the absence of several alloys with even correlation between three elements for $n = 4$, 4 elements for $n = 5$ etc.

The diagrams, built for different systems, are called equilibrium diagrams.

The common regularities for building equilibrium diagrams were the foundation for building 4-, 5- and 6-component schemes of equilibrium diagrams for components which form eutectic mixes; isothermal and polythermal sections of these diagram schemes were analyzed [45].

Krukovich's three-dimension divergent chard grid may be used for volumetric figures and dimensions with a certain divergence angle for all axis in a dispersed space.

5.3.2 Building the Equilibrium Diagram Scheme for Fe–Cr-Ni-B

Boriding of multicomponent alloys (e.g. super-alloys) and high-alloy steels (e.g. high-speed steels) are accompanied by significant difficulties in the preliminary choice of saturating mediums, i.e. their saturation ability, and the choice of saturation temperatures due to the possible risk of fusing of obtained diffusion layers. Another difficulty is connected to thermal processing of these alloys and heating at high temperatures in particular. Solving these problems is limited by the absence of corresponding multicomponent equilibrium diagrams for alloys in the form of working plans and the lack of sufficient technological developments.

In accordance with the described rules for building schemes of multicomponent equilibrium diagrams and taking into account regularities of structure forming in double and triple systems Fe-B, Ni-B, Cr-B, Cr-B-Fe, Cr-Ni-B, Ni-Fe-B [17, 18, 21, 22] and the regularities of structure forming under boron diffusion saturation of Fe-Ni-Cr alloys, the scheme for a 4-component equilibrium diagram for Fe-Ni-Cr-B alloys was created (Figs. 5.12 and 5.13) [44, 45].

The main distinction of the obtained diagram scheme is its well-defined correlation with the concentration distribution of boron due to the high affinity of Cr, Ni, Fe to boron. The liquidus surface of the 4-component system at different mutual positions of components can be seen in Fig. 5.12 (without Fe_3B and FeB_2).

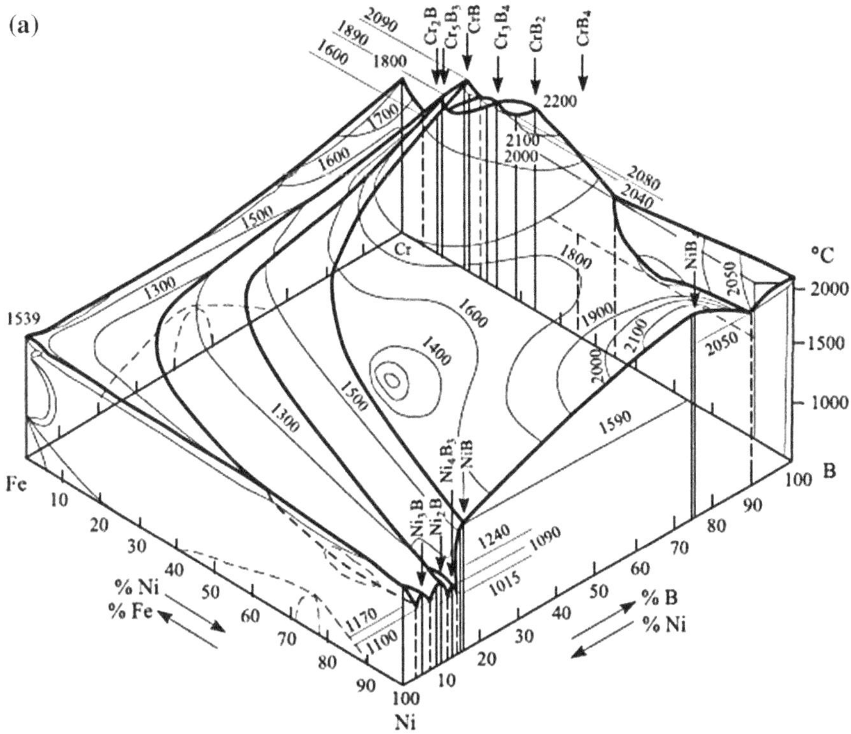

Fig. 5.12 **a** The general view of the liquidus surface in the 4-component system; a. the position of Ni-Fe–Cr-B elements [44, 45], **b** The position of Fe-Ni-Cr-B elements [44, 45], **c** The position of Fe–Cr-Ni-B elements [44, 45]

The areas of low boron concentration (3.5–5.0 mass%) is a medium for the formation of eutectic mixtures on the foundation of solid solutions and low borides in all double diagrams; consequently, the fourth diagram would also reflect the generation of the eutectic mixture. The eutectic temperature, calculated according to the temperature rule of the eutectic reaction [45, 46], is 1022–1035 °C at various positions of components and is hypothetically located in the area of the following components' correlation (mass%): 12–25% Fe, 50–60% Ni, 4–11% Cr, 4.1–4.8% B.

Fe_2B and Ni_2B borides have iso-structural crystal lattices and form a continuous range of solid solutions. Simultaneously iron and nickel are soluted in the rhombic crystal lattice of boride Cr2 (Cr_2B solutes 0.60 of mass units of Fe_2B; Fe_2B solutes 0.18 of mass units of Cr_2B at 1000 °C [17]). Therefore this interaction is reflected by iso-structural and iso-concentration lines which join melting points of Fe_2B, Ni_2B and Cr_2B. Thus the system between these phases contains mutual solid solutions which ensure the smooth change in the temperature of crystallization at the transition from one phase to another.

(b)

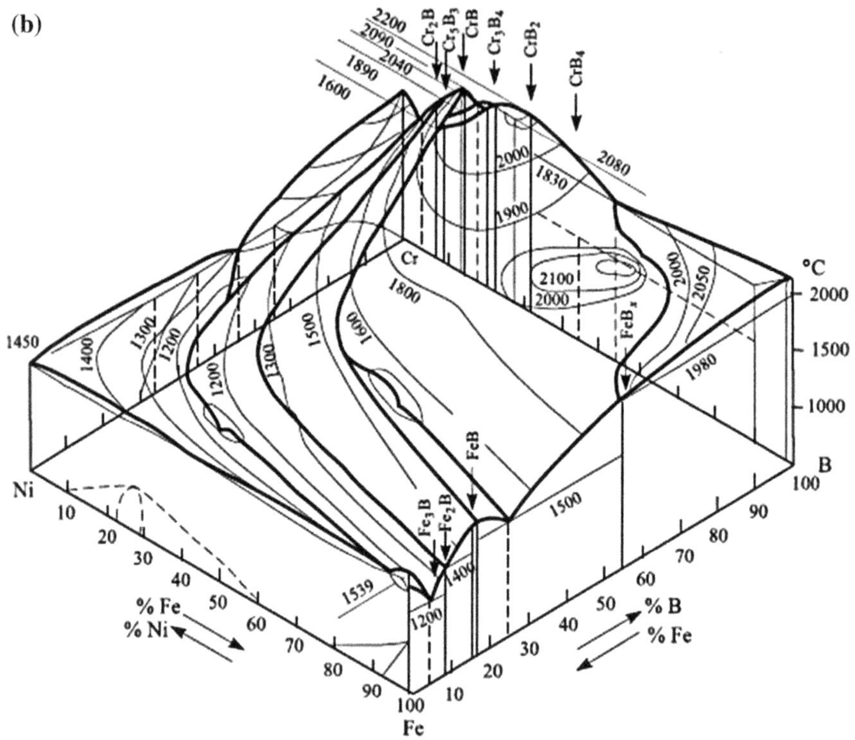

Fig. 5.12 (continued)

The similar interaction takes place in the area of boron concentrations between 15.5 and 17.22% during the formation of iso-structural borides FeB, NiB and CrB with rhombic crystal lattices. Herewith the boride FeB solutes 0.32 mass units of CrB; CrB solutes 0.36 m.u. of FeB at 1100 °C. The solubility of FeB and CrB increases up to 0.65 m.u. with the temperature rise to 1500 °C [17]. A nickel boride has also high solubility in borides FeB and CrB, i.e. borides form a continuous range of solid solutions.

Therefore the fourth diagram reflects this interaction by the iso-structural and iso-concentrative boron line on the liquidus surface that combines melting points of the phases at any position of the initial elements in tops of the concentration square.

In the fourth diagram lower borides Ni_3B and Fe_3B, orthorhombic and monoclinic borides Ni_4B_3, chrome borides Cr_5B_3 and Cr_3B_4, found in double diagrams, wedge out in the form of solid solutions of the wide homogeneity area on the basis of borides Fe_2B, Ni_2B, Cr_2B and of solid solutions on the basis of borides FeB, NiB and CrB. However, it doesn't eliminate the possibility of forming new complex borides ($Fe_3Ni_{20}B_6$ (4.61%B), $Cr_2Ni_3B_6$ (18.8%) and Cr_3NiB_6 (23.2% B) etc.), which are alloyed by the missing system's elements (Cr or Fe), or other compounds. It's worth mentioning that if formed phases are ordered solid solutions on the

(c)

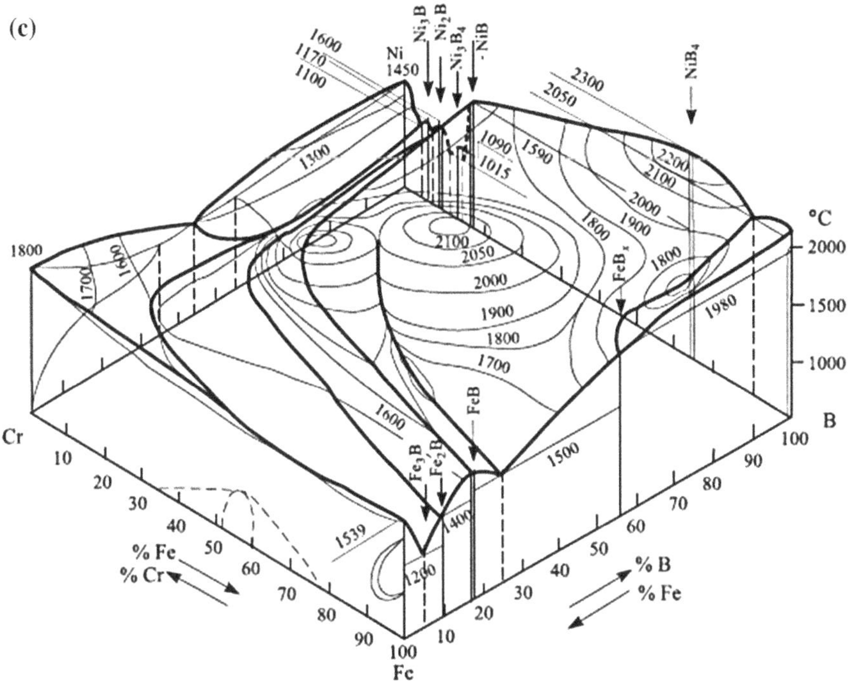

Fig. 5.12 (continued)

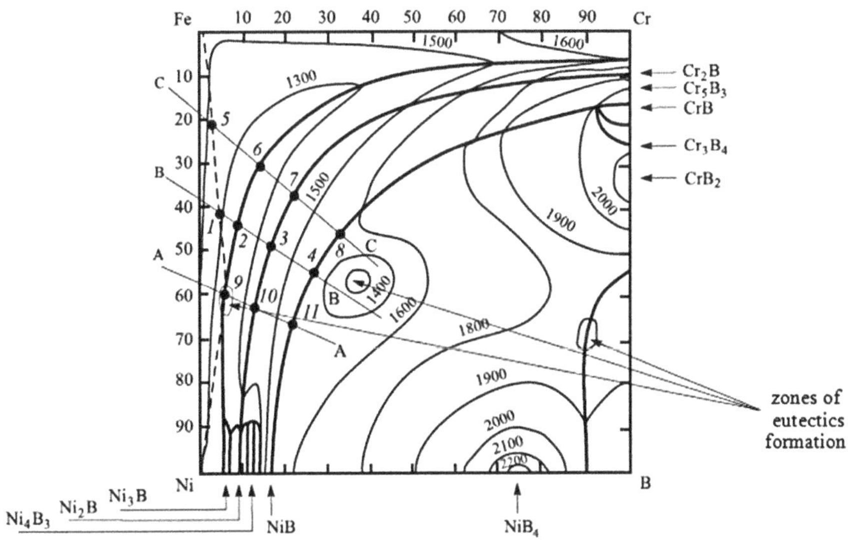

Fig. 5.13 The liquidus surdace of Ni-Fe–Cr-B equilibrium diagram scheme

foundation of borides, then their formation doesn't significantly influence the relief character of the liquidus surface.

The rise in boron concentration up to 20–50% transits the liquidus surface into the area of higher temperatures. This area is characterized by the second temperature minimum in the fourth diagram. The formation of this eutectics is the result of the interaction of eutectics in Fe-B system at ~25% B in mass at the melting point of 1500 °C and the wide area of liquid solution in the system Ni-B at the melting point of boride NiB—1590 °C. The eutectic concentration of components calculated by the temperature rule hypothetically has the following composition (in mas. %): 30.0% Fe, 30.0% Ni, 19% Cr, 21% B. The melting point of eutectics lies in the range 1240–1290 °C [45, 46].

The transition into the area of increased boron concentrations is accompanied by the formation of the third eutectic mixture as a result of the interaction between eutectics in system Cr-B (melting point 1830 °C) and phase FeB_x (melting point 1980 °C). The formed eutectics include borides NiB_4, CrB_2, FeB_x and solid solutions of system's metals Ni, Cr, Fe in boron at the following calculated elements' content (mas. %): 28–30% Cr, 6–8% Ni, 3–5% Fe, 58–61% B. The melting point of eutectics lies in the range of 1690–1755 °C.

Thus, the main characteristics of the 4-component diagram scheme are logically repeated in any position of the system's elements (Fig. 5.12) with a slight change in the relief of the liquidus surface due to the influence of a particular element on the structure formation. Consequently, the analysis of similar 4-component systems requires only one equilibrium diagram at one mutual position of elements.

The general view of the liquidus surface (Fig. 5.13) brings additional data to the description of the system elements' interaction characteristics, allows to make the analysis of particular alloys more detailed and determine the directions of polythermal sections (e.g., A-A, B-B, C-C).

The solidus surface of this equilibrium diagram scheme was built using the calculated eutectic temperatures and concentrations (Fig. 5.14). The separate representation of the solidus and liquidus surfaces eases the use of the diagram.

The built schemes of Ni-Fe–Cr-B 4-component diagram and its polythermal sections (Fig. 5.15) were used for developing saturating compounds and determining the temperature conditions for boriding of the alloys of this system [45, 47].

5.3.3 Building the Equilibrium Diagram Scheme for Cr-Mo-Fe-W-B

The scheme of Cr-Mo-Fe-W-B 5-component diagram, necessary for the analysis of the structure formation of high-speed steels during boriding, was built on the foundation of the developed principles. As for the 4-component diagram, the character of the liquidus surface is closely connected to the concentrative

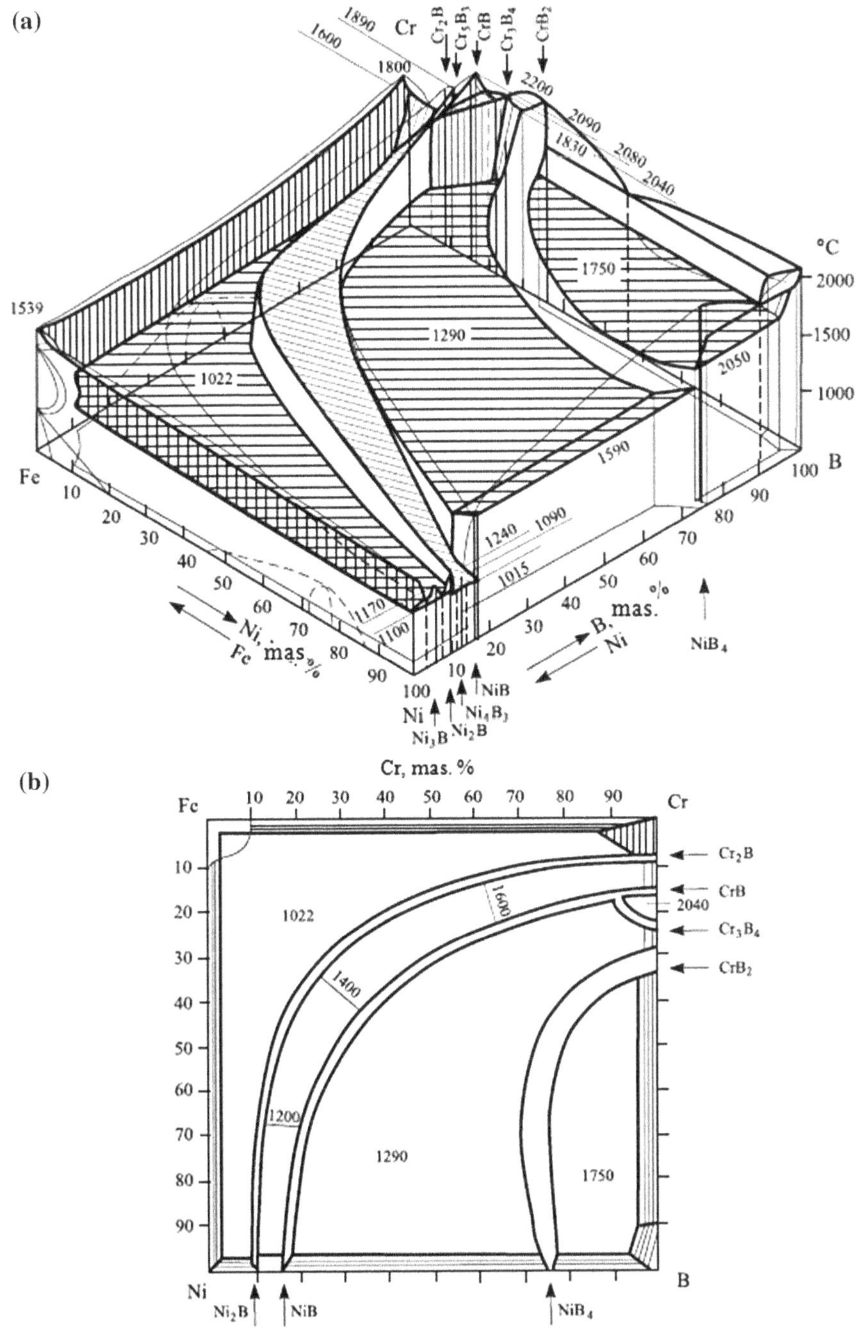

Fig. 5.14 **a** The equilibrium diagram scheme of Ni-Fe–Cr-B, the general view of the solidus surface, **b** The equilibrium diagram scheme of Ni-Fe–Cr-B, the view from above the solidus surface

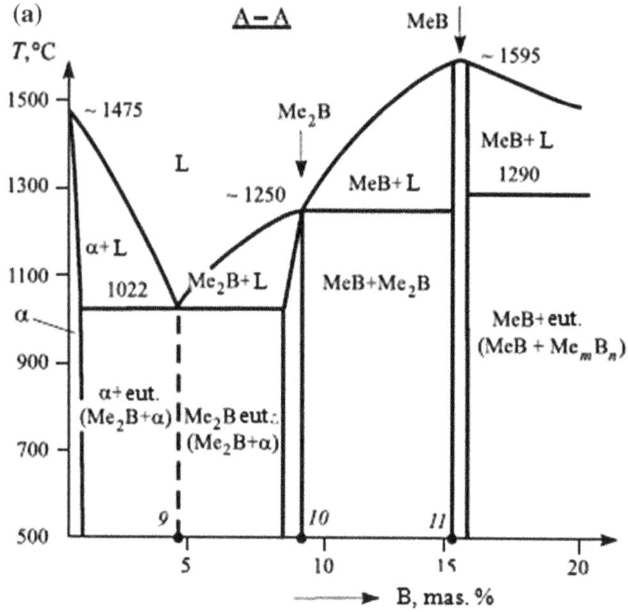

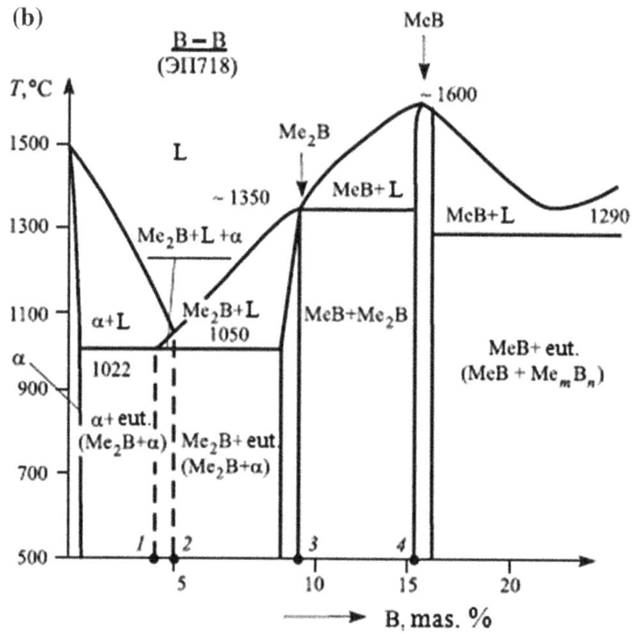

Fig. 5.15 **a** The polythermal section A-A of the equilibrium diagram scheme of Ni-Fe–Cr-B (the section is shown in 5.13), **b** The polythermal section B-B of the equilibrium diagram scheme of Ni-Fe–Cr-B (the section is shown in 5.13), **c** The polythermal section C-C of the equilibrium diagram scheme of Ni-Fe–Cr-B (the section is shown on 5.13)

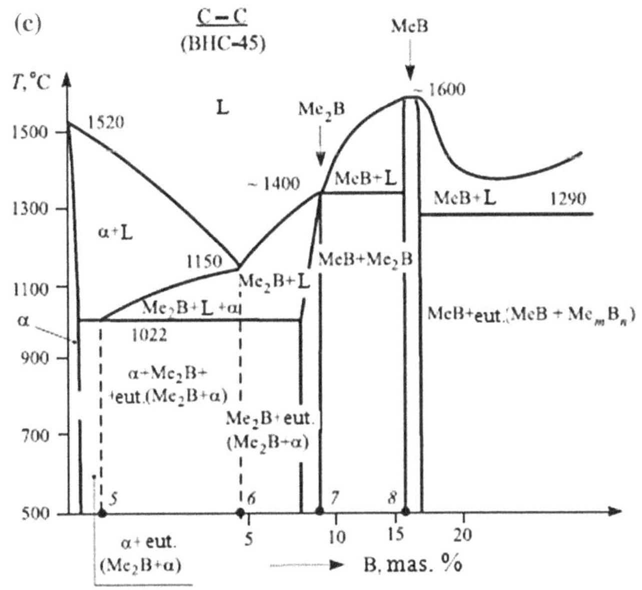

Fig. 5.15 (continued)

distribution of boron due to the presence of strong boride-generators Cr, W, Fe, Mo in steel (Figs. 5.16 and 5.17) [45].

Taking into account the fact that carbon in high-speed steels is found in a bond state—carbides, which are preserved in the structure even during the heating, it's necessary to assume that its influence on the boride-formation would be insignificant. For this reason, it wasn't taken into account while building the diagram scheme and calculating the critical points.

Another note concerns wolfram; it is the most high-melting point element of the system and a strong carbide-generator; it is found in the form of carbides in annealed steels (Table 5.1). The solid solution of steel P6M5 contains only 0.3 mass% W. Thus the calculation of the eutectic temperature which defines the temperature of boriding was made either with taking wolfram into account and without it. In the first case T_{eut} was ~1100 °C; in the second— ~1157 °C [45]. The practical boriding proved the correctness of the calculations that excluded wolfram.

Molybdenum is also preserved in the form of carbides at the boriding temperature. However solid solutions contain it in larger amounts. It's worth noting that molybdenum and wolfram are very similar in the manner of boride-generation: they form isomorphic structures and easily substitute each other in borides' lattices. Thus the calculation of melting point for eutectics under boriding requires only one element. In paper [45] only molybdenum was taken into account for the analysis.

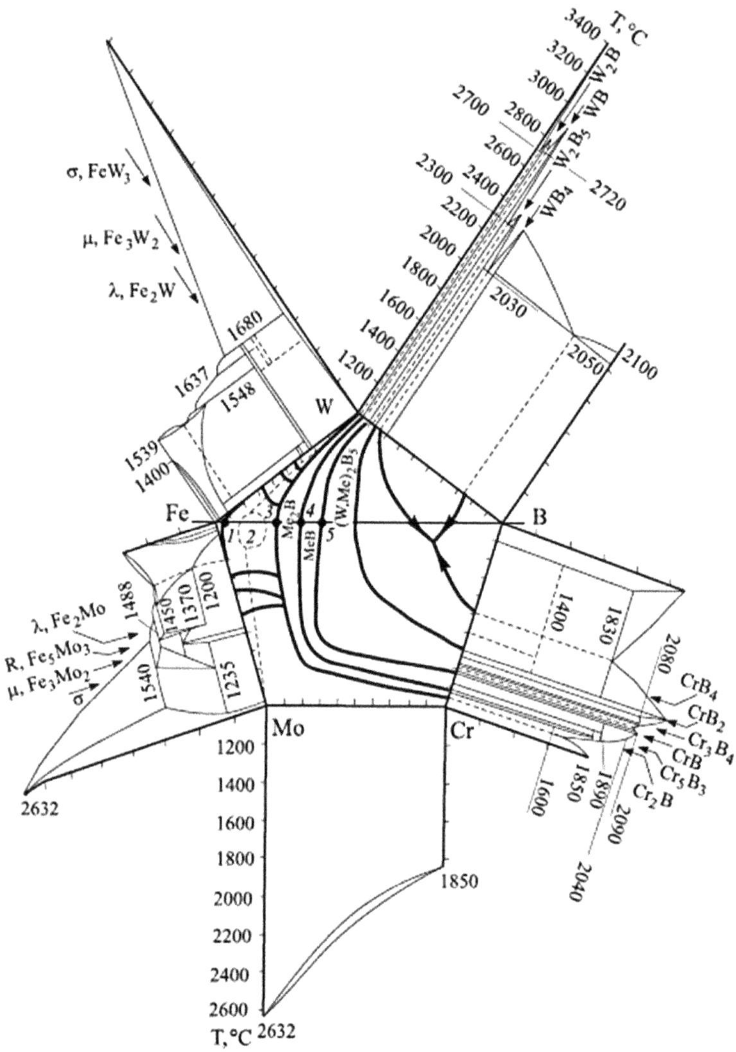

Fig. 5.16 The general view from the above of the liquidus surface of Cr-Mo-Fe-W-B diagram scheme and the view of double equilibrium diagrams

The created diagram scheme of Cr-Mo-Fe-W-B, containing the analysis of generated mutual solutions of isostructural borides Me_2B and MeB and possible interaction of eutectics, and the built polythermal section (Fig. 5.18) allow to evaluate the mechanisms of the structure formation of particular alloys under boriding.

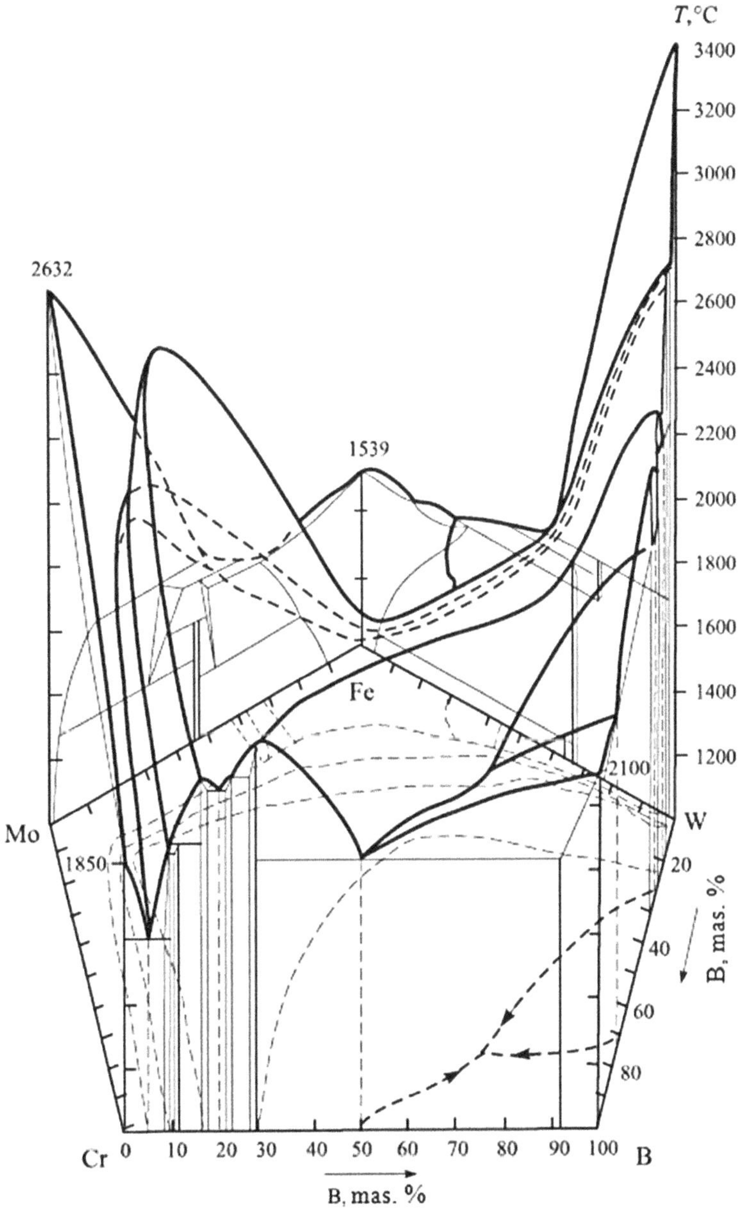

Fig. 5.17 The scheme of 5-component equilibrium diagram Cr-Mo-Fe-W-B

Table 5.1 The character of changes in the phase composition and content of alloying elements in steel P6M5 (in accordance with the type of processing) [48]

Type of processing	The phase composition		The carbide type	The content of alloying elements in solid solution			
	Solid solution	Carbides		W	Mo	Cr	V
Annealing	78	22	Me_6C, MeC, $Me_{23}C_6$	0.3	0.7	3.3	0.2
Hardening from 1225 °C	91	8.6	Me_6C, MeC	2.6	2.7	3.8	1.2

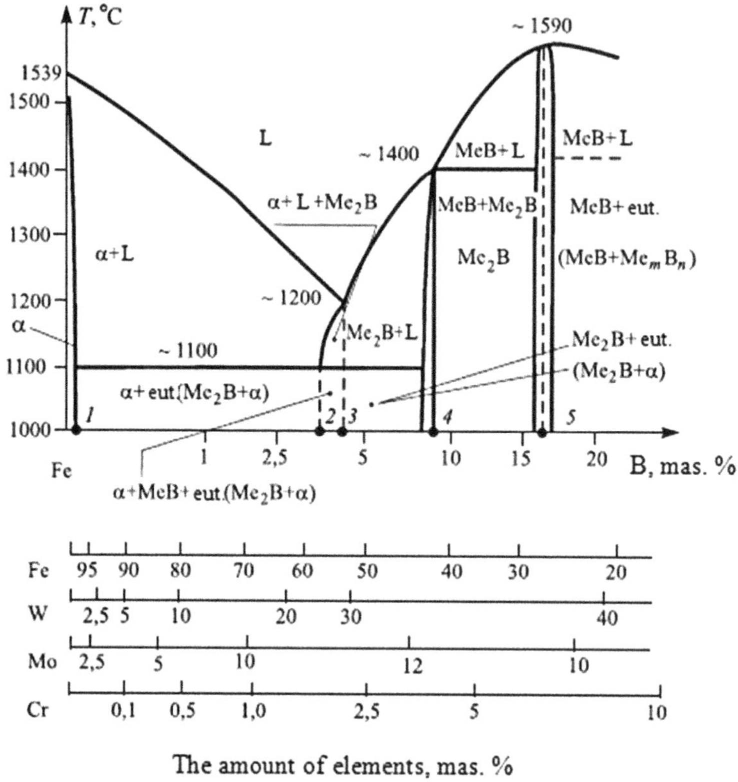

Fig. 5.18 Polythermal section of diagram scheme Cr-Mo-Fe-W-B (Section Fe-B is shown in 5.16)

References

1. Largest Collection of Ternary Phase Diagrams to be Published by year-end / Special Prepublication Offer // ASM International. The Materials Information Society, 1994. 12 p.
2. *Bannykh O.A., Dritsa M.E., ed.* Diagramy sostoyaniya dvoinyh i mnogokomponentnykh sistem na osnove zheleza. Spravochnik. Moscow, 1986. 440 p. [The equilibrium diagrams of binary and multicomponent iron-based systems]
3. *Villars P., Prince A., Okamoto H.* Handbook of Ternary Alloy Phase Diagrams / ASM International. The Materials Information Society. 1994. V. 1–10. 15000 p.
4. *Mchedlishvili V.A., Khovrin V.V., transl. Vinarov S.M., ed.* Bor, kalcii, niobii i tsirkonii v chugune i stali. Moscow, 1961. 460 p. [Boron, calcium, niobium and zirconium in cast iron and steels].
5. *Vogel R. und Tammann G.* Uber das ternare System Eisen-Bor-Kohlenstoff // Z.anorg. allgem. Chemie. 1922. V. 123. P. 225–275.
6. *Busby P.E., Warga M.E., and C. Wells.* Diffusion and Solubility of Boron in Iron and Steel // J. Met. 1953. November. P. 1463–1468.
7. *Carroll K.G., Darken L.S., Filer E.W., and Zwell L.* A new Iron Borocarbide // Nature. 1954. V. 174. P. 978–979.
8. *Nicholson M.E.* // J. Met. 1957. January. P. 1–6.
9. *Habig K.H., and Chatterjee-Fischer R.* // Tribol. Int. 1981. V. 14, №4. P. 209–215.
10. *Voroshnin L.G., Lyakhovich L.S., Panich G.G., Protasevich G.F.* Struktura splavov sistemy Fe-B // Metallovedeniye i termicheskaya obrabotka metallov. 1970. №9. Pp. 14-17. [The structure of Fe-B system smelts].
11. *Koifman I.S. et al.* Rentgenograficheskii analiz borotsementita // Metallovedeniye i termicheskaya obrabotka metallov. 1969. №2. Pp. 59–60 [X-ray analysis of boron-cementite]
12. *Stadelmaier H.H. and Gregg R.A.* // Metall. 1963. V. 17. P. 412–414.
13. *Cameron T.B. and Morral J.E.* // Met. Trans. A. 1986. V. 17A.P. 1481–1483.
14. *Borlera Lucco M. and Pradelli G.* // Met. Ital. 1967. V. 59, №11. P. 907–916.
15. *Samsonov G.V., Umanskii L.S.* Tverdye soedineniya tugoplavih metallov. Moscow, 1957. 265 p. [Solid compounds of high-melting point metals].
16. *Golutvin Yu. M.* Teplota obrazovaniya i tipy khimicheskoi svyazi v neorganicheskikh kristallah. Moscow, 1962. 380 p. [The heat of formation and types of chemical bonds in non-organic crystals]
17. *Kuz'ma Yu. B., Chaban N.F.* Dvoinye i troinye sistemy, soderzhaschie bor. Spravochnik. Moscow, 1990. 320 p. [Double and triple systems containing boron. Reference book.]
18. *Bannykh O.A., Budberg P.B., Alisova S.P. et al.* Diagramy sostoyaniya dvoinyh i mnogokomponentnykh sistem na osnove zheleza. Spravochnoe izdaniye. Moscow, 1986. 440 p. [The equilibrium diagrams of binary and multicomponent iron-based systems. Reference book.]
19. Powder Diffracting File. Inorganic Section. Set. 1–30. — Going Committed on Powder Diffraction Standards. Swarthmore, Pennsylvania, USA. 1977, Card № 19–613.
20. *Glukhov V.P.* Boridnye pokrytiya na zheleze i stalyah. Kiev, 1970. 208 p. [Boride coatings on iron and steels].
21. *Portnoy K.I., Chubarov V.M., Romashov V.M.* Diagramma sostoyaniya sistemy nikel-bor // Poroshkovoya metallurgiya. 1967. №2. Pp. 15-21. [The equilibrium diagram of the system nickel-boron]
22. *Portnoy K.I., Levinskaya M.Kh., Romashov V.M.* Diagramma sostoyaniya zhelezo-bor // Poroshkovaya metallurgiya. 1969. № 8. Pp. 66-70. [The equilibrium diagram of iron-boron]
23. *Schobel J.D., Stadelmaier H.H.* Das Zweistoffsistem Ni-B-Z // Metallkunde 1965. Bd. 56, №12. S. 856–859.
24. *Portnoy K.I., Romashov V.M.* Binarnye diagrammy sostoyaniya ryada elementov s borom // Poroshkovaya metallurgiya. 1972. №5. Pp. 48-56. [Binary equilibrium diagrams for a range of elements with boron.]

25. *Gorbunov A.E., Boduryan F.M.* // 1976. №16. C. 172–178.
26. *Telegus V.S., Kuz'ma Yu.B.* // Lviv University Bulletin. Chemistry. 1969. Iss. 11. Pp. 21–24.
27. *Pradelli G., Gianoglio C.* //Metallur. Ital. 1976. V. 68, №4. P. 191– 194.
28. *Kanaizuku Tadahito* // J. Solid State Chem. 1982. V. 41, №2. P. 195–204.
29. *Post B., Glaser F.W., Moskowitz D.* // Acta Metallurg. 1954. V. 2, №1. P. 20–25.
30. *Tai Shou-wei G.A., Yasinska G.A., Samsonov G.V.* Dopovidni AN URSR. 1960. №1. Pp. 48–50.
31. *Samsonov G.V., Kovalichenko V.S., Yasinskaya G.A.* // Academy of Science Journal. Metallurgy and Oil. 1960. №2. Pp. 115–119.
32. *Kolomytsev P.T., Moskaleva N.V., Snetkov A.Ya.* // Poroshkovaya metallurgiya. 1969. №10. Pp. 76–81.
33. *Kuz'ma Yu. B., Telegus V.S., Kovalyk D.A.* // Poroshkovaya metallurgiya. 1969. №5. Pp. 79–87.
34. *Telegus V.S., Kuz'ma Yu.B.* // Poroshkovaya metallurgiya. 1968. №2. Pp. 68–75.
35. *Kuz'ma Yu.B.* Kristallokhimiya boridov. Lviv, 1983. 41 p. [The crystal chemistry of borides]
36. *Kuz'ma Yu. B., Starodub P.K.* // Academy of Science Journal. Non-organic materials. 1973. Vol. 9, №3. Pp. 376–381.
37. *Anosov V.Ya., Ozerova M.I., Fialkov Yu. Ya.* Osnovy fiziko-khimicheskogo analiza. Moscow, 1976. 503 p. [The basics of physical-chemical analysis]
38. *Bocke H.E.* // Mineral. Petrogr. Geol. 109 (1916).
39. *Schout P.N.* // Mehrdimensional Geometri. Leipzig, 1902.
40. *Lodochnikov V.N.* Prosteishiye metody izobrazheniya mnogokomponentnyh sistem. // Izv. IFHA. 1924. Vol. 2. Iss. 2. Pp. 255-352 [Simple methods of representing multicomponent systems]
41. *Bochvar A.A.* // Academy of Science Journal, 1944. p. 851.
42. *Radischev V.P.* // SFHA Bulletin. 1936. Vol. 9. P. 203, 209; Ibid. 1938. Vol. 11. P. 5.
43. *Petrov D.A.* Diagramma sostoyaniya pyatikomponentnoi evtekticheskoi sistemy v koordinatah trehmernoi proektcii pentapopa // Diagramy sostoyaniya metallicheskih sistem. Moscow, 1981. Pp. 35-45. [Equilibrium diagram of five-component systems in the coordinates of 3-D projection of pentapope]
44. *Krukovich M.G.* Principy postroeniya mnogokomponentnyh diagram sostoyaniya v koordinatah koncentratciya - temperatura / Moscow, 1994. 24 p. [The principles of building multicomponent equilibrium diagrams in the coordinates concentration - temperature.]
45. *Krukovich M.G.* Razrabotka teoreticheskikh i prikladnykh aspektov upravleniya strukturoi i svoistvami borirovannykh sloev i ikh ispolzovanie pri proizvodstve transportnoi tekhniki. Dis. dokt. tekh. nauk. Moscow, 1995. 416 p. [The development of theoretical and applied aspects of controlling the structure and properties of boronized layers and their use in producing transport machinery. Doctorate dissertation]
46. *Krukovich M.G.* Raschyot evtekticheskikh koncentratcii i temperatury v dvukh- i mnogokomponentnykh sistemakh // MiTOM. 2005. №10. Pp. 9-17. [The calculation for eutectic concentrations and temperature in two- and multicomponent systems]
47. *Golubin A.V., Banas I.P., Krukovich M.G., Shalkevich A.B., Goncharova G.I.* Poverkhnostnoye uprochneniye austenitnykh Fe-Ni-Cr splavov borirovaniem. // The collection of MIIT papers. Moscow, 1991. Iss. 843. P. 20-36. [The surface hardening treatment of austenite Fe-Ni-Cr alloys by boriding]
48. *Gulyaev A.P.* Metallovedeniye. Moscow, 1986. 544 p [Physical Metallurgy].

Chapter 6
Plasticity as the Key to the Durability of Boronized Layers

Abstract The chapter introduces the classification of boronized layers according to the plasticity indicators such as methods of obtaining, groups of plasticity, shear stress, types of structure, phase composition, textureness evaluation, level of residual tensions, type of substrate, boron concentration on the surface. The layers with different borderlines determining the interconnection of the layer and sublayer are studied; the influence of substrate structure on the final properties of boronized layers is demonstrated. The chapter also presents the detailed descriptions of evaluation methods applied to the mechanical properties (plasticity) of boronized layers. The integral characteristics of plasticity is shear stress which covers the physical and mechanical properties of borides themselves (microhardness, plasticity modulus), phase composition, the correlation between phases, their dispersability and mutual position in the layer and on the surface, stress state of the layer and its plasticity. The recommended widths of one-phase and two-phase boronized layers in different steel brands are given in order to increase the wear-resistance of different tools and details used in different conditions.

The working capacity of boronized layers is mainly determined by the work conditions of strengthened details or tools. However, the common characteristics in all cases is a material's predisposition to the accumulation of damage at the surface or in the body of details and also the resistance to the environment's influence.

Plastic materials accumulate damage during friction for a significant amount of time, because this process is accompanied by the plastic deformation of surface layers and followed by the cold work and the loss of plasticity; for brittle materials the accumulation consumes the minimum of time and the brittle fracture of surface layers can happen even at the initial stage of a detail's exploitation.

Thus, the foundation for the durability of friction points' details with coatings is determined by a certain reserve of surface layers' plasticity; its optimal level depends on a material's properties and the conditions of friction.

The surface's resistance to the influence of an external medium during exploitation depends on its electrochemical heterogeneity, the presence of thick protective secondary structures formed under the process of friction, and the type of the

M.G. Krukovich et al., *Plasticity of Boronized Layers*,
Springer Series in Materials Science 237,
DOI 10.1007/978-3-319-40012-9_6

environment. The dominant significance of one or another factor is determined by the value and difference in electrode potentials of phases and the structural components of alloys on its surface, a type of the environment and conditions of friction.

The one-phase structure is considered to be the most adequate from the point of corrosion interaction between boronized layers and environment. In certain situations, the multi-phase structure with imminent electrode potentials' values (at minimal electrochemical heterogeneity) is allowed. The multi-phase structure, due to the existence of a softer layer component on the friction surface, facilitates the formation of screening secondary structures which decrease the electrochemical inhomogeneity of the surface.

The best qualities from the position of boronized details' wear process at abrasive and borderline rubbing friction are possessed by details with a compact structure of boride phases. The conditions of rubbing friction and rolling friction at alternating loading with the presence of grease require layers with multi-phase structures with dense and soft components and with a certain level of solidness ~1100–1400 HV as having the most appropriate qualities.

6.1 The Classification of Boronized Layers by Their Plasticity

The plasticity (ductility) of boronized layers is determined by the particular set of internal and external factors which influence their formation. The internal factors include: phase composition, a type of steel for a processed detail, a combination of its composition and thermal processing characteristics with the width and structure of the layer, a level of the layer's texturedness and the character of tension distribution in the layer and in the detail. The external factors are mostly the processing conditions: the composition of the environment, the temperature-temporal conditions and methods of electro-physical impact.

The classification of layers by their plasticity, which reflects the diversity of boriding conditions and forming structures, is given in Fig. 6.1 (colour inset); it is done to give a more complete understanding of the interconnection of layers' properties with peculiarities of the process and acquirement of methods that are used to control the plasticity (ductility) of boronized layers.

The traditional methods and ways of boriding include those which engage the traditional heating equipment and usual technologies. These are boriding processes in gas, liquid and solid mediums. The untraditional methods are rare methods of processing which are now used more often. They presuppose the use of plasma, electro-beam, luminescent and other methods of heating, the combination of processing with electromagnetic impact and additional treatment of boronized layers obtained with other methods.

The properties of boronized layers are divided into groups that are characterized by low, medium and high levels of ductility. Each group corresponds with a certain

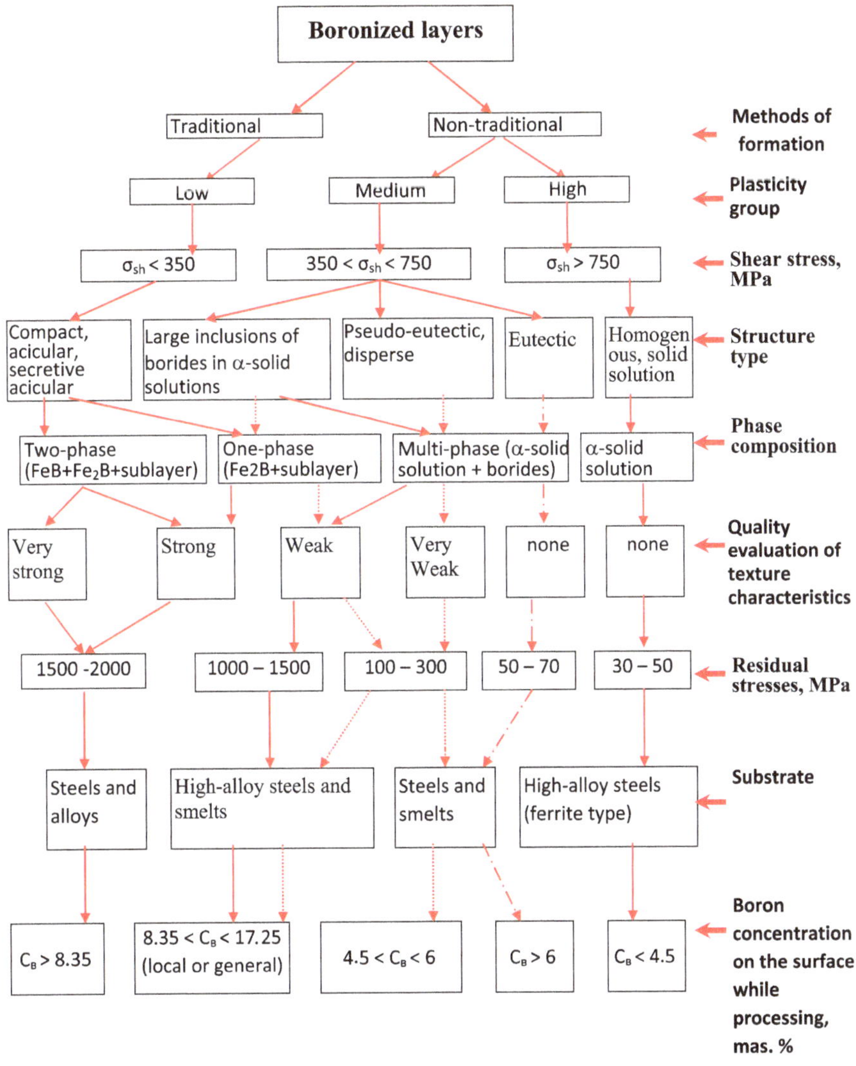

Fig. 6.1 The classification of boronized layers by their plasticity

interval of fracture stress values. The fracture stress is an integral value closely connected to the type of structure and phase composition. Minimal fracture stresses are common for compact acicular boronized layers and layers which have smoothed needles and a sublayer due to the withdrawal of alloy elements. They form the group of low plasticity. The group of medium plasticity with the intervals of fracture stress values between 350 and 750 MPa includes boronized layers with disconnected borides with different levels of dispersibility. The group of high plasticity is formed by the layers on the basis of α-solid solution or with a different type of dominant structure with boride residues on the borders of grains.

The quality evaluation of texturedness is determined by the intensity of roentgenogram lines. The X-ray diffraction analysis of obtained layers showed a good combination of texturedness levels and compactness of borides' layers. It was the result of the regularity of borides' growth which in the end defines the level of residual tension.

Residual tensions significantly depend on the type of processing; they can reveal themselves in the phase composition and the structure of a layer. They also depend on the brand of steel substrate, layer's width and types of final thermal processing. The classification ends with determining boron concentration on the surface under saturation which directly connects the technology of boriding with the plasticity of boronized layers.

The border between the boronized layer and the matrix and a type of the substrate also play important roles in forming the plasticity. The substrate determines the interconnection of the layer with the matrix and the type of residual voltages distribution. E.g., the formation of a continuous substrate on the basis of α-solid solution of Si in Fe is accompanied by spontaneous peeling of boride areas on the processed surface as a result of the big difference in coefficients of thermal expansion of the solid solution and matrix's borides.

The following types of sublayers are found in boronized layers:

based on α-solid solution of boron in Fe;

- based on a zone with increased carbon content which was obtained as a result of its replacement with the growing boride layer (the carbon content is increased in comparison with the matrix which contains a small quantity of boron);
- based on a layer of α-solid solutions of alloy elements of the matrix (Si, Ni etc.) in Fe obtained as a result of their replacement with a growing boride layer;
- based on a mixture of borides and solid solutions of alloy elements obtained by simultaneous and consecutive saturation or as a result of changing the character of saturation during heating or soaking. Divided inclusions of borides can be found on the surface, in the middle part or on the border "layer-matrix".

Thus, the solidness and plasticity of these areas in a boronized layer is determined by phase properties and their numerical correlation. In particular, layers which contain 15–60 % of solid solution should be perceived as a group of medium plasticity; layers with higher content—a group of high plasticity.

The formation of boronized layers either with a sublayer or without it depends on the technology of their production; it leads to the necessity of type classification which will reflect the interconnection between layers' exterior, technological conditions and plasticity (Fig. 6.2, colour inset).

In some cases boronized layers without the sublayer are formed outside of the matrix as a result of physical (PVD) or chemical (CVD) deposition. However, they are more likely to have a little diffusion zone and can be found in a matrix.

Boronized layers with the sublayer obtained with traditional methods lie outside the matrix only partially. This happens due to the diffusion of matrix atoms (iron and alloy elements' atoms) which has the counter direction to boron, and the growth

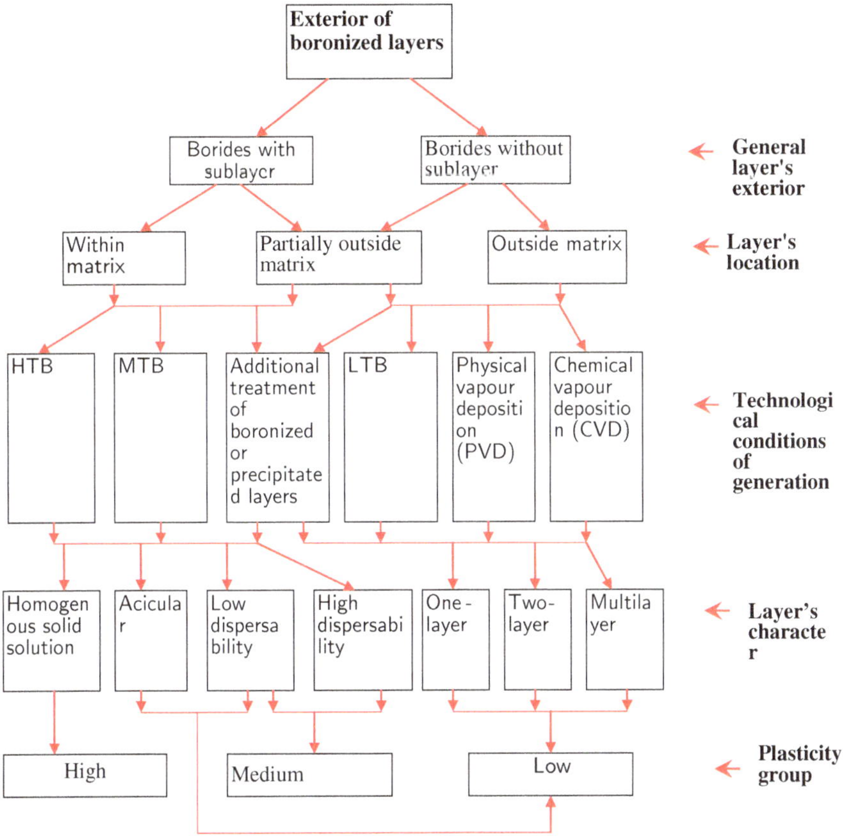

Fig. 6.2 The classification of boronized layers by their exterior

of a layer towards saturating medium. The increase in the matrix size (detail) in this case doesn't exceed 15 % of the boride layer's width.

The formation of layers within the matrix boundaries is ensured by the generation of homogenous solid solutions with a high level of plasticity. The group of medium plasticity level contains layers on the basis of different quantitative correlation between borides and solid solutions. The group of low plasticity consists of precipitated (PVD, CVD) layers and boride-based layers.

Simultaneously the maximum width of boronized layers can be obtained for a particular steel and type of its thermal processing, when they have proper connection with the matrix and have a certain plasticity reserve (until the natural brittleness of borides hasn't expressed itself). Moreover, some conditions of details and tools' exploitation allow to use the width values of boronized layers with tolerable defectiveness (Table 6.1). These recommendations allow to use boronized details in a wide range of working conditions.

Table 6.1 The recommended widths of boronized layers for different working conditions

No.	Details' working conditions	Steel brand	The recommended widths of boronized layers (mm)			
			One-phase layers		Two-phase layers	
			Post-hardening	Post-normalizing	Post-hardening	Post-normalizing
1	High pressures and impact loading (forming, coining etc. stamps)	40X	0.08–0.11/0.12–0.15	0.15–0.18/0.18–0.20	0.05–0.07/0.08–0.10	0.12–0.15/0.16–0.18
		X12 M	0.04–0.06/0.06–0.08	0.05–0.07/0.07–0.09	0.03–0.05/0.05–0.06	0.03–0.05/0.05–0.06
		ШХ15	0.08–0.10/0.10–0.12	0.10–0.12/0.12–0.15	0.04–0.06/0.08–0.10	0.10–0.12/0.11–0.13
2	Details exploited under abrasive wear processes	20	0.15–0.20	0.15–0.20	0.13–0.15	0.15–0.18
		40	0.13–0.16	0.15–0.18	0.12–0.14	0.15–0.17
		ШХ15СГ	0.10–0.12	0.12–0.15	0.08–0.10	0.11–0.13
		У8, У10	0.10–0.12	0.12–0.15	0.08–0.10	0.10–0.12
		55ХМА	0.12–0.14	0.13–0.15	0.10–0.12	0.12–0.14
3	Details of techonological rigging	65Г	0.06–0.08/0.13–0.15	0.10–0.12/0.15–0.17	0.04–0.06/0.11–0.13	0.10–0.12/0.13–0.15
		40	0.08–0.11/0.14–0.17	0.15–0.20/0.18–0.22	0.06–0.08/0.12–0.15	0.12–0.15/0.16–0.18
		ШХ15	0.08–0.10/0.12–0.14	0.10–0.12/0.14–0.17	0.04–0.06/0.10–0.12	0.10–0.12/0.13–0.15
4	Details and tools working at increased temperatures	ШХ15	0.10–0.12	0.12–0.15	0.08–0.10	0.11–0.13
		У8	0.12–0.15	0.15–0.18	0.10–0.12	0.12–0.14
		3Х2В8	0.08–0.10	0.08–0.10	0.06–0.08	0.06–0.08
		Х12Ф	0.06–0.08	0.07–0.09	0.05–0.06	0.05–0.06
		5ХНВ	0.12–0.14	0.13–0.15	0.10–0.12	0.10–0.12
		7Х3	0.12–0.14	0.11–0.13	0.10–0.12	0.10–0.13
		Р18	0.05–0.07	0.05–0.07	0.03–0.05	0.03–0.05

The spontaneous fracture of certain layer's parts can happen due to the high residual tensions; the reason for that is the excess of the optimal combination between the boride layer width and matrix's material. The process can also depend on the speed of cooling after boriding and the type of thermal processing of the matrix.

6.2 Evaluation Methods Applied to the Plasticity (Brittleness) of Boronized Layers

Many mechanical properties of materials are closely interconnected therefore the quantitative evaluation of some properties may function as comparative characteristics for other properties. Such characteristics include brittleness, plasticity and toughness. These terms reflect opposite properties of materials: 'brittleness' versus 'plasticity'/'toughness'. Therefore methods of evaluating brittleness can be used as methods of evaluating plasticity for a wide range of materials.

A large number of scientists mention that boride layers are characterized by a significantly high hardness and increased brittleness [2–5]. Simultaneously this fact is normally presented as qualitative characteristics; quantitative data on brittleness are not given. The authors of [1], relying on their own rich experience and the results of many researchers in boronized layers' properties, state that the high brittleness of boronized layers is exaggerated because it presents itself when technological recommendations are violated.

Thus, it becomes possible to systematize existing methods of evaluating plasticity or brittleness of boronized layers, discover the most efficient one and show the comparative data on different boronized layers.

The most popular methods for evaluating brittleness of boronized layers are methods of determining microbrittleness by measuring microhardness and hardness with the Vickers test (Diamond Pyramid) [6]. Later these methods were spread on to hard alloys [7] and high-melting compounds [8].

The evaluation of the brittleness grade

Brittleness is a property of hard bodies to break without residual deformation expressed macroscopically [9].

The brittleness grade is a qualitative characteristics of boronized layers; it is carried out by the method of brittleness evaluation used for nitrated layers with a 4-point scale (Fig. 6.3, colour inset) [10].

The brittleness grade is determined by the appearance of the diamond pyramid imprint in accordance with the brittleness scale (VIAM) which grades imprints by the number and the character of cracks and fractures. There are 4 brittleness grades:

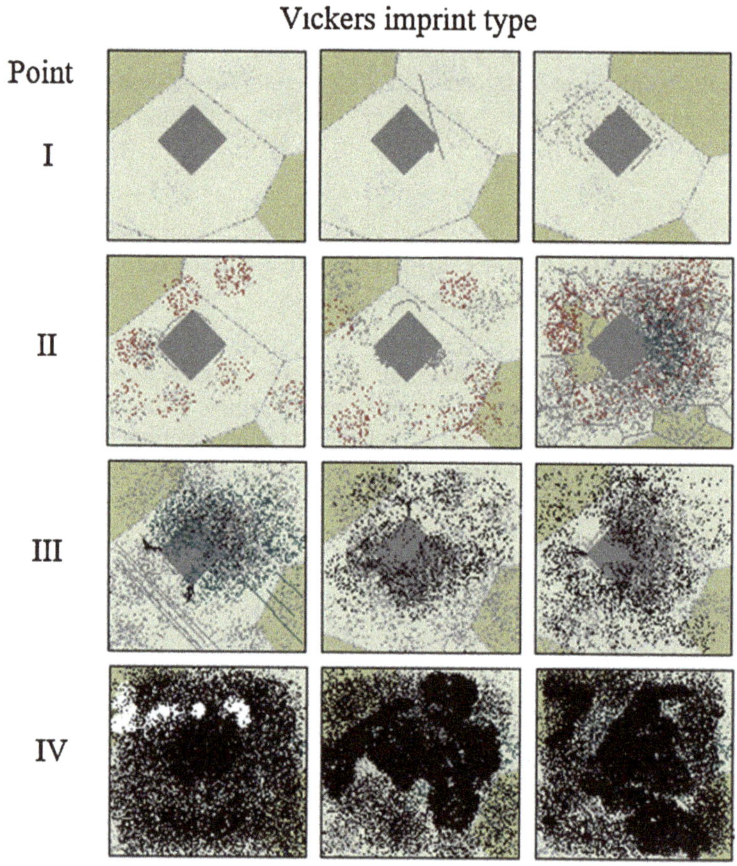

Fig. 6.3 The scale of brittleness point estimation for boronized and nitrated layers

I. crack-free non-brittle layers with non-deformed imprints;
II. slightly brittle layers without any significant deformation of imprints; 1–2 minor cracks positioned along the imprint's length on a certain distance or beginning and locking on one side of the imprint;
III. brittle layers with deformation and cracks; cracks beginning from the imprint's top (1–2 cracks), or double cracks positioned along all the imprint's sides, or the combination of the described cracks;
IV. exceedingly brittle layers with multiple cracks (3–5); cracks, positioned on a significant distance along the imprint's sides, or cracks, beginning from all tops of the imprint, or the combination of the described cracks.

Grades I and II coincide with operable ductile diffusion layers, grade IV indicates an inadmissible brittleness of layers and rejects such details as defective.

Table 6.2 Conventional scale of brittleness grade evaluation

Brittleness grade	The character of imprint
0	Without visible cracks and fractures
1	1 minor crack
2	1 crack which doesn't coincide with the course of the imprint's diagonal. 2 cracks in adjacent corners of the imprint.
3	Two cracks in the opposite corners of the imprint
4	More than 3 cracks. 1–2 fractures at the imprint's sides
5	The destruction of the imprint's form

The quantitative evaluation of the brittleness grade

The comparative evaluation for zones of chemical compounds (a boride zone of a general boronized layer, in particular) can be done with the method developed by Ikornikova and Pikunova [6]. It is based on the evaluation of microbrittleness with the microhardometer. The foundation of this method is based on determining the following two results:

- the change in number of imprints with definite cracks under the increase of loading;
- the change in number of cracks near one imprint.

After imprints are made, they are closely analyzed focusing on the character of the cracks that were formed under loading. After that, each imprint is given a certain brittleness grade in accordance with the developed scale (Table 6.2). Despite the subjective approach, the character of the formed cracks describes properties of any analyzed brittle material with a high level of reliability.

The total brittleness grade of a layer is calculated according to the formula:

$$Z_p = 0 \times n_0 + 1 \times n_1 + 2 \times n_2 + 3 \times n_3 + 4 \times n_4 + 5 \times n_5 \tag{6.1}$$

where n_0, n_1, n_2, n_3, n_4, n_5—relative numbers of imprints from the total amount with a certain brittleness grade (0, 1 … 5).

Normally the total number of imprints in from 25 to 100.

To take into account the tempo of elastic strain growth before the brittle fracture due to the growth of loading P, it is necessary to calculate the correlation between total brittleness level growth and loading growth $\Delta Z/\Delta P \approx (\delta Z/\delta P)_p$.

Knowing the total brittleness grade, it is possible to determine the relative index of the material's (layer's) brittleness Υ_p [11]:

$$\Upsilon p = Z_p \left(\frac{\delta Z}{\delta P}\right)_P. \tag{6.2}$$

Toughness under the diamond pyramid indentation

Toughness is the ability of a solid body to absorb energy while being plastically deformed before being fractured [9].

The calculating method, introduced by Palmquist, presupposes the determination of work (S_{kp}), that is done with the force P_{kp} during the indentation of a diamond pyramid. It is defined as a ratio between the loading (P_{kp}) (at 50—there are cracks) and the value of hardness which correlates to this loading. The author called this work "toughness". Using the pyramidal contact point with a square base and top angle 136°C and knowing the hardness (kgs/mm^2) which is correlated to the loading P_{kp}, it is possible to calculate the 'toughness' (ductility) of a layer by the formula:

$$S_{\mathrm{Kp}} = 2 \cdot 10^{-4} \sqrt{\frac{P_{\mathrm{Kp}}^{3}}{HV}} \tag{6.3}$$

Fracture toughness Fracture toughness is one of the major characteristics in the fracture mechanics. It describes a beginning moment of the over-critical development of a crack, when its length reaches a critical value. It is indicated as K_c in a flat strained condition; K_{1c} in a flat deformed state [9].

The fracture toughness is calculated according to the formula:

$$K_C = \frac{X \cdot P}{C_0^{\frac{3}{2}}} \tag{6.4}$$

where P—loading; C_0—radial line of a crack; X—intensity of residual crack advancing.

In an ideal toughness-ductile situation, when the process of indentation can be represented as a broadening cavern, the parameter X can be calculated the following way:

$$X \approx \left(\frac{E}{HV}\right)^{\frac{1}{2}} \tag{6.5}$$

where E—the Young's modulus; HV—hardness.

The constructive foundation for this idealization is a statement about a ductile part of deformation having a constant volume of radial transient process for a definite strained layer's state [12]. In particular, the formula for a one-phase boride layer is:

$$K_C = \frac{A \cdot P}{C_0^{\frac{3}{2}}} \tag{6.6}$$

where $A = 0.02_8 \left(\frac{E}{HV}\right)^{\frac{1}{2}}$.

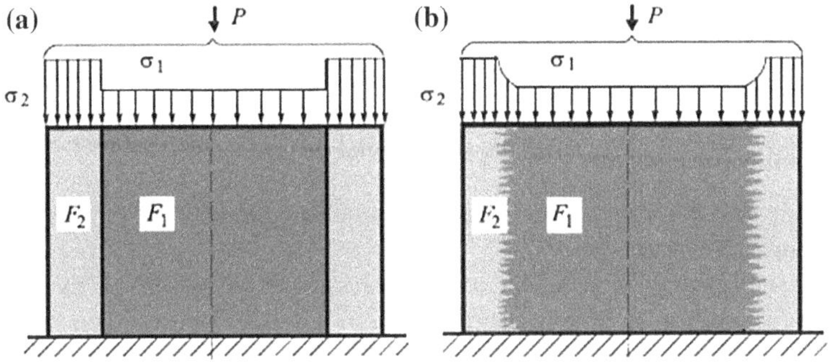

Fig. 6.4 The diagram of cross-cut end tensions of section homogeneous models in conditions of the linear stress condition: **a** for homogenous levels with a regular borderline; **b** for boronized layers

The evaluation of durability and plasticity of diffusion layers under compression

The durability is a material's property to resist deformation, plasticity is a property of solid bodies to change their sizes and forms under stress without fracturing [9].

The evaluation of mechanical properties of boronized standard models involve the material's heterogeneity and ends in getting approximate results which poorly reflect the substrate and layer's properties. The properties of the substrate and layer's various phases, defined separately, don't give the full picture of the properties which they have when put together.

To determine the properties of boronized details which can be used in engineer calculations, it is necessary to exploit the methods introduced in work [13]. The method involves the calculation of true tensions at any given moment of the deformation under compression in conditions of linear stress and homogenous contralateral strain (Fig. 6.4).

The total force is calculated by the formula:

$$P = \sigma_1 F_1 + \sigma_2 F_2 \tag{6.7}$$

where σ_1 and σ_2 are the true tensions which occur in the core and diffusion layer of the model correspondingly; F_1 and F_2 are the areas of lateral contraction of the core and the diffusion layer correspondingly.

The known core's material properties (e.g., σ_1) allow to calculate the properties of the diffusion layer at any loading in accordance with the formula

$$\sigma_2 = (P - \sigma_1 F_1)/F_2 \tag{6.8}$$

The obtained results are presented in the form of the curve of material's diffusion layer strengthening. The given method involves several stages:

- building curve "stress—absolute deformation" for a model with a diffusion layer;
- building curve "stress—absolute deformation" for a model without a diffusion layer;
- building the curve for a layer's material by substracting the second curve values from the first curve values;
- building the curve of deformation strengthening with the use of the indicator curve for a diffusion layer's material;
- the calculation of durability and plasticity properties of the diffusion layer's material using methods described in works [14, 15].

The experimenting process should be carried out according to the following recommendations:

- the compression of models should be done according to the scheme of linear stress conditions;
- the models which are characterized by the axis perpendicular to mutually parallel flanks should be used;
- the absolute deformation of the model should be fixed with the precision <0.002 mm [16];
- all the models should be processed in equal temperature modes under the process of chemical-thermal working.

Specific elongation
One of the ways to evaluate boronized layers' plasticity is to analyze their predisposition towards delamination while determining their specific elongation [17]. The delamination of boronized layers in most cases begins with general specific elongation or compression over 1.5 % [18, 19]. Two-phase boronized layers ($FeB + Fe_2B$) begin to delaminate at the plastic deformation 1.5—3 %, one-phase (Fe_2) at 3–45 %. The plasticity of two-phase boronized layers is 2 %. The predisposition towards delamination increases with the increase of the layer's width [1].

Bend tests on the formation of cracks
The additional (comparative) evaluation of boronized layers' plasticity can be done by carrying out a bend test in accordance with State Russian standard 1419-90. It allows to determine the limiting plasticity. Bend tests are carried out until the first crack appears on the surface of boronized models. Comparing the value of loading, it is possible to carry out a comparative analysis of plasticity for the boronized layers obtained by different methods of boriding.

The fracture stress
The fracture stress (σ_{fr}) is a minimal tension in a model which initiates the layer's fracture. It is the result of cutting σ_{cut} and shear τ impacts. The stress on the cut is seemingly bigger than τ and doesn't exceed its critical value during tests. The

fracture of a layer happens due to the increased τ which exceeds limit values. The calculation of τ and σ_{cut} based on the properties and loading conditions taken from the table is not efficient as it is impossible to take into account the structure factors of a layer and its stress condition.

The multiple diamond pyramid tests revealed that the loading which creates τ impacts the area which can be approximately calculated as $(2 \cdot l + c) \cdot l$. It determined the outlook of the final formula (6.9).

The fracture stress is an integral characteristic of a layer's brittleness which spontaneously involves physical-mechanical properties of borides (microhardness, plasticity modulus), phase composition, phase correlation, their dispersibility and mutual position in the layer and on the surface, stress condition of the layer and its plasticity. The brittleness decreases with the rise of the value of σ_{fr}; therefore the layer is more plastic (ductile) [20]. The fracture stress is determined while indenting the diamond pyramid into the surface of a section according to the formula [21]

$$\sigma_{fr} = \frac{0.174 \cdot P}{2 \cdot l^2 + C \cdot l} \tag{6.9}$$

where P—loading on the indenter while measuring microhardness; l—minimal distance from the centre of the imprint to the model's edge which doesn't cause the fracture of the boronized layer; c—the diagonal of the imprint.

The limiting deformation

The limiting deformation ε_{lim} is a relative deformation (limiting ductility) which is calculated from the correlation between the imprint's diagonal (which characterizes the value of plastic deformation of the material) and the length of the crack between the imprints (which reflects the material's predisposition to brittle fracture) [22].

The limiting deformation can be measured by Vicker's hardness tester or microhardness tester with a pyramid indenter. Two imprints are made on a tested surface; the distance between the imprints is optional. Then additional imprints are made in such a manner that approaching to the first imprint leads to the crack formation between them. The crack forms while indenting the pyramid in the corners of the imprint—where the stresses are concentrated. The value of the limiting deformation and therefore the plasticity of the material depends on the distance between the imprints at the crack formation (small distance means better plasticity).

The limiting deformation ε_{lim} is calculated after measuring the crack length (the distance between its edges) and the imprint's diagonal in the moment of crack formation:

$$\varepsilon_{lim} = \frac{D_{imp}}{L_{cr}} \tag{6.10}$$

where D_{imp} is the imprint's diagonal; L_{cr} is the crack's length between imprints.

The critical scratch length

The mechanical properties of coatings are determined with the sclerometric method by measuring the critical scratch length $L_{cr\ max}$. The method is based on a material being continually loaded, deformed in elastic and elastic-plastic areas and later fractured by relocating an indenter horizontally after positioning on a certain depth [23, 24]. Sclerometric tests are carried out in the conditions of a permanent depth of the indenter's intrusion for the purpose of obtaining comparable results. A standard Vicker's diamond pyramid is used. Parameters of a scratch (its width and depth) are highly stable [25]. Cracks are formed not only due to a material's fracture but its plastic deformation [26].

The process of the material's deformation with a diamond indenter consists of two parts:

- preliminary cold work of a material during the indenter's vertical intrusion on the given depth;
- following deformation of the processed material by a side surface of the indenter during its horizontal motion.

This horizontal motion of the indenter causes the gradual increase of a roll collar, which reaches its maximum height at the maximum denting stress $P_{d\ max}$. It is accompanied by a local brittle fraction of the metal on the top of the roll collar. As a result, a scratching stress decreases; it can be presented on the diagram as a smooth transition of a curve or as a slight fall (depending on the material's surface layers' properties) (Fig. 6.3). The following motion of the indenter slightly raises the scratching stress; then it stabilizes. The obtained diagrams of scratching allow to determine the range of mechanical properties of the material which reflect its ability to resist fracture. As shown in [27], the critical scratch length $L_{scr\ max}$ directly depends on the material's ductility. It can be explained by the fact that both

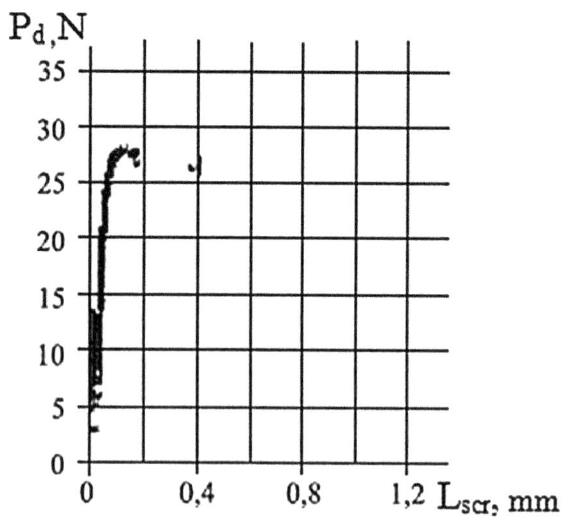

Fig. 6.5 The scratching diagram of the boronized layer [28]

extension and scratching at the moment of fracture are accompanied by the exhaust of material's plasticity reserve. The scratching diagram (Fig. 6.5) clearly shows the moment of the exhaust of a boronized layer's ductility reserve. This method is used for comparing ductility properties of different boronized layers.

References

1. *Voroshnin L.G., Lyakhovich L.S.* Borirovanie stali. Moscow, Metallurgiya, 1978. 239 p. [Voroshin L.G., Lyakhnovich L.S. Steel Boriding]
2. *Minkevich A.N.* Khimiko-termicheskaya obrabotka metallov i splavov. Moscow, 1965. 491 p. [Chemical-thermal processing of metals and alloys]
3. *Glukhov V.P.* Boridnye pokrytiya na zheleze i stalyah. Kiev, 1970. 208 p. [Boride coatings on iron and steels]
4. Tekhnologiya termicheskoi obrabotki stali. Perevod s nem. Moscow, 1981. 608 p. [The technology of thermal steel processing]
5. *K.H. Habig, R. Chatterjee-Fischer* // Tribol. Ing. **14**(4), 209–215 (1981)
6. *Ikornikova N.Yu., Pikunova L.A.* // Microhardness Proceedings. 21-23.11.1950. Moscow, 1951. pp. 226–233
7. *S. Palmquist* // "Jernkontorets Annaler". Sweden **141**(5), 300–307 (1957)
8. *Samsonov G.V., Neshpor I.S., Khrenova L.M.* // FMM. 1959. Vol. 8. Issue. 4. pp. 622–630
9. *Kershenbaum V.Ya., Prusakov B.A., ed.* Mnogoyaychnyi tolkovyi slovar' 'Metally'. Moscow, 1999. 710 p. [Multi-language explanatory dictionary 'Metals']
10. *Lakhtin Yu. M., Arzamasov B.N.* Khimiko-termicheskaya obtabotka metallov. Moscow, 1985. 256 p. [Chemical-thermal metal processing]
11. *Glazov V.M., Vigdorovich V.N.* Mikrotvyordost' metallov. Moscow, 1962. 224 p. [The microhardness of metals]
12. *A.H. Ucisik, C. Bindal,* Fracture toughness of boride formed on low-alloy steels / Surface and Coatings Technology. **94–95**, 561–565 (1997)
13. *Ryzhkovich R.L.* Metod opredeleniya prochnosti i plastichnosti diffuzionnyh sloev // Metallurgy. Iss. 8. 1976. pp. 62–67 [The method of determining hardness and ductility of diffusion layers]
14. *Timoschuk L.T.* Mekhanicheskiye ispytaniya materialov. Moscow, 1971. 225 p. [Mechanical testing of materials]
15. *Gubkin S.I.* Plasticheskaya deformatciya metallov. Volume 2. Moscow, 1961. 255 p. [Plastic deformation of metals]
16. Severdenko I.P., Ryzhkovich R.L. Ustroistvo dlya izmereniya deformatcii pri osadke // Metallurgy. Physical metallurgy and chemical-thermal processing. **1**, 37–41 (1970) [The device for measuring deformation]
17. *Lyakhovich L.S.* Khimiko-termicheskaya obrabotka metallov i splavov. Spravochnik. Moscow, 1981. 424 p. [Chemical-thermal processing of metals and alloys. Reference book]
18. *H. Ornig, O. Shaaber, Harterei* // Tech. Mitt. **17**(3), 131–140 (1962)
19. H.H. Neumann // Ind. Anr. **88**(33), 697–698 (1966)
20. *Krukovich M.G.* Razrabotka teoreticheskikh i prikladnykh aspektov upravleniya strukturoi i svoistvami borirovannykh sloev i ikh ispolzovanie pri proizvodstve transportnoi tekhniki. Dis. dokt. tekh. nauk. Moscow, 1995. 416 p. [The development of theoretical and applied aspects of controlling the structure and properties of boronized layers and their use in producing transport machinery. Doctorate dissertation]
21. *Grigorov P.K., Kathanov B.B.* Metodika issledovaniya khrupkosti borirovannogo sloya // Povysheniye nadezhnosti i dolgovechnosti detalei mashin: NIITM Proceedings. Rostov, 1972. Iss. XVI. pp. 97–99

22. Patent RF №2085902. The method of evaluating the ductility of hardened metals / Skudnov V.A., Grigoryev I.N., Evdokimov S.V., Gavrilov L.A. // Bulleten' izobreteniy. 1997. № 18. p. 21
23. *Matyunin V.M., Volkov P.V., Saidakhmedov R.Kh. et al.* Opredeleniye mekhanicheskih svoist i adgezionnoi prochnosti ionno-plazmennyh pokrytii sklerometricheskim metodom // MITOM. 2002. №3. pp. 36–39. [The determination of mechanical properties and adhesive hardness of ionic-plasma coatings with a sclerometric method]
24. *Matyunin V.M., Semenov V.F., Volkov P.V. et al.* Avtomatizirovannii nerazrushayuschii ekspress-kontrol mekhanicheskikh kharakteristik metalla // Informacionnye sredstva I tekhnologii: Conference proceedings. Moscow, 1998. Volume 2. pp. 20–27
25. *Markovets M.P.* Opredeleniye mekhanicheskikh svoistv po tverdosti. Moscow, 1979. 192 p. [The determination of mechanical properties by hardness]
26. *Borisov V.G.* Issledovaniye zavisimosti mezhdu kharakteristikami tverdosti i plastichnosti metallov: Candidate of Science Dissertation. Moscow, 1968. 134 p. [The study on the dependence between hardness characteristics and the ductility of metals]
27. *Volkov P.V.* Metod lokalnoi ekspress-otsenki mekhanicheskih svoist poverhnostnyh sloev mashinostroitelnyh materialov: Candidate of Science Dissertation. Moscow, 2000. 166 p.
28. *Sizov I.G.* Mekhanicheskiye svoistva boridnykh sloev posle elektronno-luchevogo borirovaniya // Polzunovskii almanakh. 2005. №4. pp. 42–45. [Mechanical properties of boride layers after electron-beam boriding]

Chapter 7
The Formation Conditions for Boride and Boronized Layers and Their Influence on the Layers' Plasticity

Abstract The chapter presents the study of the formation mechanism of saturating boron atoms in ionic liquid, gaseous and powder mediums and their mass transfer to the processed surface in self-organizing mode. The authors have come to the conclusion that the dissipative compounds of self-organization are represented by boron subions, i.e. boron irons of lowest valency. The reduction of subions goes according to the electrochemical laws. In the mediums which include boron in atomic state, saturating atoms are formed due to the mechanical interaction between medium and substrate both in solid and liquid states of interacting components. The conditions for mass transfer are also analyzed. Diffusion boronized layers are generated according to the diffusion or diffusion-crystallization mechanisms. In the first case the saturating object is solid, in the second—in liquid-crystal state. The forming stages of boronized layers generated in these mechanisms are analyzed. Acicular structures of boronized layers are characteristic for the first mechanism; pseudo-eutectic and eutectic layers of heterogeneous structure based on solid solutions and boride inclusions—for the second mechanism. The saturation parameters which determine the control over the processing and properties of obtained boronized layers and are singled out.

The properties of borides are determined by bond types between atoms in their crystal lattice and their structures (Table 7.1). However, the properties of boronized layers, apart from intrinsic boride properties, are influenced by the peculiarities of a layer's structure which emerge as a result of diffusion processes during boriding. The formation of boronized layers' structure begins with heating when the surface temperature reaches a certain point; it depends on the level of the surface's contamination, the speed of boron precipitation on the surface, the speed of boride crystallites' formation and their growth during soaking; in ends at the stage of cooling after the hardening process and the final thermal or mechanical processing of the details.

Some changes in plastic properties of boronized layers can be seen during exploitation while a wear process is taking place in a boronized layer or under the influence of other external conditions.

M.G. Krukovich et al., *Plasticity of Boronized Layers*,
Springer Series in Materials Science 237,
DOI 10.1007/978-3-319-40012-9_7

Table 7.1 The dependence of boride properties on their crystal structure

No.	Type of compound	Type of crystal lattice	Structure type	Density (t/m^3)	B content (mas. %)	Melting point (°C)	Microhardness (HV)	The coefficient of thermal expansion (10^{-6} K^{-1})
1	2	3	4	5	6	7	8	9
1	Fe_2B	Tetragonal	$CuAl_2$	6.93–7.43	8.82–8.84	1389; 1410;	1260–1960	7.85
2	FeB	Rhombic	FeB	6.47–6.8	16.25	1540; 1650;	1760–2340	23
3	Ni_3B	Rhombic	Fe_3C	7.5	5.78	1170;	1190	–
4	Ni_2B	Tetragonal	$CuAl_2$	7.9	8.43	1240; 1125;	1430	–
5	Ni_4B_3(o)	Ortho-rhombic	o-Ni_4B_3	7.55	12.13	1580; 1025;	1480	–
6	Ni_4B_3(m)	Monoclinal	m-Ni_4B_3	7.55	12.14	1580; 1031;	1480	–
7	NiB	Rhombic	CrB	7.13	15.55	1546; 1035;	1550–1700	–
8	Cr_2B	Rhombic	Mg_2Cu	6.07–6.24	9.09–9.65	1870	1350	14.2
9	Cr_5B_3	Tetragonal	Cr_5B_3	6.03–6.12	10.45–11.1	1890	1420–1520	13.7
10	CrB(ξ-CrB)	Rhombic	CrB	6.05	16.1–17.21	2090; 2050;	2100–2200	12.3
11	$Cr_1 - xB$	Tetragonal	α − MoB	6.05	17.21–17.22	2090	2100–2200	12.3
12	Cr_3B_4	Rhombic	Ta_3B_4	5.22–5.76	21.72–25.7	2100	1800–1850	11.8
13	CrB_2	Hexagonal	AlB_2	5.22–5.6	28.8–32.7	2200	2130–2340	10.5
14	$Fe_3Ni_{20}B_6$	FCC	$Cr_{23}C_6$	–	4.61	–	–	–
15	$Cr_2Ni_3B_6$	Orthorhombic	V_5B_6	–	18.8	–	–	–
16	Cr_3NiB_6	Orthorhombic	V_2B_3	–	23.2	–	–	–
17	Co_2B	Tetragonal	$CuAl_2$	7.9–8.33	8.44	1400	1150	–
18	CoB	Rhombic	FeB	7.25	15.51	1500	1150	–
19	CoB_2	Rhombic	–	–	26.86	–	2575	–
20	TiB	Rhombic	FeB	5.09	18.43	2400	3700–2800	–

(continued)

Table 7.1 (continued)

No.	Type of compound	Type of crystal lattice	Structure type	Density (t/m^3)	B content (mas. %)	Melting point (°C)	Microhardness (HV)	The coefficient of thermal expansion (10^{-6} K^{-1})
21	TiB_2	Hexagonal	AlB_2	4.5	31.12	2980	3370	8.1
22	V_3B_2	Tetragonal	U_3Si_2	5.6	12.4	2070	–	–
23	VB	Rhombic	CrB	5.44	17.52	2250	2300	–
24	V_3B_4	Rhombic	Cr_3B_4	5.35	22.07	2350	2400	–
25	VB_2	Hexagonal	AlB_2	5.28	29.81	2400	2800	7.5
26	NbB	Rhombic	TaB	7.6	8.2	2380	2195	–
27	NbB_2	Hexagonal	AlB_2	6	–	3000	2600	7.9–8.3
28	Mo_2B	Tetragonal	$CuAl_2$	9.1	5.34	2140	2500	–
29	Mo_2B	Tetragonal	MoB	8.2–8.3	10.13	2350	2500	–
30	MoB-ß	Rhombic	TaB	8.2	10.13	2180	2250	–
31	MoB_2	Hexagonal	AlB_2	7.78	18.4	2100	2200	–
32	Mo_2B_5	Rhombohedron	Mo_2B_5	7.01	22.01	2100	2350	–
33	TaB	Rhombic	TaB	14	5.85	2430	3130	–
34	TaB_2	Hexagonal	AlB_2	11.7	13.5	3100	2500	5.12
35	W_2B	Tetragonal	$CuAl_2$	16	2.86	2770	2420	–
36	WB-α	Tetragonal	MoB	15.3	5.56	2400	3700	–
37	W_2B_5	Hexagonal	W_2B_5	11	12.82	2300	2660	–

Thus it is necessary to analyze the influence of the mentioned stages on the structure and plastic properties of boronized layers to be able to control and effectively use them during the whole technological process of hardening and exploitation.

According to the given classification (Fig. 2.1) there are several boriding processes: in liquid, gaseous, solid and plasma mediums. The common feature for all saturating mediums is the excessive presence of boron in ionic and atomic states. The mass transfer of boron to the processed surface takes place due to the ionic exchange in saturating mediums or the direct contact between a processed surface and a boron-containing material.

7.1 The Formation Mechanism for Saturating Boron Atoms in Ionic Melted Mediums

Ionic mediums include liquid or solid salt electrolytes which are characterized by the ionic type of conductivity. The process of boriding involves the regular use of oxide, oxyhalogenide and halogenide electrolyte systems. They are tetraborates and monoborates of alkaline and alkaline earth metals diluted by halogenides and not diluted, tetrafluoroborates etc.

Boriding in melted salt mediums can be accompanied by electrolysis (with forced electrocuting from an exterior source) or can be carried out without it (by a spontaneous redox-reaction).

Oxide systems (commonly used tetraborates of alkaline metals are among them —$Na_2B_40_7$, $K_2B_4O_7$, $Li_2B_4O_7$) are a mixture of alkaline metal oxides and boron oxide (B_2O_3). Bonds of ions which determine the structure and properties of the melted electrolyte are different for oxide and halogenide systems.

Thus, the presence of ion and covalent bonds in oxide systems allows to divide their cations into two groups. The first group has cations Na^+, Mg^{2+}, Ca^{2+}, Fe^{3+}— their bond with anions O^{2-} is of mainly ionic character. The second group consists of cations Si^{3+}, P^{5+}, Al^{3+}, B^{3+} etc. which have a covalent bond with anions O^{2-}. The second group of cations is a grid-forming group. The covalent bond causes the formation of compound closed anions $(B_4O_7)^{2-}$, $(SiO_3)^{2-}$, $(PO_4)^{3-}$ etc. [1].

The concept of micro-heterogeneity which covers peculiarities of each ion helps to get the better picture on the behavior of melted oxide systems. According to this notion, the oxide system is a micro-heterogenic electrolyte which consists of simple cations from group I, oxygen anions and compound complex anions. Therefore, the cations from group I arc less bound; this determines their electrical conductivity in oxide systems. Their increased mobility leads to the initial reduction during cathode processes under electrolysis and non-electrolysis saturation. It is proved by the fact that smelts on the basis of cations from group II have a rather low electrical conductivity ($7 \times 10^{-5}\ \Omega^{-1}\ cm^{-1}$) [1] due to the strong covalent bond between their cations and anions in closed structure complexes.

One of the most energy-efficient elements in the melted halogenide structure of alkaline metals are groups of highly polarized ions around counter ions with lower polarization; the most stable elements are symmetrically build tetrahedral and octahedral complexes. All vacancies and other ions (cations of group I) are positioned between complexes. The particles of complex ions interact with each other more actively than with the particles out of complexes. This means that, even in halogenide systems, alkaline and alkaline earth metals are the most mobile ones which effects their ability to control the electrical conductivity.

Sumultaneously it is worth mentioning that the presence of group I cations in oxyhalogenide systems weakens strong covalent bonds between group II cations and oxygen which creates conditions for an increased mobility.

Therefore, the regulation of proportion between cations from group I and II in salt smelts for boriding allows to control the mass transfer and manage the technological process during boriding (smelt's toughness, boron concentration on a processed surface and the speed of details' cleansing from the sticked smelt after the end of the hardening process).

For the sake of the mass transfer under the electrolysis of borax ($Na_2B_4O_7$) and other oxide systems, it is better to follow the theory which defines group I cations as primary, i.e. the ones that reduce on a cathode (Li, Na, K, Ca etc.) [2–4].

Finally, taking into account everything mentioned, the most probable formation mechanism for boron saturating atoms at the electrolysis of $Na_2N_4O_7$ can be divided into several stages:

(1) the thermal dissociation:

$$Na_2B_4O_7 \rightarrow 2NaBO_2 + B_2O_3; \tag{7.1}$$

(2) the electrolytic dissociation of monoborate:

$$NaBO_2 \rightarrow Na + (B_2O_4)^-; \tag{7.2}$$

(3) the precipitation of sodium on the cathode:

$$Na^+ + e \rightarrow Na; \tag{7.3}$$

(4) the discharge of boron-containing anions on the anode:

$$(B_2O_4)^- \rightarrow B_2O_3 + 1/2O_2 + e; \tag{7.4.}$$

(5) the interaction of metallic sodium with the smelt:

$$B_2O_3 + 2Na \rightarrow Na_2O + B_2O_2; \quad (7.5)$$

$$B_2O_3 + 2Na \rightarrow Na_2O + 2BO; \quad (7.6)$$

(6) the reduction of boron subions on the cathode:

$$B^{2+} + 2e \rightarrow B; \quad (7.7)$$

(7) the discharge of oxygen anions on the anode:

$$O^{2-} \rightarrow O + 2e \quad (7.8)$$

The peculiarity of the proposed mechanism isn't a direct sodium reduction of boron from its oxides to the atomic state, but a gradual reduction in accordance with the formula (7.5, 7.6) which leads to the break of closed boron-oxygen complexes and weakening of B-O bonds. The obtained compounds—B_2O_2, BO, as well as B_2O_3—can enter the compounds of broken chains or form their own complexes. Saturating boron atoms are generated mostly on the processed surface; they take part in the formation of the diffusion layer [5].

It is necessary to admit that the interaction of new boron-oxygen complexes with metallic Na cab be prolonged; it will be accompanied by the formation of new subions or boron atoms at a certain distance from the processed surface. However the generated atoms won't be able to take a full part in the formation of a boronized layer due to their remoteness from a near-cathode zone of the surface.

The accumulated Na-atoms are gathered in the form of separate particles, transform into a steam and emerge on the surface of a bath burning with the generation of yellow sparks in the near-cathode area. It is well seen during low-temperature boriding when the floating of sodium gas bubbles is impeded due to the high toughness of the smelt. At the end of the processing, Na burns in the air during the extraction of cathode residue and generates frequent sparks. The constantly changing correlation between boron-oxygen complexes and Na oxides in the near-cathode area allows to assume the possibility of their simultaneous reduction in different zones of the processed surface or a consecutive reduction in the same zone. The last assumption correlates with the study results in [44].

It's important to mention a significant difference in the amount of boron atoms which are used to form a diffusion layer and the amount of boron atoms which may be reduced in accordance with the electrical charge passing through the cathode without using Na. This difference is about 10 times, i.e. it takes only 0, 1 of boron atoms from their hypothetical total number to form a diffusion layer. The established fact indirectly proves that the boron reduction is gradual and it leads to the losses and it also shows the existence of possibilities of increasing the efficiency of the process of electrolytic saturation. This was reflected in the technological recommendations on electrolytic boriding with the use of reverse and asymmetric electrical currents, the decrease of its density, the 1.3–1.5X acceleration of layer's growth or putting additives into electrolyte which leads to the generation of subions and increase the effectiveness of the mass transfer [6–13].

The oxy-halogenide systems are characterized by a more intense weakening process of boron-oxygen covalent bonds due to the influence of additional number of group I cations and halogenide anions; they are partly replaced by bonds B-F and B-Cl. The heterogeneity of the system causes the formation of one-valent boron subions in the smelt; they reduce on the cathode with smaller energy losses.

The non-electrolytic process in a liquid medium makes boriding in borate and other smelts with the addition of solid powder-like bonds (electro-chemical reducing agents), similar to oxygen, to follow the electrochemical theory of these processes [8, 14–20].

The nature of this theory is that the mass transfer is carried out by a certain spontaneous precipitation of a more electro-negative metal on the surface of a less electro-negative metal in its salt smelt or by generation and effective spontaneous functioning of galvanic elements during saturation.

Saturated metal (−) (cathode)	Smelt (ionic conductor)	Electrochemical reducing agent (+) (anode)

The polarity of materials of a galvanic element is caused by processes on their surfaces. The predominance of a particular mechanism depends on types of electrolytes and powder additives. It seems that the first mechanism is just a specific type of the second one.

In fact, if two electrochemically different substances (a saturated metal and a more electronegative powder additive) are dipped into a melted electrolyte, the generation and work of galvanic elements begin due to the contact between two different substances; the process of this interreaction is divided into anode and cathode stages. The interaction is accompanied by the migration of electrons from a more electronegative electrochemical reducing agent to the surface of a less electronegative saturating metal; this causes the disbalance on the border between them and smelt (Fig. 7.1).

The anode stage is carried out on the surface of the electrochemical reducing agent (powder additive). It aims to establish a balance and compensate the number of lost electrons by giving new cations to the smelt (by self-melting). The obtained cations form complexes with smelt anions, leaving a sufficient number of electrons on the surface; these electrons will migrate from the place of contact to the processed surface with fewer electrons.

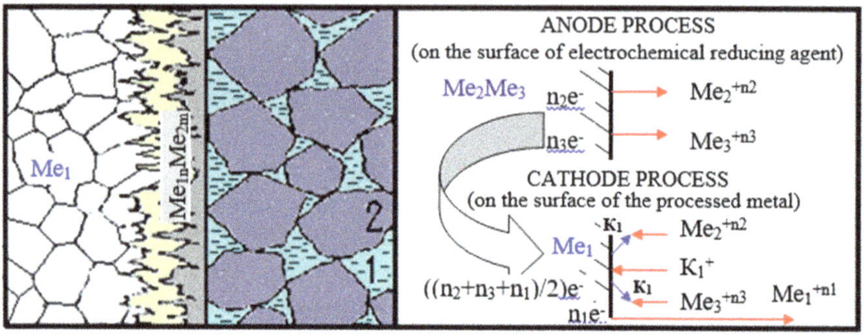

Fig. 7.1 The scheme of anode and cathode processes during non-electrolysis saturation ($n_2 \geq n_3 \geq n_1$). *1*—electrolyte (salt smelt); *2*—electrochemical reducing agent; Me_1, Me_1^{+n1}—saturated metal and its cations; Me_2, Me_2^{+n2}—saturation element (metal or non-metal) and its cations; Me_3, Me_4, Me_3^{+n3}, Me_4^{+n4}—components of electrochemical reducing agent and its cations; K^{1+}—smelt cations of the first group. Technological variants for compositions of salt smelts and electrochemical reducing agent in relation to the saturation element: **I**—*1*—$K^1{}_XO_Y \cdot Me_2O_Z$; *2*—nMe_2mMe_{III}; **II**—*1*—$K^1{}_XO_Y \cdot Me_2O_Z$; *2*—nMe_3mMe_4. **III**—*1*—$K^1{}_XO_Y \cdot K^1F_Y \cdot K^1Cl_Y$; *2*—$nMe_2mMe_{III}$, where n, m, X, Y, Z—coefficients

The cathode stage is carried out on the surface of the processed material; as a result of the excess of electrons, which migrated from the powder additive's surface in the places of contact, it is accompanied by the reduction of cations which compose the electrolyte in accordance with reactions (7.3)–(7.7). Thus, the balance with the smelt is again spontaneously established on this border. The reduced cations may be the cations that were formed in the smelt during the anode dissolution of the powder additive (electrochemical reducing agent).

The speed of cathode and anode stages depends on the difference between equilibrium (or stationary) electrode potentials of powder additives and the processed material, the level of polarizability of their surfaces and the composition of the melted salt and the temperature. The processed surface produces the phase or provides the boron concentration which are less electronegative than the processed surface during the process. The boron mass transfer is controlled by the speeds of its subions generation in the smelt and by their direct reduction on the processed surface during the work of short-circuit galvanic elements [3, 8, 18–20] etc.].

The following three technological compound variants of the boron content in smelts and electrochemical reducing agents are known:

- According to the first variant, both smelts (e.g. $Na_2B_4O_7$) and electrochemical reducing agents (e.g. B_4C) contain boron. The generation of boron atoms may be carried out by the electrochemical reduction in subions under the work of short-circuit microgalvanic elements or by the reactions of subions disproportioning on the processed surface. Boron subions spontaneously appear in the smelt due to the fact that borax is a salt of a boron-containing powder additive [2–5]. The predominance of a certain method depends on the size of electrochemical reducing agent's particles and its chemical composition. In this case the surface concentration of boron is determined by the anode and cathode stages of the process.

- According to the second variant, boron can be found only in a smelt. The saturation is limited by the cathode stage therefore the maximum speed of the process is recommended. It can be done by choosing the composition of the electrochemical reducing agent with the highest electronegativity. It is advisable that the composition of the electrochemical reducing agent should not contain the elements that could negatively affect the process of the cathode stage or change the composition and toughness of the smelt.

During boriding, saturating boron atoms are formed only when the smelt subions are reduced at the work of short-circuit galvanic elements. Boron subions may appear in the smelt due to the chemical interaction of electrochemical reducing agent's cations and smelt's cations, i.e. these cations are capable of breaking boron-oxygen bonds as they are quite similar to oxygen. The metal-thermal reduction of boron into the atomic state is also possible. A cathode stage of a spontaneous redox-reaction process between bath components that have high resemblance to oxygen leads to the formation of Na atoms which may cause the generation of subions.

- According to the third technological variant, saturation is carried out in neutral halogenide salt systems without boron but in the presence of solid boron-containing powders. Boron ions with lowest valency appear in the smelt after some time since the bath begins to function; the smelt becomes a boron salt due to its enrichment with halogenides and/or boron oxides while bath components interact with each other and the furnace atmosphere. The mass transfer in these mediums is controlled by the formation speed of boron ions with the lowest valency, the speed of their transfer to the processed surface and the speed of reduction which can be spontaneous or connected to the work of short-circuit galvanic elements.

All the mentioned technological variants propose the generation of boron phase or the establishment of boron concentration on the surface, both of which make the saturated processed surface more electronegative than the original one.

One of the boriding variants in melted salts is a protector method which is based on the electrochemical nature of saturation in melted salts. Saturating boron atoms are generated on the processed surface as the result of the work of a customized galvanic couple: the processed details and the electrochemical reducing agent. They are connected mechanically on some surface or with the help of a point contact conductor.

The electrochemical protector must be more electronegative that the processed material. During saturation, it must be in a solid (compact or powder) or liquid state (e.g. aluminum and its alloys). In each case certain technological methods of boriding are used. The formation intensity of saturating atoms depends on the difference between electrode potentials of the couple's elements in the used salt smelt and on the correlation between the area of the protector's surface and the area of the detail's surface. The process of boriding is accompanied by the protector's dilution as a result of the anode stage of redox-reaction. A bigger area of the protector's surface signifies a smaller speed of the anode stage. Therefore the working period for a compact protector is limited; if a prolonged working period is necessary, the replacement of protectors is needed.

The distribution of a boronized layer on a processed surface depends on the type of the contact between a detail and a protector, the number of protectors and the correlations of detail and protector's areas. The formation mechanism for boron saturating atoms applicable for the protector method is similar to the electrolysis boriding in boron-containing salt smelts [5].

Thus, the control of boronized layers' phase composition, the level of their alloying and their plasticity during boriding in liquid mediums is carried out through the choice of:

- a composition of the melted salt;
- a type of powder additive with an appropriate chemical composition and grain size;
- a type of powder additive with an appropriate electronegativity;
- an appropriate current density during electrolysis;
- a valid material of hardened details;
- optimal temperature-temporal parameters of the process.

7.2 The Formation Mechanism for Saturating Boron Atoms in Ionic Gaseous Mediums

Any gaseous medium used for boriding can be perceived as a strongly diluted electrolyte, i.e. a substance with ionic conductivity and slow interaction between ions and molecules. This notion doesn't contradict the basic laws of gas mechanics. Therefore the interaction between a gas and a solid body (a saturated surface) may be described by electrochemical effects (the exchange of electrons and ions at the electrochemical heterogeneity of a solid surface).

According to the laws of electrical chemistry, the cathode and anode stages of the redox-reaction process may be spatially divided or may run alternately on the same area of the surface [21], i.e. the speed of disproportioning and exchange reactions can be imposed upon or accelerated by electrochemical effects. In fact the saturation of undiluted, thermally stable BCl_3 in gaseous mediums is accompanied by a saturation process but at a minimal speed. It may be explained by the electrochemical effects on the surface of iron which significantly increase the possibility of exchange reactions on the processed surface, e.g.

$$2BCl_3 + 3Fe = 3FeCl_2\uparrow + 2B; \tag{7.9}$$

$$BCl_3 + Fe = FeCl_3 + B. \tag{7.10}$$

Table 7.2 Isobaric-isothermal potentials of possible reactions in gas and solid mixtures

No.	Reaction type	Temperature values of isobaric potentials		
		1073 K	1173 K	1273 K
1	$B_4C + 4NaF + 2O_2 = 4BF + CO_2 + 2\ Na_2O$	−70	−85	−112
2	$3BF = 2B + BF_3$	−210	−232	−250
3	$2BF = B + BF_2$	−100	−160	−200
4	$2BF_3 + 3Na_2O = 6NaF + B_2O_3$	−138	−146	−152
5	$BCl_3 + Fe = FeCl_3 + B$	+42	+40	+37
6	$2BCl_3 + 3Fe = 3FeCl_2 + 2B$	+27	+20	+12
7	$BCl_2 + Fe = FeCl_2 + B$	−50	−75	−100
8	$2BCl + Fe = FeCl_2 + 2B$	−60	−93	−131
9	$2B2O_3 + 3Si = 3SiO_2 + 4B$	+47	+5	−45
10	$2B2O_3 + Si = SiO_2 + 4BO$	+110	+75	+50
11	$2B2O_3 + Si = SiO_2 + 2B_2O_2$	+30	+12	−70
12	$B_2O_3 + 2Al = Al_2O_3 + 2B$	−120	−200	−250
13	$B_2O_3 + 3CaSi_2 = 6Si + 3CaO + 2B$	+70	+7	−95
14	$B_2O_3 + CaSi_2 = 2Si + CaO + B_2O_2$	+40	−5	−130
15	$4B_2O_3 + 9Mn = 3Mn_3O_4 + 8B$	+80	+50	+20
16	$B_2O_3 + Mn = MnO + B_2O_2$	−60	−100	−150
17	$3BF_2 + 3B_2O_2 = 2(BOF)_3 + 3B$	−160	−210	−250
18	$3B_2O_2 = 2B_2O_3 + 2B$	−70	−120	−170
19	$3BO = B_2O_3 + B$	−85	−130	−190

The process significantly slows down during the formation of $FeCl_3$ on the surface (7.10) due to the fact that this iron chloride is preserved on the surface at boriding temperatures. It may be the explanation for the slow speed of saturation. The progress of chemical reaction (7.9), evaluated by the value of isobaric potential (Table 7.2), is quite probable and become real after applying electrochemical effects.

The dilution of BCl_3 with hydrogen leads to the partial interaction between gases and the reduction of BCl_3 to BCl_2 with the generation of separate bonds B-H, H-Cl or even the reduction to BCl. Therefore, the interaction leads to the formation of lower boron and chlorine compounds, i.e. ions of lower valencies (subions).

Saturating boron atoms in this case appear on the processed surface due to the exchange reactions (Table 7.2) between BCl and BCl_2 or the electrochemical interaction of the subions and the surface by disproportioning reactions:

$$2B^{2+} + Fe = Fe(B) + B^{3+} \tag{7.11}$$

The main rule for the spontaneous mass transfer according to the electrochemical laws is the electronegativity of a saturating element (boron) in relation to the processed material. The quality evaluation of boron electronegativity in relation to iron in the medium of BCl_3 and H_2, carried out by measuring stationary electrode potentials, showed that the electronegativity of this pair completely correlates with the general rule. The boron electronegativity in relation to iron in chloride and halogenide salt smelts on air is determined in the study [22].

The interaction between hydrogen and the processed detail's surface and its activation are not ruled out. Thus, there is a high possibility for the formation of boron subions in a saturating medium during saturation of iron with the clear gases of BCl_3 or the mix $BCl_3 + H_2$. The insufficient dilution of BCl_3 (with the content <0.06% V) leads to the formation of a multiple B^{n+} subions on the processed surface which causes the emergence of boron residue which prevents the following electron exchange and mass transfer.

The use of diborane (B_2H_6), less stable than BCl_3, leads to the mass transfer accompanied by the thermal dissociation of B_2H_6 into elementary boron and boron-hydrogen compounds BH and B_2H_2. It may be assumed that the presence of hydrogen in the diluent causes a saturating gaseous mixture to consist only of boron hydrides, i.e. boron compounds of different valencies, and provokes the generation of saturating boron atoms on metallic surfaces according to the disproportioning reactions. The possibility of boron reduction involving the use of electrochemical effects should not be eliminated.

The control of boronized layers' phase compositions and their plasticity during gas processes is carried out the following way:

- a choice of a boron-containing gas;
- a choice of a gas-carrier and a level of saturating medium dilution;
- establishing a certain pressure and speed of a gas flow;
- using electrical-physical methods of impact;
- a reasonable choice of the material of a hardened detail or the type of preliminary processing;
- a choice of optimal temperature-temporal parameters of the process.

7.3 The Formation Mechanism for Saturating Boron Atoms in Solid Mediums

Solid mediums is the name for saturating mixtures on the basis of powder, granular or composite materials which partly or fully stay in a solid aggregate state during the process of saturation. These mediums are mixtures of materials with ionic (a salt compound) and electron (boron compounds) conductivity. The result of their chemical and electrochemical interaction at an increased temperature is the formation of a gaseous saturating medium which determines the boriding process. The continuity of the process is provided by the gas exchange with the external medium and the ionic and electron exchange with the processed surface.

The mass transfer is done by boron subions during boriding in a gaseous medium obtained at the interaction of boron-containing substances (B_4C, B, ferroboron, MgB_{12}, CaB_6 etc.) with activators (KBF_4, $NaBF_4$, NaF, AlF_3, NH_4Cl etc.) without their direct contact. Gaseous mediums keep them in the form of the following compounds: BO, B_2O_2, BCl, BCl_2, BF, BF_2 [23–25].

The spontaneous reaction progress is explained by the system transition from a less stable state into a more stable state of heterogeneous surfaces which are found in one closed volume (e.g. the surface of particles B_4C and the processed surface).

The possibility of reactions is evaluated by a value of isobaric-isothermal potential (ΔGo):

$$\Delta G^0 = \Delta H_T^0 - T\Delta S_T^0, \tag{7.12}$$

where ΔH_T^0—the change in the enthalpy as a result of the reaction at temperature T, K; ΔS_T^0—the change in entropy.

The condition for equilibrium is $\Delta G^0 = 0$, the reaction can go from left to right only at its negative values.

When the saturation process ends or an "idle" heating mode of the container without processed details is on, the gas formation slows down; in case of the complete decapsulation (e.g., heating in a sealed retort), the gas formation can spontaneously stop as there will be no conditions for reaction products to leave a closed medium ($\Delta G^0 = 0$). The process of gas formation should be attributed to self-organizing processes which are characterized by the balance of two stages of the redox reaction process in a gaseous medium—the formation of boron atoms with lowest valency and their reduction and obtaining boron atoms on the saturated surface. The surface of the metal in this case is less electronegative than boron. The products of boron atoms formation reactions are gas boron compounds with higher boron valency (Table 7.2).

The formation of B_2O_3, which precipitates on the surface in the form of glass-like films, leads to the deceleration of the saturation process. The negative impact of B_2O_3 is neutralized when a gas-forming mixture of aluminum-containing activators is added.

A gaseous medium for contact-free boriding is also acquired during the scavenging of a powder or a granular mixture by gas BF_3. The process involves fluoric activators containing decelerators of decomposition reaction for BF_3 [26]. The authors used silicon carbide as a decelerator; when gas BF_3 went through it, the gas saturating medium's activity decreased due to the formation of unstable compounds $(BOF)_3$ (Borudif Technology). The unstable compounds formed complex dimensional constructions. These compounds, in our opinion, can't provide a directed boron mass transfer to the processed surface in a saturating medium. Taking into account the similarity between a mass transfer process and oxygen deficit in most cases of gas saturation, it is necessary to conclude that the boron mass transfer is provided by boron subcompounds (boron ions of lower valencies) under the process of boriding according to "Borudiff" technology [26]. In this case $(BOF)_3$ in saturating medium decomposes into simple compounds which have different boron valencies—they play a transport role for the mass transfer in saturating mediums.

The activity of gaseous mediums may be determined by measuring stationary electrode potentials of the studied metallic samples in reliance to the control electrode (graphite or carbon-oxygen). The largest difference of potentials between them will define the activity of boron in the saturating medium during the change of the gas flow, i.e. "boron potential" of the medium; the comparison of different samples will show the boron activity in studied metals.

Saturation in solid powder mediums (contact way of boriding in gaseous mediums) doesn't change the procedure of mass transfer—it goes according to the given regularities of saturation in gaseous mediums. However the role of electrochemical effects will even rise due to the contact of the electronegative powder component of the mixture and less electronegative processed surface. Thus, the mass transfer will be determined by the number of more mobile boron subcompounds in a gaseous medium. Some portion of mass transfer is ensured even in the areas of direct contact between the processed surface and the boron-containing material, but this process isn't very relevant for the formation of boronized layers.

The regularities of the mass transfer during chemical-thermal processing with the use of ions of lower valency are quite fully described in the literature and proved from the point of thermodynamics. The evaluation of isobaric potentials during the formation of boron compounds of low valencies show rather a high possibility of their emerge (Appendix 6.2) and their spontaneous reduction on the processed surface. Electrochemical regularities or different electrical-physical impact on the surface may influence the speed of mass transfer, as described for other instances.

Pastes, coatings or saturating models, which stay in a solid state during saturation, are defined as compact materials (composites); in cases when these materials are used, the realization of gas contact or partially non-electrolysis mass transfer in liquid medium is possible. In both cases the mass transfer is caused by the formation of boron compounds.

When boriding is carried out in the presence of glow-discharge, ionic processing, corona discharge and other processes, the saturating medium near the processed surface would be in an ionized state and contain all the subions which determine the saturation process. Normally saturation is carried out in vacuum or a controlled atmosphere.

These mediums can be used for boron mass transfer by the direct contact of boron-containing material and the processed surface. However it isn't primary.

The control of the phase composition of boronized layers and their plasticity during saturation in solid mediums can be done the following way:

- choosing a saturation method;
- choosing a composition of a saturating mix and a level of its density;
- controlling the level of a container's capsulation;
- applying electrical-physical and/or mechanical methods of impact;
- applying scavenging of saturating medium by a gas-carrier;
- choosing an appropriate detail's material or preliminary processing;
- choosing optimal temperature-temporal parameters for the process.

7.4 The Mechanism of Mass Transfer in Mediums Containing Atomic Boron

When boriding is carried out in mediums which contain atomic boron, it takes place in the following combinations of the processed surface and saturating medium's states.

- Solid + solid: the mass transfer is carried out slowly due to the absence of substances which would facilitate the activity of this exchange. This process can take place in solid mediums; the contact between clear surfaces of the processed material and the saturating element is strong; it takes place in vacuum. This method is also used for annealing galvanic, electrochemical and other precipitated surfaces. This method also involves additional processing of boronized layers in solid states if they were obtained by other methods to make them more plastic or to generate specific structures. The mass transfer in these cases is done in the areas of contact between heterogeneous materials due to the diffusion exchange.
- Solid + liquid: the process is carried out in melted metals where the saturating element is in a dissolute or dispersed state. The mass transfer is provided by the gradient of concentrations on the border between the processed metal and the saturating medium.
- Liquid + solid: the process of generating boronized layers on foundries takes place while pouring a liquid metal into a mold with the coating made of a boron-containing mixture. The mass transfer is caused by the melting accompanied by the dissolution of solid boron-containing additives or by the combination of alloying with the mass transfer through a gas phase which is formed in the saturating mixture put into a porous ceramic side of the form. Partial alloying is observed during saturation in powder mixtures while fusing the processed surface.
- Liquid + liquid: the process also takes place while pouring a liquid metal into a mold which has a coating that partly melts during filling and gets soaked with the dominant metal of casting.

The control of the phase composition of boronized layers and their ductility during saturation in mediums that contain atomic boron is carried out by making consecutive choices of:

- a method of saturation with a certain combination of a processed surface and saturating medium's states;
- a composition of saturating mixture;
- a material of a hardened detail and preliminary processing;
- an optimal temperature for filling which provides the formation of a quality molding and a boronized layer.

7.5 The Stages of Boriding and the Control Over the Speed of Their Progress

Boriding as any other chemical-thermal process consists of three interconnected basic stages.

- The formation of saturating boron atoms on the processed surface. The formation speed depends on the medium's state of matter, the composition of the saturating medium, the character of interaction between the medium components and with the processed surface, the circulatory processes etc. The amount of atoms, formed in a unit of time, characterizes the power of the diffusion source.
- The adsorption of the generated boron atoms by the surface of the processed details, i.e. the formation of strong interatomic bonds (ion, covalent, metallic, molecular (Van-der-Vaals)) with the surface. A particular type of bond is determined by the character of interaction between the boron and the metal in the expected phase. The strongest bond is ionic (hetero-polar) and chemical; the weakest—molecular.
- The diffusion of boron into the depth of the metal with the generation of a boronized layer. The formation of a diffusion layer is possible at the constant adsorption of boron atoms by the surface and the presence of a diffusion flow directed towards the surface. The speed of the layer's formation in solid solutions depends on the boron diffusion coefficient in solid solutions; the boride formation depends on the coefficient of boron diffusion in borides.

The thermodynamic system at the first stage of chemical-thermal processing in a saturating medium can be analyzed as a system which transforms the external heat energy into the energy of chemical interaction of the components of the mixture and the solid body. The regularities of this transformation are determined by the processes' kinetics and the type of open thermodynamic systems where the exchange of energy and substance with the environment goes in accordance with the first and second laws of thermodynamics. These systems in certain conditions of interactions between energy flows and substance are places for processes of matter reordering that correspond to the entropy fall and the formation of self-organizing dissipative structures. The formation of such structures is possible if:

- the system is thermodynamically open, i.e. exchanges the energy and substance;
- the dynamic equations of the system are non-linear;
- the deviations from the equilibrium excess critical values;
- macroscopic processes take place cooperatively (synergistically).

In accordance with the thermodynamic theory of structure, stability and fluctuations [27], any system in chemical equilibrium ensures the generation of compensatory processes which try to weaken any changes occurred in one of the factors controlling this equilibrium. This system organization is the foundation of Le Chatelier-Braun's principle (the Equilibrium Law). This principle describes the system's reaction to spontaneous fluctuations. At a certain stationary system state, characterized by minimum entropy, these fluctuations decrease as in the thermodynamic equilibrium. In this case the principle is also valid.

The compensating processes that counteract the influence of changes in the system lead to the formation of dissipative structures (compounds of elements) which appear and stay due to the exchange of energy and substance with the external environment in metastable conditions. The dissipative structure is a temporal and dimensional material ordering in open systems outside the area of thermodynamic stability.

The specific field of interest is instabilities which break the symmetry (equilibrium); they lead to the spontaneous self-organization of the system in its dimensional order and its functions. In this case the dissipative structure correlates with low levels of entropy in this open system. Outside the equilibrium, different chemical reactions may compensate the diffusion influence and cause the formation of ordered structures on a macroscopic scale. The flow of energy may regulate systems and increase or decrease the entropy production. It can change the entropy production by adding a new mechanism of dissipation.

Self-organization is observed in any saturating medium in special conditions for each case. The energy analysis of the mass transfer process (Fig. 7.2) reveals the three levels of activation—insufficient, normal and excessive. Each level correlates with a certain combination of system reactions; only normal activation is appropriate for the purposes of mass transfer process control. System self-organization accompanied by the generation of dissipative structures is expected on other levels. However the results of these kinds of self-organization don't meet technological requirements of the process.

The boron subcompounds which are found in saturating mediums are dissipative structures (compounds) of self-organization; they guarantee the mass transfer and disappear at the end of the saturating process.

The energy, necessary for saturation, i.e. for the formation of subcompounds (G_{sc}), is defined as a sum of its composites that guarantee the mass transfer. They include the energy of subcompounds formation ($G^{form}{}_{sc}$), the energy of compounds transportation ($G^{trans}{}_{sc}$), the energy of their reduction ($G^{red}{}_{sc}$) and the energy for the withdrawal of the reaction products from the processed surface ($G^{wit}{}_{pr}$). Each of these energies is a function of internal system parameters (Fig. 7.2): the temperature (T), the medium composition or the saturating element content (C), the area of gas-forming surface or the size of the material's particles (Sr), the internal pressure of a gaseous medium (P), a type of a processed material (M), the speed of circulation or medium movement (V), the electrophysical impact (EPI) and the temperature of the metal during the electrophysical impact (Tm).

The sum of these energies defines optimal conditions for the mass transfer that would provide the maximum (or set) concentration (C_n) of a saturating element on the surface. The hypothesis is that each constituent can be calculated and evaluated from the thermodynamic parameters of interacting medium components.

The quantitative evaluation of energy $G^{form}{}_{sc}$ and $G^{red}{}_{sc}$ can be done using the value of isobaric-isothermal potential (ΔG^{o}) of correlating reactions at the contact

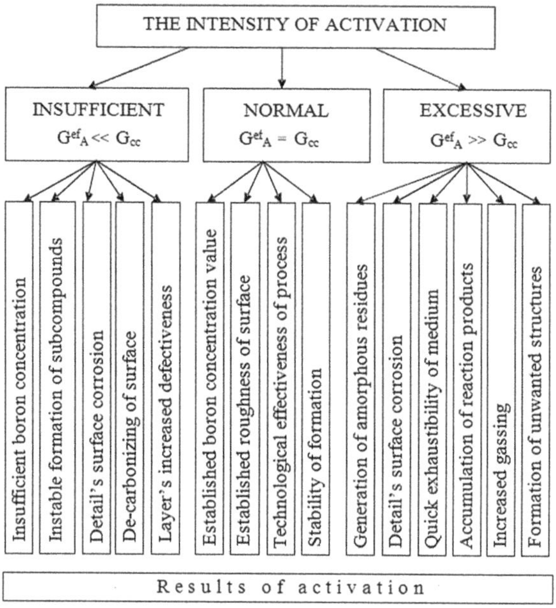

Fig. 7.2 The energy classification of mass transfer in a saturating medium. G_A^{ef} the energy of activation that provides mass transfer; G_{cc} energy necessary for mass transfer $G_{cc} = G^{form}{}_{sc} + G^{form}{}_{sc} + G^{red}{}_{sc} + G^{wit}{}_{pr}$; $G^{ef}{}_{A} = (T, C, Sr, M, EPI)$; $G^{form}{}_{sc} = (T, C, P, Sr, EPI)$; $G^{form}{}_{sc} = (V, P)$; $G^{red}{}_{sc} = (T_M, M, V, EPI)$; $G^{wit}{}_{pr} = (T_M, M, V, EPI)$

and contact-free methods in gaseous mediums. Electrolysis and non-electrolysis methods in liquid mediums demand the evaluation on the basis of the density of a used current (i) and the difference in potentials of a short-circuit galvanic element ($\Delta\varphi$).

The evaluation of the energy needed for the withdrawal of reaction products ($G^{wit}{}_{pr}$) should take into account the state of reaction products and the value of the energy of their evaporation or dissolution in the environment. It may also be the partial pressures of reaction products and their possible interaction with the saturating and external mediums.

The energy of transportation $G^{trans}{}_{sc}$ in circulatory and direct-flow methods can be measured on the basis of the energy used for the formation of the flow at the balanced distribution of compounds.

The effective energy of activation $G^{ef}{}_{A}$ is a part of external energy used for establishing the conditions for a saturation process. It can be evaluated on the basis of the total value of energy used for the hardening process without the energy used for the preliminary heating of a saturating medium, details and heating equipment and the energy losses during saturation etc. It is the energy used for the compounds formation, their transportation, reduction and withdrawal of reaction products. It determines the temperature of the process at a set value C_p, the speed of gas

Energetic conditions	Kinetic conditions	Temperature and physic-chemical parametres
$G^{ef}{}_A = \Sigma\, G_{sc}$ $K_{st} = G^{red}{}_{sc} / G^{ef}{}_A$ Minimization of stocked energy, K_{st} -> min	The cyclicity of formation, transportation of subcompounds and withdrawal of reaction products Coordination of process speeds $mv^{form}{}_{sc} = nv^{trans}{}_{sc} = kv^{with}{}_{pr}$ $t = mt_1 + nt_2 + kt_3$	The temperature of mass carry process -> opt The stability of subcompounds -> min The valency of subcompounds -> min The circulation of medium -> opt The boron concentration on the surface C_n->C_{set}

The indicators of self-organization
The speed of surface formation -> max
The stability of saturation process in time -> max
The state of processed surface -> opt
The layer's phase composition -> to set parameters

Fig. 7.3 The structural-energetic complex of conditions and parameters of self-organizing processes in saturating mediums

formation, the circulation and pressure of gases, the level of electrophysical influence etc.

In cases when the temperature is a set parameter, one can change the medium composition, its circulation speed and internal pressure to control the mass transfer to provide the optimal level of self-organizing.

The quality correlation of self-organization parameters gives a structural-energetic complex which combines energy and kinetic conditions and physical-chemical and temperature parameters (Fig. 7.3).

The main energetic condition is the tendency of a stored energy to the minimum ($K_{st} \rightarrow$ min). The most important kinetic conditions are the necessity of coordination between the process speeds of subcompounds' formation ($V^{form}{}_{sc}$), their transportation to the processed surface ($V^{trans}{}_{sc}$) and the withdrawal of reaction products ($V^{wit}{}_{pr}$). Simultaneously the optimal factors should be provided for certain conditions: the medium's temperature, the metal's temperature and the electrophysical impact (T, T_m, EPI), boron concentration (C_p) on the surface—all of these factors lead to the maximum speed of boronized layer's growth. It is proved by the development of methods and composites which ensure the maximum saturation speed; the use of asymmetrical and reverse currents; the use of alternate electrolyze periods and exposures etc. Coefficients *m, n* and *k* take into account the difference in speeds, subcompound losses and volume changes of substances after acquiring reaction products. The saturation in closed volumes without gas-carriers equalizes these coefficients: the process of gas formation is carried out spontaneously at the right choice of a gas-educing mixture; when a self-organizing mode is established, the number of newly formed subcompounds correlates with the number of compounds that reacted with the surface. The total time of the cycle should be optimal:

it is supposed to relate to the time needed for the generation of adhesive bonds between boron atoms and a metallic substrate and the transfer of these atoms from the surface into deeper layers.

The continuity of boron atoms transfer into the metal requires at least two atomic layers: adhesive layers of atoms with bond B-Me and an atom layer with bond B-B ('free layer'). When the concentration on the surface is set, the coordination of speeds of the bond generation and their continuity should ensure this concentration by decreasing the number of boron atoms in bond B-B, i.e. decreasing the number of boron adatoms (atoms, adsorbed by the saturating surface) in the "free layer".

The physical-chemical and temperature parameters include additional conditions for providing a set or maximum concentration of boron on the surface.

The indicators of self-organization are the maximum speed of saturation, the process stability, the state of the processed surface and a certain phase composition of the layer.

Thus, the inflow of boron ions to the processed surface with the purpose of obtaining the maximum speed of saturation needs to be correlated with the maximum available number of atoms which are necessary to go from the processed surface to the deeper layers of substrate in certain conditions during the formation of the layer of a particular composition.

As a result of the electron exchange of boron subcompounds (ions of lower valencies) with a metallic surface, boron atoms are formed. *The second process stage*—the adsorption of formed atoms—goes intensely and is similar to the chemisorption process in its character and intensity. The first stage of the interaction is hypothetically the formation of a covalent bond B-Me which later transforms into a covalent-metallic type. When subsequent atoms of boron go through the process of adsorption, previously adsorbed boron atoms also take part in it which significantly increases the share of covalent bonds.

The described regularities are the reason for the following processes: if the amount of boron atoms on the surface is sufficient, boriding is accompanied by the quick accumulation of adsorbed atoms and the generation of α-solid solution of boron in iron and Fe_2B boride. If the amount of boron atoms is insufficient for forming Fe_2B, we can witness the formation of a layer based only on α-solid solution of boron in iron. The excessive amount of boron atoms (more than necessary for the formation of FeB phase) leads to the formation of an "amorphized" boron layer which is often found after electrolysis and non-electrolysis saturation in liquid mediums, after saturation in gaseous mediums and sometimes after saturation in solid (powder) mediums.

The formation of such "amorphized" layers causes the deceleration of saturation process as it impedes the electron exchange between medium subions and the surface without any significant drops in the electrical conductivity of inter-electronic areas. This leads to the conclusion that the "amorphized" layer is the mix of ions (adions) and atoms (adatoms) of the boron adsorbed by the surface. The subions' adsorption takes place due to the formation of bonds with some adatoms, i.e. the formation of B-B bonds. Thus, the adions provide the necessary conditions for the electronic exchange between the processed surface and the saturating medium.

The diffusion stage of boriding involves the volume boron diffusion (in α-solid solution of boron in iron and in borides), the surface boron diffusion (on the borders of grains, sub-grains and blocks of the substrate and on the border between borides and α-solid solution), the self-diffusion of iron atoms (increasing diffusion) and the diffusive redistribution of alloying elements in the substrate, sublayer and borides. All these processes are somehow involved in the formation of properties: in particular, the plasticity of boronized layers.

The volume diffusion largely depends on the crystal structure of the material and the presence of defects—vacancies and dislocations. The main role in the diffusion of solid solutions is played by non-equilibrium defects (vacancies); their number depends on the external impacts: rapid electro-heating, cold plastic deformation, preliminary quenching etc. The formation of non-equilibrium vacancies is accompanied by the refining of metal grains' structure and substructure; it improves the role of surface diffusion in the heterophase system. If the boron content is higher than the limit of dissolubility in the metal's structure, boride compounds form on the grain borders. Such formation of separate boride compounds is often seen during boriding in the sublayer of boride zone. Therefore, the speed of formation of solid solution of boron in iron and the number of centres for forming boride crystals are the instruments of controlling different types of preliminary or simultaneous processing of a processed metal's surface.

The analysis of regularities for boride needles' growth, based on the predominance of surface boron diffusion on the border between borid and solid solution, allowed to mathematically explain the predominant growth of borides in the form of cylinders [28, 29] (Fig. 7.4, colour inset).

The obtained results of calculating a parabolic boride growth satisfactorily corresponds to the experimental results. However, due to the made assumptions, this interpretation doesn't explain the reasons for a boride growth in the form of

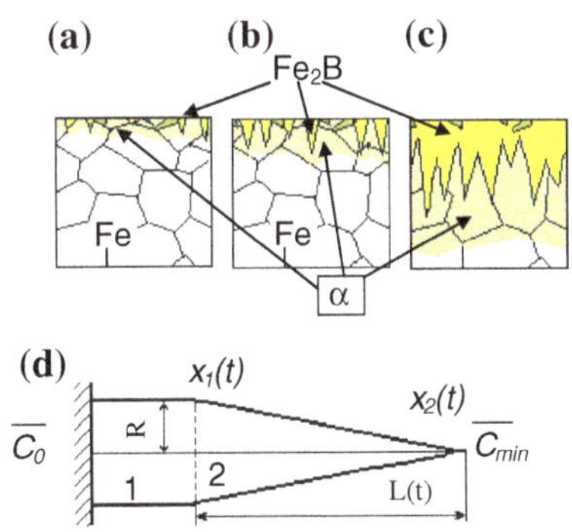

Fig. 7.4 The scheme of growth pattern for boride needles and pinching out of unfavorably positioned crystals [28]: **a**, **b**, **c** consequent stages of boride crystals growth; **d** schematic presentation of separate boride needle; 1, 2—cylinder and conical parts of boride crystals; α—*solid* solution of boron in iron; *yellow color* boride crystals favorably positioned towards the diffusion front; *gray color* boride crystals unfavorably positioned to the diffusion front

needles and the change in the level of acicularity while boronizing the same material in dependence with the saturating medium, i.e. in dependence with surface events and the crystal structure of borides themselves.

A mathematical approach to the analysis of the parabolic growth of boride needles, used in [30, 31], with the multiple references to the experimental data showed the predominant role of volume diffusion in forming boride needles and allowed to create a mathematical model which explains the reason for the growth of needles positioned perpendicularly to the processed surface.

At the same time a certain role is played by the surface diffusion. However the mathematical analysis of both individual and combined influences of the surface and volume diffusion of boron didn't show any reasons for changing the character of acicularity (a correlation between conic and cylindrical boron compounds) that characterize the boride layers depending on the external conditions.

The volume boron diffusion in borides is mostly connected to their crystal structure. The presence of varied bonds in borides leads to the formation of structural elements outside the boron atoms in their crystal lattices (Fig. 7.5).

The presence of boron chains determines the generation of through channels in boride crystal lattices or whole boron-free planes [32–35]. The quality and quantitative anisotropy of boride crystals according to the density of atoms (a low level of symmetry) provides the anisotropy of a diffusion coefficient and anisotropy of boride crystals growth during boriding.

Therefore, the acicular structure of boride layers is determined by the anisotropy of boron diffusion coefficient in the crystal lattice of borides. The presence of through channels and boron-free planes in directions [001] and [100] [36, 37] leads to the predominant growth of Fe_2B crystals in direction [001] parallel to the diffusion flow. Plane [002] in this case is parallel to the saturated surface. The formation and growth of FeB crystals is accompanied by the crystal-graphic orientation of the Fe_2B matrix lattice and formed FeB boride, i.e. FeB crystals are also oriented in the crystal-graphic direction [001]. This influences the formation of residual tensions and the plasticity of the boronized layer.

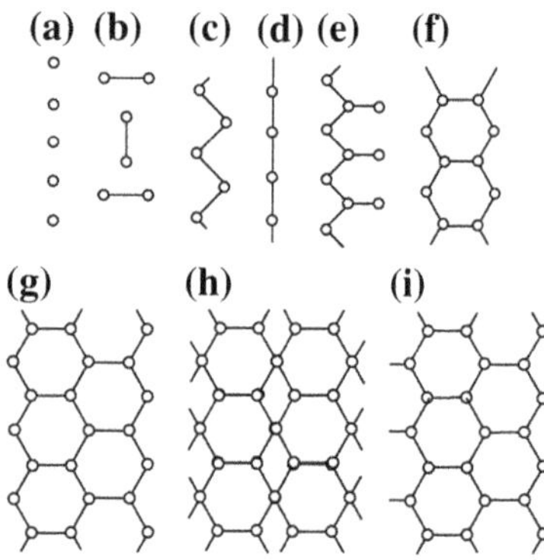

Fig. 7.5 The structural elements of boron atoms in borides [33–35]: **a** isolated atoms; **b** pairs of atoms; **c** zigzag chains; **d** direct chains; **e** branched chains; **f** double chains; **g** triple chains; **h** crimpled flat grids; **i** flat packed grids

The increasing diffusion of iron atoms and diffusive redistribution of alloying elements in the substrate, sublayer and borides are caused by the similarity of these elements to boron. When iron atoms and alloy elements come to the surface, they take part in the formation of boride layers and provide the growth of the detail's size (<15% from the boride layer's width).

In most cases the layer's growth is controlled by diffusion processes. Therefore the acceleration of these processes can be done by different methods of electrical-physical impact: electromagnetic fields, ionization of the environment, electrical or glow currents, cathode-ray processing, the use of electrolysis for creating a directed flow in the environment, laser processing, high frequency current heating etc.

7.6 The Formation Mechanism for Diffusive Boronized Layers in Steels

The boron concentration on the surface determines the phase composition of the diffusion layer according to the equilibrium diagram. When iron is boronized in a medium which provides a boron concentration on the processed surface at the value no less than 8.4 mas.%, layers, based on Fe_2B + α-solid solution, are formed; at boron concentration from 8.34 to 16.25%—layers based on FeB + Fe_2B + sublayer. Layers on the basis of α-solid solution may be formed on iron at concentrations less than 0.02 mas.% B.

Each of obtained structures has its advantages in certain exploitation conditions and may have a significant effect. Two-phase layers with micro-hardness up to 2200 HV (FeB + Fe_2B) have advantages in abrasive wear processes in the absence of impact stresses; one-phase layers with micro-hardness up to 1700 HV provide a maximum wear resistance in the presence of impact stresses and cyclic loading due to its plasticity reserve. Layers on the basis of α-solid solution of boron in iron in medium-carbon steels after hardening increase hardenability, durability, wear resistance (by generating boron on the grains borders) and can ensure the high work capacity for a wide range of details working in the presence of grease.

Simultaneously the best effect is provided by boride layers with a certain plasticity reserve and by layers with a composite structure. The appearance and growth of borides and the formation of a composite structure of a boronized layer may take place in solid, liquid-crystal and liquid states of the processed surface.

Normally two formation mechanisms are studied: diffusion and diffusion-crystallization.[1]

[1]The term "diffusion-crystallization mechanism" is proposed in [5] instead of using the term "liquid-solid mechanism" [39] which reflects only an aggregate state of the zone where saturation takes place. The new term reflects both the aggregate state of this zone and the way the process of forming a new layers goes.

7.6.1 The Diffusion Formation Mechanism for Boronized Layers

The formation of borides in accordance with the diffusion mechanism takes place in a solid state of the substrate: it may be accompanied by the primary formation of heterogeneous solid solution of boron in iron or spasmodic changes in boron concentrations along the width of the diffusion layer which characterizes a reactive diffusion [38]. The insufficient boron concentration for boron formation leads to the formation of α-solid solution of boron in iron alone.

The initial period of saturation is characterized by the difference of the phase layer's composition and borderline phase concentrations from the equilibrium values. Taking into account this fact and the unevenness of boron distribution in metals and alloys [40], an assumption can be made: a primary formation of solid solution, uneven in concentrations, is highly possible. In this case boron is concentrated on grains and blocks' borders, in places where dislocations go through the surface, on package defects, in places where single-point defects are concentrated etc. These areas, due to the boron's horophility, provide a maximum solubility of boron in iron and a level of concentration which is sufficient to form a Fe_2B boride.

These statements are proved by the results of electron metallography and the X-ray study of surfaces in samples of ARMCO-Fe after short-term boriding. In particular, the formation of separate areas of Fe_2B phase during boriding in borax smelt with 40% B_4C becomes fixed in 5 min [24], during electrolysis boriding in a borax smelt—in 10 s [41]. The most of the surface in these cases is α-solid solution of boron in iron.

Thus, if the limit of boron solubility in iron is exceeded, the surface is covered first with areas and then with the whole layer of Fe_2B phase accompanied by the formation of a new interphase border Fe_2B—α-solid solution of boron in iron in accordance with the equilibrium diagram of alloys. The further soaking leads to the rise in boron concentration on the surface to maximum values, which are determined by the equilibrium conditions of the surface and the saturating medium, while the composition of Fe_2B and α-solid solution of boron on their borderline stays the same. The intensity of Fe_2B boride layer growth is provided by the boron concentration on the surface and the speed of boron diffusion in its crystal lattice without involving changing concentration conditions in the sublayer. The formation of FeB phase on the surface (if boron concentration on the surface is sufficient) is accompanied by the emergence of a new phase border FeB-Fe_2B and one more diffusion limiting factor—the speed of boron diffusion in FeB boride lattice.

The formation of a layer according to the diffusion mechanism is met by the diffusion of iron atoms; iron atoms are used in the formation of borides when they appear on the surface. Their diffusion increases details' size by 10–15% of the boride layer width.

In general, the formation of a boronized layer begins in non-isothermal conditions during heating. The process of boron atoms adsorption can be detected even at 500 °C and is accompanied by the appearance of α-solid solution. The following heating of the processed surface causes the formation of a layer according to the described pattern.

Thus, the formation of a layer according to the diffusion mechanism goes through several stages:

- the formation of phases in non-isothermal conditions;
- the formation and growth of the boronized layer's phases (borides and a sublayer) in isothermal conditions at constant and unstable concentrations of boron and alloy elements on the surface;
- the redistribution of the layer's phases and the formation of a sublayer during the process of slow cooling or further thermal processing.

7.6.2 The Diffusion-Crystallization Formation Mechanism for Boronized Layers

The diffusion-crystallization mechanism is used during saturation when the processed surface is in a liquid-crystal or liquid state. In both cases the saturation process is accompanied by the diffusion of boron and alloy elements and the crystallization of a liquid solution; these processes are detected during isothermal soaking and also during heating or cooling. The presence of a liquid phase significantly increases the diffusion mobility of atoms which leads to the formation of wider boronized layers with a heterogeneous structure and improved plasticity.

The diffusion-crystallization formation mechanism for boronized layers requires a creation of special conditions for generating a liquid-crystal state of the processed surface with the correlation of liquid and crystal phases which provide the preservation of the detail's form and coarseness of the detail's surface. According to the general laws, the amount of liquid phase shouldn't exceed 30%. A solid phase must be presented by unmelt areas of matrix solid solution or crystal-forming borides that make up a solid carcass. The temperature and concentration conditions for this equilibrium are determined by the multicomponent equilibrium diagrams of smelts.

The diffusion-crystallization mechanism can also be used while forming a liquid-crystal state in a boride sublayer by forcing out or redistributing alloy elements. In this case the amount of liquid phase should also be preserved at the same level (<30%).

The diffusion-crystallization mechanism involves the following stages to form a boronized layer:

- the formation of a diffusion zone and borides in real conditions of heating in production;
- the transfer of a generated layer and its areas into a liquid-crystal state at reaching a certain processing temperature; it is accompanied by a sharp increase in the total boron mass transfer;

- the further development of the layer at isothermal soaking (boride crystallization from liquid solution and simultaneous melting of new matrix areas);
- the crystallization of liquid residues during cooling;
- the secondary crystallization (the isolation of disperse borides from oversaturated solid solution) during further cooling or additional aging or tempering.

The formed structure of a boronized layer consists of separate inclusions of the borides of different dispersibility distributed in a comparatively soft solid solution. Such a structure in some way creates conditions for following the Sharpie principle during a wear process and is good for friction points. Moreover, the control over the phase correlation, their dispersibility, the form and mutual position allows to change the properties of boronized layers in wider areas. This creates the foundation for expanding the usage of boriding. The formed structures are called pseudo-eutectic in comparison with the eutectic structures which are obtained after the complete melting of the layer (Fig. 7.6.) or during the melting of composite electrolyte surfaces.

The diffusion-crystallization mechanism is used when a metal is poured into a melting form, the interior walls of which are covered with a certain coating which contains, e.g., boron carbide. The layer formation goes through certain stages [42]:

- the partial destruction of the coating layer by the liquid metal and the mechanical agitation of separate boron carbide particles with the liquid;
- soaking of B_4C particles and their dissolution in the liquid;
- the convective agitation and flattening diffusion of boron into a certain depth of the mold's surface;
- the crystallization of the layer and the mold; the final diffusion formation of the layer in the solid state during cooling.

The structure of the layer changes during the crystallization of molds with different sections. In particular, if the crystallization's speed is high (the mold's wall

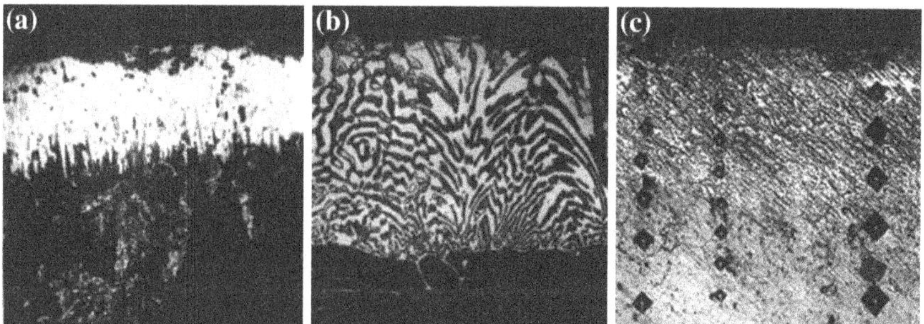

Fig. 7.6 The microstructures of boronized layers obtained by the diffusion (**a**) or diffusion-crystallization (**b**, **c**) formation mechanisms: **a** steel 45, × 320, 950 °C, 4 h, electrolysis saturation at j = 0.15 A/cm^2; **c** the eutectic layer, × 100 (saturation from a coating under high frequency current heating); **b** a pseudo-eutectic boronized layer × 100, smelt BHC-45, 1050 °C, 1.5 h, powder saturating mix containing B_4C

is thin) then the agitation is slight; the result is the increased boron content on the mold's surface. Thus, thin-walled molds get boronized layers with a high boron content and boride eutectics. If the wall is wider, the agitation of boron-containing substances and metal is witnessed and a boronized layer grows into the metal. The boron content along the layer's depth is evened out which allows to obtain an even structure. The speed of crystallization is relatively low, boronized layers are wider but contain less boron.

Saturation from ceramic melting forms (saturating mix is part of it) leads to the formation of boronized layers which goes in several stages [25, 43]:

- the formation of a layer during molding by the direct contact of the metal with the form's surface and agitation;
- the boron diffusion from a coating material without its melting during mold's metal crystallization;
- final formation of boronized layer at isothermal soaking or without it and following cooling.

Boron from a porous ceramic fire-proof material goes to the surface though a gaseous phase which is formed as a result of B_4C interaction with a saturation activator.

The diffusion-crystallization mechanism of boronized layers' formation is also realized during saturation in solid coatings or linings made in saturating mixtures during high frequency current heating [44, 45]. The best results are provided by a close control of the temperature. A set temperature mode is achieved by a multiple thermo-circulation: heating up to 1100–1150 °C during 1–2 min; soaking (light cooling) for 1.5–2.0 min (the inductor is turned off) [45].

The formation of a boronized layer goes through the following stages:

- the formation of a layer under heating according to the diffusion mechanism in a gaseous medium;
- the formation of a layer in a liquid-crystal state;
- the crystallization of liquid residues in the surface layer during cooling;
- the final diffusion formation of a layer in a solid state during cooling.

The obtained composite boronized layers with disperse structures are characterized by the increased plasticity.

7.7 Gas Formation Patterns in Boriding

The technology of boriding in gaseous phases includes the use of gas, solid and liquid original substances. Currently the most productive method is the use of solid and liquid substances as it completely eliminates the toxicity and explosiveness during the process of boriding. In all instances the sources of saturating atoms are: B_4C, B_2O_3, $Na_2B_4O_7$, H_3BO_3, KBF_4, $NaBF_4$, boron, borides of metals, alkyl-borides etc.

A saturating gas medium is obtained as a result of redox-reactions between mixture components in heterogeneous concentration conditions in a saturated volume. It ensures the formation of boron subcompounds. These subcompounds are oxide, halogenide or hydrate boron compounds which are preserved in gas states at boriding temperatures and cause the process of the boron mass transfer by reducing to an atomic state on the processed surface in an electrochemical way or according to the reactions of exchange, disproportioning. They include B_2O_2, BO, BF_2, BF, BCl, BCl_2, H_2B_2 etc. These subcompounds determine the process of mass transfer during the saturation in powders. However, due to the shortness of the transportation way of boron compounds and the electron exchange between powder particles with the working surface in places of contact, the boron mass transfer is carried out more intensely and thus provides a more intense saturating ability.

In the situations when the number of electrons on the working surface is excessive (the inclusion of details in the chain of electrical current as a cathode, spontaneous work of short-circuit galvanic elements at the contact method), a subion electrolysis reduction is possible. Formed gas reaction products spontaneously tend to an even distribution in the whole volume which causes their withdrawal from the working surface. The reaction products are balanced near the processed surface; they interact with a boron-containing mixture compounds and form new boron subcompounds in the saturated volume. This explains the cycle character of the gas formation.

The thermodynamic evaluation of the redox-reaction ability of the medium components was done by measuring an isobaric potential of possible reactions according to L.P. Vladimirov's method[2] for several temperatures (Table 7.2). The minimal negative value of isobaric reaction potential reflects a higher possibility of its progress.

The temperature values of isobaric potentials show that the formation of saturating boron atoms goes gradually. Boron gas subcompounds are generated at initial periods; later they are disproportioned on the working surface. The oxy-halide systems, during formation of boron subcompounds with oxygen and fluorine, provide the possibility of interaction of these subcompounds between each other (reaction 17); this creates an additional source for boron atoms.

References

1. *Esin O.A., Geld P.V.* Fizicheskaya khimiya pirometallurgicheskikh processov. Moscow, 1966. 352 p. [Physical chemistry of pyro-metallurgy processes].
2. *Eytel W.* Fizicheskaya khimiya silicatov. Moscow, 1962. 362 p. [Physical chemistry of silicates].

[2]Vladimirov L.P. Termodinamicheskiye raschety ravnovesiya metallurgicheskih reaktcii. Moscow, Metallurgy. 1970. 330 p. [Thermodynamic equilibrium calculations for metallurgy reactions].

3. *Dolmanov F.V.* Issledovaniye odnovremennogo nasyscheniya zheleza i stali nekotorymi elementami 3, 4 i 6 grupp. Dissertaciya kandidata tekhnicheskikh nauk. Minsk, 1968. 158 p. [The study of simultaneous saturation of iron and steel with elements of 3, 4 and 6 groups. Candidate of Science dissertation].
4. *Appen A.A.* Temperaturoustoichivyye neorganicheskiye pokrytiya. Leningrad, 1967. 158 p. [Temperature-resistant non-organic coatings].
5. *Krukovich M.G.* Razrabotka teoreticheskikh i prikladnykh aspektov upravleniya strukturoi i svoistvami borirovannykh sloev i ikh ispolzovanie pri proizvodstve transportnoi tekhniki. Dis. dokt. tekh. nauk. Moscow, 1995. 416 p. [The development of theoretical and applied aspects of controlling the structure and properties of boronized layers and their use in producing transport machinery. Doctorate dissertation].
6. *Pasechnik S. Ya., Korotkov V.D.* Issledovaniye processa borirovaniya reversivnym tokom. // Zaschitnye pokrytiya na metallah. Kiev, 1967. Issue 1. pp. 40–45. [The study of boriding carried out by a reverse current].
7. *Korotkov V.D., Stasov A.A.* // Author certificate №773149, C23C. 1980, № 39.
8. *Krukovich M.G.* Issledovaniye zhidkostnyh bezelektrolyznykh processov khimiko-termicheskoi obrabotki. Dissertaciya kandidata tekhnicheskikh nauk. Minsk, 1974. 298 p. [The study of liquid non-electrolysis processes of chemical-thermal processing. Candidate of Science dissertation].
9. *Voroshnin L.G., Sharif A., Turov Yu. V., Krukovich M.G.* // Author certificate № 876773, C23C. 1981, №40.
10. *Sokolova V.A., Vorobyev G.M., Klimenko F.K., Zelenskiy G.V.* // Author certificate № 964023, C23C. 1980, №37.
11. *Afanasyev A.A., Venevtcev Yu.M., Korotkov V.D.* // Author certificate № 768854, C232C. 1980, № 36.
12. *Spiridonova I.M., Bodrikov V.P.* // Author certificate № 1661245, C232C. 1991, № 25.
13. *Kuskov V.N.* // Patent № 2004616 Russia, C23C. 1993, № 45-346.
14. *Anfinogenov A.I., Smirnov M.V., Ilyuschenko N.G., Belyaeva G.I.* // Trudy instituta elektrokhimii UFAN USSR. Iss. 3. Sverdlovsk, 1961.
15. *Smirnov M.V., Loginov N.A.* // Siberian Branch Russian Academy of Science Bulletin. 1962. 4. pp. 64–71.
16. *Ilyuschenko N.G., Anfinogenov A.I., Kornikov N.I.* Perenos metallov v rasplavlennykh solyakh i opredeleniye aktivnosti metallov v splave metodom EDS. // Trudy 3-go Vsesoyuznogo soveshaniya po fizicheskoi khimii i elektrokhimii rasplavlennykh solei i shlakov. Sverdlovsk, 1966. pp. 115–120 [The transition of metals in melted salts and the determination of metal activity in the smelt by EMF method].
17. *Ilyuschenko N.G., Anfinogenov A.I., Chernov Ya. B.* Tekhprotcess zhidkostonogo borirovaniya metallov v rasplavlennykh solyakh. Sverdlovsk, 1989. 45 p. [The technological process of liquid boriding of metals in melted salts].
18. *Dolmanov V.F., Krukovich M.G.* K voprosu o mekhanizme bezeletroliznogo nasyscheniya iz rasplavov solei // Mashinostroeniye i stroitelstvo. Tallin, 1970. pp. 23–27.
19. *Lyakhovich L.S., Kosachevskiy L.N., Dolmanov F.V., Krukovich M.G.* Nekotorye teoreticheskiye i prikladnye aspekty zhidkostnykh protcessov nasyscheniya. // Progressivnye metody termicheskoi obrabotki. Moscow, 1971. pp. 84–86. [Several theoretical and applied aspects of liquid saturation processes].
20. *Lyakhovich L.S., Kosachevskiy L.N., Dolmanov F.V., Krukovich M.G.* Zhidkostnye bezelektrolyznye protcessy khimiko-termicheskoi obraborki. MiTOM, 1972. №2. pp. 61–62. [Liquid non-electrolysis processes of chemical-thermal processing].
21. *Skorceletti V.V.* Teoreticheskiye osnovy korrozii metallov. Leningrad, 1973. 264 p. [Theoretical basis of metal corrosion].
22. *Belyaeva G.I., Ilyuschenko N.G., Plotnikova A.F., Anfinogenov A.I.* Vzaimodeistviye bora v rasplavlennykh solyakh // Fizicheskaya khimiya i elektrokhimiya rasplavlennykh solei i tverdykh elektolitov. Sverdlovsk, 1973. Part II. pp. 48–49. [The interaction of boron and melted salts].

23. *Samsovnov G.V., Epik A.P.* Tugoplavkiye pokrytiya. Moscow, 1974. 400 p. [High-melting point coatings].
24. *Voroshnin L.G., Lyakhovic L.S.* Borirovaniye stali. Moscow, 1978. 240 p. [Boriding of steel].
25. *Bel'skii E.I., Sitkevich M.I., Pankratin E.I., Stefanovich V.A.* Khimiko-termicheskaya obrabotka instrumentalnykh materialov. Minsk, 1986. 247 p. [Chemical-thermal processing of tool materials].
26. *Gœriot P., Thevenot F., Driver J.H., Laurent A.* Borudif: Nouvelle technique industrielle de boruration des aciers // Traitement thermique. 152–81. pp. 21–28.
27. *Glensdorf P., Prigozhin I.* Termodinamicheskaya teoriya struktury ustoichivosti i fluktuatcii. Moscow, 1973. 280 p. [Thermodynamic theory of the structure of sustainability and fluctuation].
28. *Voroshnin L.G., Khusid B.M.* Diffuzionnyy massoperenos v mnogokomponentnykh sistemakh. Minsk, 1979. 256 p. [Diffusion mass transfer in multicomponent systems].
29. *Prusakov G.M.* Matematicheskiye modeli i metody v raschetakh na EVM. Moscow, 1983. 144 p. [Mathematical models and methods of computer calculations].
30. *Yuryev V.A., Kotov A.P., Yuryeva M.V.* Kinetika rosta boridnykh faz v armko-zheleze // Voronezh State Technological University bulletin. Material Studies. 1999. №6. Pp. 98-99. [The kinetics of growth of boride phases in ARMCO-iron].
31. *Isakov G.M., Prusakov G.M., Sherbedinskiy G.V.* Issledovaniye kinetiki rosta boridov v sisteme Fe-B i Fe-B-C // USSR Academy of Science Bulletin. Metalls. 1987. №1. pp. 186–190. [The study of growth kinetics in borides in the system Fe-B and Fe-B-C].
32. *Belov N.V.* Struktura ionnykh kristallov i metallicheskikh faz. Moscow, 1947. 155 p. [The structure of ionic crystals and metal phases].
33. *Samsonov G.V., Umanskiy L.S.* Tverdye soedineniya tugoplavkikh metallov. Moscow, 1957. 265 p. [Solid compounds of high-melting point metals].
34. *Samsonov G.V., Markovskiy L.Ya., Zhigach A.F. et al.* Bor, ego soedineniya i splavy. Kiev, 1960. 590 p. [Boron, its compounds and smelts].
35. *Glukhov V.P.* Boridnye pokrytiya na zheleze i stalyakh. Kiev, 1970. 208 p. [Boride surfaces in iron and steels].
36. *Kunst H., Shaaber O.* / Harterei-Techn, Mitt. 1967. Bd. 22, H. 1. S. 1-25.
37. *Minkevich A.N., Rastorguev L.N., Yusfina L.I.* // MiTOM. 1967. №3ю pp. 36–38.
38. *Popov A.A.* Teoreticheskiye osnovy khimiko-termicheskoi obrabotki stali. Sverdlovsk, 1962. 120 p. [Theoretical grounds of chemical-thermal processing of steel].
39. *Dubinin G.N.* Diffuzionnoye khromirovaniye splavov. Moscow, 1964. 450 p. [Diffusion chromizing of smelts].
40. *Krishtal M.A., Ivanov L.I., Grinberg E.M.* // MiTOM, 1970. №8. pp. 74–76.
41. *Turov Yu.V.* Issledovaniye osobennostei formirovaniya i svoistv boridnykh pokrytii na stalyakh. Minsk, 1974. 25 p. [The study of specifics in formation and properties of boride coatings in steels].
42. *Akulinichev E.V.* Mekhanism obrazovaniya i strukturnye osobennosti borirovannykh sloev, poluchennykh pri kristallizatcii stalnykh otlivok. // MiTOM, 1979. №1. pp. 5–7. [The formation mechanism and structural peculiarities of boronized layers obtained during crystallization of steel castings].
43. Bel'skii *E.I., Liventcev V.E., Sitkevich M.V.* Issledovaniye zakonomernostei formirovaniya i svoistv boridnykh pokrytii, poluchennykh v processe litya // MiTOM, 1983. №10. pp. 51–54. [The study of patterns in formation and properties of boride coatings obtained in the process of casting].
44. *Shalya M.A.* Snizheniye khrupkosti borirovannogo sloya // MiTOM, 1968. №8. pp. 36–37. [Reducing the brittleness of a boronized layer].
45. *Afanasyev A.A.* Elektroliznoye borirovaniye reversirovannym tokom konstruktcionnykh stalei. Voronezh, 2001. 311 p. [The electrolysis boriding of construction steels with a reverse current. Doctorate dissertation].

Chapter 8
Methods of Reducing the Brittleness of Boronized Layers: The Parameters of Boriding Technology Aimed at Determining the Plasticity of Boronized Layers

Abstract The chapter deals with the technological methods developed in order to decrease the brittleness of boride phases in layers. The general interconnected methods of decreasing the brittleness of boride phases are their optimal micro-alloying and creating a favourable strain state. The role of sublayers in boronized layers in relation to the formation of final properties of boronized objects is shown in the chapter. The general parameters of technological processes and compositions of saturation mediums which guarantee the formation of boronized layers of adequate plasticity are determined. The authors study boriding processes: in the mixture of gases under flow or circular gas movement; in powder and grained mediums or in lined containers where the mass transfer goes through gaseous medium; in metal-thermal mixtures using the effect of self-spreading high-temperature synthesis; in liquid mediums on the base of melted salts or metals under electrolysis or non-electrolysis methods; while using the electrochemical reducing agents, protector processing method etc. Each case is provided with the description of optimal conditions for obtaining plastic boronized layers in steels and iron casts. The structures of boronized layers and schemes of used equipment are given.

The brittleness of boronized layera is determined by many factors. They are: the electron structure of phases, the correlation between boride phases (FeB, Fe_2B) in a layer, the morphology of borides, their stress state etc. It also depends on the continuity of a borides' layer on the processed surface. Following the condition of balance between the amount of a saturating element that goes to each area of the surface and to deeper layers of the metal helps to obtain a continuous width-regular layer. An optimal saturating medium provides the continuity of the process of delivering saturating element's subions to the processed surface. If the concentration continuity is broken, the width of the layer would be uneven; other results might be a fragmented structure of certain boride zones or a layer with disconnected boride needles. In this case the intervals between borides would be filled with a solid solution of boron in the matrix material or solid solutions of saturating

M.G. Krukovich et al., *Plasticity of Boronized Layers*,
Springer Series in Materials Science 237,
DOI 10.1007/978-3-319-40012-9_8

elements competing with boron and/or alloy elements. These elements have a certain impact on the boron concentration in the surface, the type of adatoms and adions' bonds with the surface and between each other; this influences the formation of a layer and determines its structure, phase composition, hardness and plasticity.

The study and practical use of mentioned regularities allows specialists to control the brittleness of boronized layers.

Boronized layers normally have a sublayer or a transition zone which is located on some distance below borides; the depth depends on the chemical composition of the matrix material, the type of the saturating medium and the temperature-temporal conditions of the process. The transition zone is much softer than the boride layer; its hardness is comparable to the matrix's hardness. However it takes part in determining the stress state of boronized layers and exploitation conditions for boronized details.

The width of a boronized layer is normally a distance from the surface to the end of boride needles which is considered as an effective width of a boronized layer.

8.1 Methods of Reducing the Brittleness of Boride Phases in Layers

The nature of the increased brittleness and hardness of borides is determined by the formation of stable electron configurations, i.e. the energetic stability and statistical weight of configurations sp^3 and d^5 [1]. The high stability of electron configurations explains the ability of boronized layers to preserve hardness at high temperatures (up to 800 °C), low friction coefficients and low wear process intensity during the work in friction points.

Alloying treatment of boride layers by elements with high stability of electron configurations (e.g., W) leads to the increase in brittleness and hardness. In particular, the microhardness tests of the boride layer in steel P6M5 showed several values at the level 2400 HV, which corresponds the microhardness of W_2B boride.

Alloying treatment of borides by metals with a smaller amount of stable electrons (e.g., Ni, Mn) leads to the decrease in microhardness and brittleness, i.e. unlocalized electrons cause the decrease in the brittleness of boride layers. Simultaneously the combined influence of the phase composition and structure's state normally dominates the change in properties due to the electron composition.

The general interconnected ways of reducing brittleness of boride phases are their optimal microalloying and the creation of favourable stress state conditions. These ways are ensured by the following technological steps:

- microalloying of a layer from a processed material or from a saturating medium;
- reducing the content of high-boron FeB phase in a layer;
- preliminary surface processing;
- setting optimal temperature-temporal conditions;

- setting a graded mode for temperature, electric and technological parameters;
- using concentrated energy sources;
- setting the width of a boride layer which coincides with the recommended value for a particular steel brand (Table 6.1);
- setting a speed of cooling after saturation;
- additional processing of obtained boride layers;
- final thermal and mechanical processing.

The decrease of stresses in a boride layer can be achieved (apart from the mentioned) with the following methods:

- obtaining a layer with disconnected boride needles;
- providing an optimal correlation between boride phases;
- regulating a level of textureness of the boride layer;
- regulating the width of a continuous layer and the level of boride acicularity (layer's density and the level of boride needles' sharpness).

In particular the disconnection of boride needles, i.e. obtaining a fragmented structure, is achieved, for instance, after carbonizing and subsequent boriding in the mixture which contains 60% B_4C, 10% graphite, 10% Fe powder, 20% Na_3AlF_6 [2]. The same effect is reached during saturation from grained mixtures with grains larger than 5 mm and a high content of boron carbide (>65%) or by the preliminary creation of alternate areas on the processed surface (these areas are protected from boriding). Obtaining a composite fragmented or disperse structure can be also achieved by high frequency current, laser or electron beam heating.

It is necessary to make the following remark: in many cases one-phase layers on the basis of Fe_2B are more plastic and durable in the conditions of dry friction and in the conditions of a borderline lubricant at high unit pressures. There are several ways of controlling the structure formation during boriding in order to obtain one-phase layers.

- Boron concentration on the working surface insufficient for the formation of a continuous layer of a high-boron FeB phase.

For iron and steels this concentration is in the range of 8.5–10 mass%. The necessary concentration can be obtained by regulating a diffusion source capacity or creating conditions for an accelerated withdrawal of boron atoms from the surface into the deeper layers of the processed detail.

The regulation of the diffusion source capacity in gaseous processes is carried out by setting concentrations of boron-containing gases and gas-carriers, providing the pressure into the saturating container and by setting a certain speed of passing the gaseous medium. Contact and contact-free saturation in gaseous mediums (obtained from powder or compact saturating mixtures) require the choice of a particular speed and type of gas formation reactions and providing a pressure of gases in the container. For instance, the saturation in the refractory-lined equipment requires the pressure of gases in the range of 0.5–2.0 kPa to obtain one-phase layers.

The accelerated withdrawal of atoms into the detail's depth is reached by creating a significant amount of flaws in the crystal structure, vacancies and dislocations in surface layers of a metal. This is achieved by the mechanical impact or thermal circulation in the process of saturation.

One of the ways to increase the number of vacancies and dislocations in a significant part of a metal during boriding is preliminary diffusion sherardizing. Boriding initiates the process of zinc sublimation from the diffusion zone which increases the boride width (1.5X). The process of sherardizing may be combined with boriding. In this case, $ZnCl_2$ (3–5%) is added to the initial saturating mixture for boriding. Then isothermal soaking is carried out for 0.5–1.0 h at sherardizing temperatures (300–500 °C). The consequent heating during boriding (speed 100–120 °C/min) causes the zinc sublimation from the formed layer.

The same effect is achieved by the introduction of Sn, Sb, Ni and their oxides in a saturating medium. For instance, Ni and NiO in the saturating mixture prevents the wirthdrawal of iron atoms from the processed surface which leads to the increase of its adsorbent capacity.

- The graded mode of diffusion source capacity by conducting an interrupted process with diffusion soaking or by decreasing the capacity in the final stage of the process

For instance, this graded mode is achieved under the electrolysis saturation by reducing a working current density on the processed surface to 0.04 A/cm^2 or its complete switching off which should be done in periods.

For non-electrolysis saturation, the graded mode of saturating capacity is provided by putting processed details into a neutral salt bath.

For saturation in gaseous mediums, it can be achieved by reducing the amount of the boron-containing gas or by increasing the amount of the diluent gas. Further repeated heating for hardening boronized details in weak saturating mediums also causes a partial dilution of the high-boron phase.

- Alloying of a diffusion layer by metals stabilizing Fe_2B phase (e.g. Mn, Ni). These elements form Mn_2B and Ni_2B borides, which are isomorphic to Fe_2B, which means that their mutual solubility is unlimited. While alloying, atoms of Mn and Ni replace Fe atoms in the boride lattice and increase the amount of the metallic compound. The growth speed of these one-phase boride layers is higher than two-phase by 20%. Alloying can be carried out both in the saturating medium and the processed material.
- The simultaneous and consecutive saturation of the metallic surface by boron or partly soluble elements. The elements which are only partially soluted during boriding are: C, Si, Al, S, Cu. Their adsorption by the surface is accompanied by several bonds of the surface being replaced; they also decrease its adsorption capacity which leads to the fall in boron concentration on the surface.
- The acceleration of boron atoms withdrawal from the surface by creating a liquid-crystal state at the initial moment of saturation on the surface followed by the diffusion front. The transition to a liquid-crystal state during boriding can be

achieved by the usage of high speed current and laser heating, preliminary application of nickel or nickel-phosphorous electrochemical coatings. A liquid-crystal state in the sublayer is obtained by alloying elements being forced out by a growing boride layer in the sublayer or by redistribution of these elements in the processed material and their diffusion into the sublayer. Carbon, nickel, chrome, aluminum and serum encourage this process. The formation of a liquid-crystal state accelerates the speed of the layer's growth (2X).

The process of boriding in the presence of a liquid phase must be carried out in rather strict temperature-concentration-temporal conditions to preserve the detail's form and working surface's smoothness. These conditions are determined by the multi-component equilibrium diagrams (Chap. 5) or experimentally.

The brittleness of boride phases can be reduced by controlling the level of the layer's acicularity and density which have a significant influence on its stress state. It can be done under high temperature boriding by isothermal soaking during heating (~700 °C)—at the initial stage it causes the formation of multiple Fe_2B phase centers. The other way is to decrease the general temperature of the process or to increase the current's density during the electrolysis saturation.

Some authors propose the term "the critical growth speed" in relation to borides. If the speed is less than critical, the growth of borides goes in different directions, i.e. not necessarily in the crystal-graphic direction [001] [3]. It is necessary to mention that the same effect is witnessed in the following situations:

- the normal boriding conditions;
- the formation of boride layers <20 mkm;
- the competing processes of chemical-thermal processing (in particular, in the Borudif process) [3];
- the liquid-crystal state of the processed surface during saturation;
- the treatment process of high-alloy materials with the effect of heredity of crystal or structure composition;
- when boride phases are moderately wide (~10–15 mkm) and have been obtained at temperatures higher than 1000 °C (before the natural selection of predominant growth of boride crystals).

Moreover, the surface after boriding often contains a disconnected layer ~10 mkm.

8.2 The Formation Peculiarities of the Transition Zone and Its Influence on the Properties

The auxiliary properties of boronized layers are determined by a complex of boride and sublayer's properties. The sublayer is significantly important for thermal processing of boronized layers when a solid substrate is created and it protects a boride layer from deformation; it also takes part in the formation of strain state of a detail.

In this case the optimal complex of the detail's properties must be obtained by correcting the heating temperature for quenching in regards to the changed chemical composition of the sublayer. A thermally hardened sublayer affects the wear process. The increased hardness and the presence of carbon-borides leads to the improvement of the detail's wear resistance in comparison with the wear resistance of the core after boride layers wear off. The sublayer also influences the development of thermo-fatigue cracks on boronized details and tools as it prevents the formation of cracks.

The formation of the sublayer, which serves as a transition zone from borides to the structure of main metal, takes place as a result of saturation as well as cooling and further thermal treatment. First it is connected to the diffusion of boron in front of the boride layer and the growing borides forcing out carbon as a result of its low solubility at saturation temperatures higher than the critical point for steels A_{c1}. The level of its content under boride needles exceeds the initial concentration in steel 1.3–1.5 times [4]. The carbon concentration smoothly decreases to the normal content for a particular kind of steel with the increase of the distance from the surface.

When the saturation temperature is lower than critical temperature A_{c1}, the redistribution of carbon doesn't happen; the diffusion zone of boron solid solution is generated. The formation is more intense during low-temperature boriding.

The width of the zone with a changed carbon content is connected to the boride layer width and exceeds it by 5.5–6.5 times. Despite the fall in carbon concentration, its structure doesn't change. In particular, steel 35 is characterized by a perlite structure in all areas of the sublayer after boriding and slow cooling (Fig. 8.1). However, the autoradiography data shows the change of carbon content in this zone from 0.7% in mass to 0.35% in this steel.

The reason for a perlite structure formation is the combined influence of boron, carbon, silicon and manganese on a steel in a shifting eutectoid concentration. Silicon, as carbon, is replaced by the boride layer and saturates the sublayer [5]. Manganese has a tendency to be soluted in borides due to its high resemblance to boron, but carbon prevents its diffusion. Thus, manganese also saturates the sublayer. Given that 1% Si shifts S point of Fe-Fe_3C diagram by 0.1% C to the left and 1% Mn—by 0.05% C [6], this explanation seems to be the most plausible.

The formation of the carbon-rich sublayer has its advantages as it provides a solid substrate for the boride layer after quenching even on steels containing 0.25–0.3%. It creates an enhanced hardened zone, smoothens the hardness differential from the layer to the core and ensures the high mechanical properties of boronized details. The width of the sublayer may be increased by preliminary carbonizing or simultaneous saturation of steel by boron and carbon.

The structure of the sublayer significantly depends on the content of alloy elements in steel. For instance, boriding of medium-carbon steels, containing Si > 1.5%, Cr > 5%, Ni > 5% and Al > 1%, leads to the formation of α-solid solution of these elements in the sublayer (Fig. 8.2); these elements have a reduced level of hardness and cause the formation of cracks and dents in the layer of borides. The comparison of properties of a similar boronized layer in steel 38XC

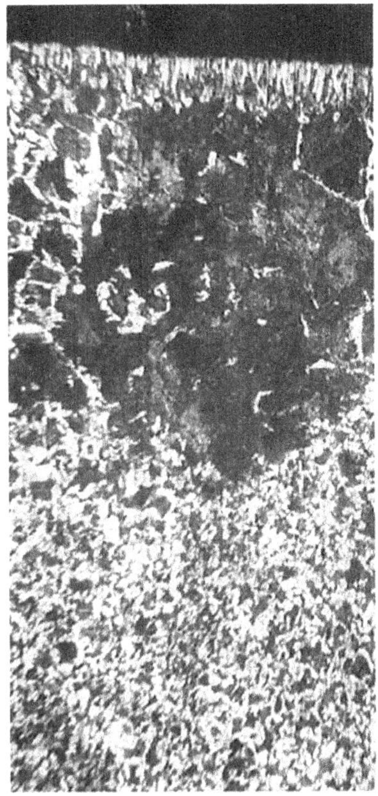

Figure 8.1 The microstructure of the sublayer in steel 35 after saturation at 850 °C for 4 h in the grained mixture (×100)

(with α-solid solution of silicon in the sublayer) and a boronized layer on steel 40X (without α-solid solution) showed an unconditional advantage of the latter one [7].

The generation of a developed zone of α-solid solution is typical for silicone steels: 40C, 40C2, 40C3, 38XC (with silicon content in the higher levels of steel), 30XГC, 9XC, 60C2 etc. At the same time a zone of α^1-phase is found in the layer of α-solid solution; due to a big difference in the coefficients of thermal expansion with steel and borides of these two phases, their combination leads to the spontaneous chipping in the boride layer. The conclusion is that the silicon content in steels under boriding shouldn't exceed 1.0%.

The influence of boron on the structure of the sublayer is more noticeable in alloy steels with low and medium carbon content. The transition zone is characterized by the significant growth of grains (from number 7–10 to 1–4) which is caused by the influence of boron and carbon (Fig. 8.1). The complete restoration of a grain size is impossible after one-time annealing. In particular, the annealing of steel 45 at 850 °C during 30 min (the steel was boronized at 950 °C) led to the fall in the grain size from number 1–2 to number 4 [8].

The growth of grains in the sublayer by no means influences the mechanical properties of boronized steels; specifically it reduces the level of fatigue hardness.

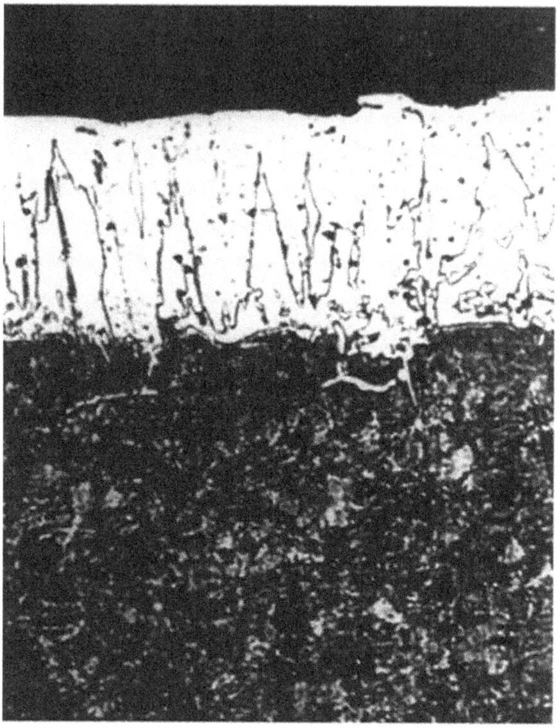

Figure 8.2 The microstructure of the boronized layer in steel 38XC after the liquid non-electrolysis saturation in the bath with boron carbide at 950 °C for 6 h (×200)

Double or triple annealings partly allow to correct large-grain structure of the sublayer. In this case heating, soaking and cooling must be accompanied by the protection of the boride layer from oxidation and deboriding. The repetitive heating leads to a slight decrease in the amount of high-boron phase in the layer's structure due to its resolution. Finally it leads to the increase of ductility of the boronized layer in general.

The level of alloying in steel determines the distribution of carbon and the structure of the sublayer. It is revealed in narrowing of the sublayer and the decrease of carbon concentration. Carbid-generating elements—chrome, molybdenum and titanium—significantly slow down the diffusion of boron atoms in the sublayer and prevents the growth of grains. For instance, 0.5% Ti completely neutralizes the influence of boron on the grain growth tendency.

Silicon in the amount of ~1% either doesn't change or increases the austenite grain growth tendency. If a steel contains more than 2% Si, grains of the transition perlite zone in a medium-carbon steel begins to shrink [8]. However, a developed zone of α-solid solution of silicon in iron begins its formation between the boride layer and the perlite zone.

Boriding of alloy steels carried out below boride layers leads to the redistribution of alloy elements which results in the formation of a zone in which the content of alloy elements is initially increased and then falls down. The width of this zone is

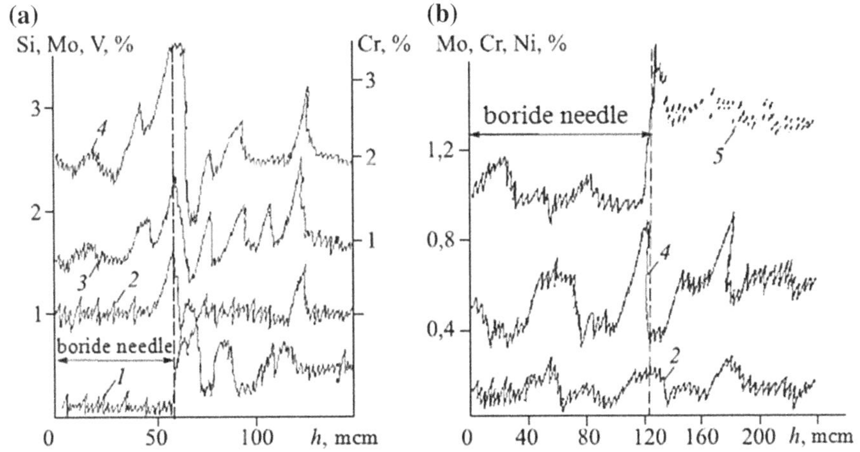

Figure 8.3 The distribution of silicon (*1*), molybdenum (*2*), vanadium (*3*), chrome (*4*), nickel (*5*) in the boronized layer of steels 5X3B3MΦC (**a**) and 5XHM (**b**) [9]

50–100 mkm. The character of elements' redistribution depends on the steel's alloying, carbon content and treatment conditions. The presence of such a zone is often accompanied by the drops in hardness and influences the formation of residual tensions and final exploitation properties of boronized details [4, 5, 8, 9]. In particular, the boriding of tool steels is characterized by the redistribution (Fig. 8.3) which significantly changes the structure and hardness of the sublayer (Fig. 8.4) [9].

A certain influence on the strain state, plasticity and wear resistance is made by precipitates of the boron-carbide phase which either adjoin boride needles in the form of feather formations or occur on the borders of ex-austenite grains in the forms of plates or sphere compounds. Feather formations are formed in the process of saturating high-carbon steels; plate formations—in the process of cooling caused by the decrease in boron and carbon solubility in the austenite.

The formation of sphere compounds with increased carbon segregation was observed in study [10] during an interrupted process of boriding which was accompanied by decreased residual tensions.

The precipitation of disperse boride inclusions happens in the transition zone and when the chane in the cooling speed. The different valucs of borides and the transition zone elasticity module leads to the complex relaxation interaction between various zones of the boronized layer [11]. It provokes the process of slow cooling to reduce the tensile stress in the transition zone and in FeB boride zone or increase the decompression stress in Fe_2B zone. In some cases two processes can happen simultaneously which may lead to the prolongation of samples. Moreover, the secretions of Fe_2B inclusions in the transition zone increases α-phase volume and improves the work of tensile stresses and the prolongation of samples [12]. Thus, quick cooling is supposed to reduce the processes of Fe_2B secreting from

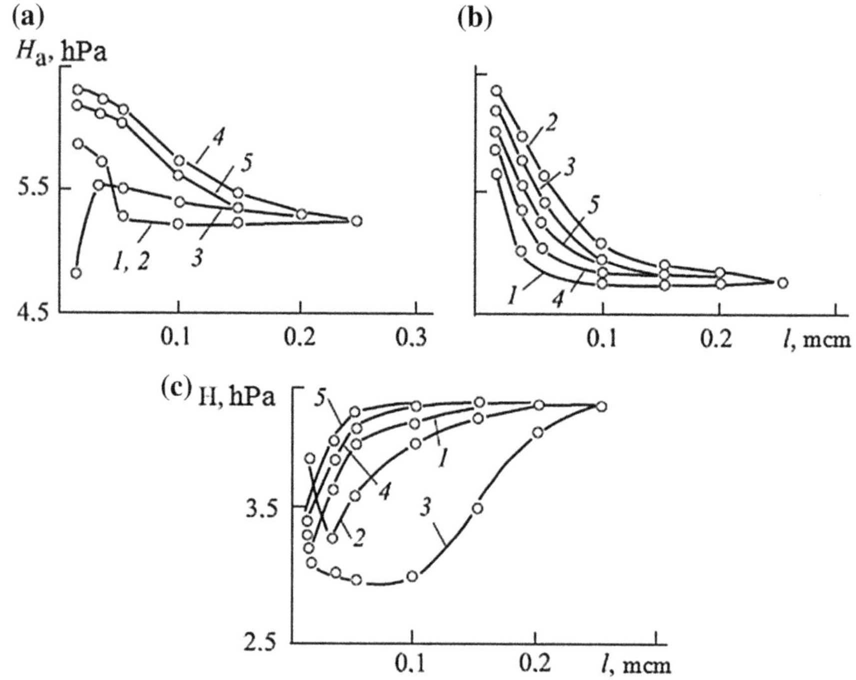

Figure 8.4 The hardness of transition zones (HV) in steels 5XHM (**a**), 7X3 (**b**) and 5X3B3MΦC (**c**) under boron-nitrating (*1*), boron-coppering (*2*), boron-siliconing (*3*), boron-chroming (*4*) and boriding (*5*) [9]. *l*—the distance from the boride needles' ends

α-phase of the transition zone and stop the unfavourable distribution of tensions in the boronized layer; by doing so, it leads to the reduction of boronized layers' deformation. It is experimentally proved by the quick cooling of containers with boronized details in a tap water from the temperatures which are equal or slightly exceed the quenching temperatures and following tempering at 200 °C; any changes in sizes of details are not observed.

The formation of boron-carbides of different chemical composition depends on the level of alloying in boride steels. For instance, medium-carbon steels' composition can change from $Fe_3C_{0.6}B_{0.4}$ to $Fe_3C_{0.2}B_{0.8}$. Boron-carbide type $(Fe, Cr)_3(B,C)$ might be formed in chrome steels [13, 14].

The hardness of the sublayer depends on the steel or smelt's brand and the speed of cooling or thermal processing that follow. After quenching a boronized detail made from carbon steels, the sublayer provides a smoother transition from a boride layer to the core. The thermal processing of high-alloy steels and smelts makes the sublayer additionally hardened due to the processes of secondary crystallization, i.e. the precipitation from solid solutions of borides and carbon-borides of different

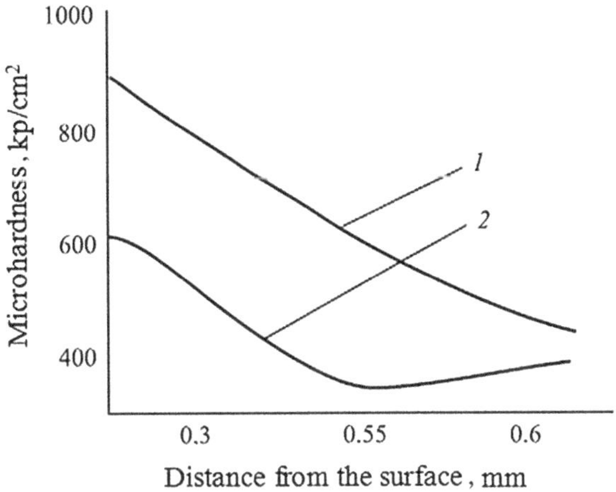

Figure 8.5 The character of hardness change in the sublayer in alloy ЭП-718 with a pseudo-eutectic boronized layer after thermal processing: *1*—after boriding at 1050 °C for 3 h; *2*—after boriding at 1050 °C for 3 h and double ageing treatment at 800 °C for 5 h and at 650 °C for 12 h

forms. This fact is illustrated by the example of changing hardness of the sublayer in a superalloy with a pseudo-eutectic layer after thermal processing (Fig. 8.5).

Thus, the change of amount, chemical composition and form of carbon-borides' inclusions in the sublayer may be done by choosing a steel with a lower content of carbon, by accelerating a cooling speed after saturation, by setting an appropriate mode of interrupted saturation to increase the plasticity of the layer and decrease the level of tension in a boronized detail.

The reduction of the boronized layer's brittleness is also achieved by using concentrated energy sources. The studies show that the use of luminescent, laser and electron beam heating during boriding allows to improve the boronized layer's plasticity; these results have led to the increased interest to these types of boriding.

8.3 The Parameters of the Technological Process Which Determine the Plasticity of Boronized Layers in Steels

The external conditions of the formation of boronized layers are dominant instruments of controlling the phase compositions, structure and plasticity of layers. A boronized layer is a composition of a boride layer on a processed surface and a sublayer which is formed by simultaneous saturation of boron and other elements or due to the shift or redistribution of alloy elements provoked by the growth of the boride layer. Thus, the properties of boronized layers are generally determined by the properties of the layer structure composites and their interaction.

8.3.1 The Connection Between the Composition of Gaseous Mediums and the Plasticity of Boronized Layers

The influence of gas-based saturating mediums on the plasticity manifests itself in the phase composition and the structure of boronized layers. The phase composition in ionic gaseous mediums can be controlled by regulating (1) the correlation between a boron-containing gas and a gas-carrier; (2) the speed of gas mixture going through the saturation medium; (3) the pressure.

The boriding of metallic surfaces in gaseous mediums is carried out by the reduction of boron subcompounds on these surfaces; boron subcompounds are the result of a partial decomposition of gaseous compounds: diborane B_2H_6, boron trichloride BCl_3, boron tribromide BBr_3, trimethylborane $(Ch_3)_3B$, triethylborane $(C_2H_5)_3B$ [15, 16]. A gas mixture is fed to a working space of a furnace through an injector mixer. The correlation between a boron-containing gas and a gas-carrier may vary in a wide range (from 2×10^{-5} to 2×10^{-1}). Gas-carriers (H_2, Ar, N_2, He and others) function as a transport medium of a boron-containing gas and delay its decomposition.

The insufficient dilution in the medium leads to the formation of an excessive amount of boron subcompounds; it causes the generation of amorphous boron residue on the processed surface which decelerates the formation of a boronized layer. Formed gaseous reaction products should go into the exhaust system and go through the range of catchers filled with water or go through a special equipment [17].

The optimal correlation for gases B_2H_6/H_2 is from 0.02 to 0.01; for gases BCl_3/H_2—from 0.06 to 0.05 [8]. The total consumption of the gas mixture must be at the level ~40–50 l/h. The first composition requires the process of boriding to go with a noticeable speed at 550 °C; the second—at 750 °C. The maximum speed of boron saturation is achieved at the pressure setting of 26.7 kPa in the reaction camera (200 mm Hg). The speed of boride layers growth is 0.04 mm/h at 3 h soaking; 0.03 mm/h at 6 h soaking. These conditions lead to the formation of two-phase boronized layers. The temperature rise which is caused by the increase of boron-containing gases decomposition and the increase in the number of subions in the medium determines the necessity of reducing the amount of boron-containing gas and/or accelerating the speed of its transmission.

The improved plasticity of boronized layers achieved by the reduction of the high-boron phase amound was accompanied by the discontinuation of feeding gas saturating mixture into the furnace in the end of soaking. The 1–2 h stop during 4–10 h soaking didn't influence the width of the common layer and boride layer in comparison with the uninterrupted feeding of the mix. At the same time the discontinuation significantly increased the economic efficiency of the hardening process. The saturation was carried out at 850 °C in the mix of gases B_2H_6 and H_2 at their correlation 1/25 and the feeding speed 75 l/h [8, 18].

Boriding in the medium of boron trichloride and hydrogen is accompanied by the corrosion of the working surface due to the low withdrawal speed of $FeCl_2$

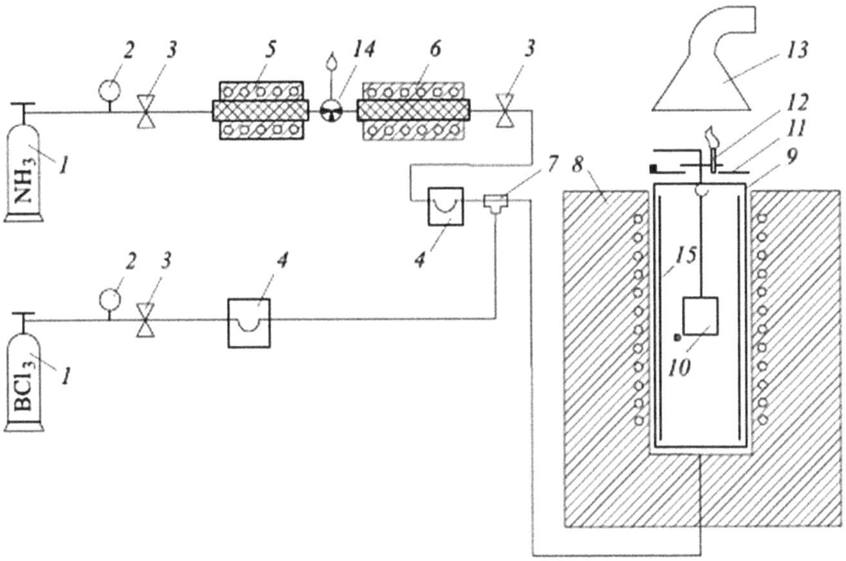

Figure 8.6 The principal setting scheme for boriding in gaseous mediums: *1*—gas containers with initial gases; *2*—control manometers; *3*—doser taps; *4*—float air rheometers; *5*—dissociator; *6*—camera for cleansing and drying of gases; *7*—mixer; *8*—heating furnace; *9*—converter; *10*—processed details; *11*—converter's lids; *12*—torch; *13*—extraction ventilation; *14*—control torch; *15*—screen

reaction products. The graded temperature mode allows to eliminate this negative effect. The first stage presupposes the use of temperature <780 °C, the second—825–1050 °C.[1]

The phase composition of the layer is controlled by the varying correlation of gases and temperature-temporal parameters. In particular, the correlation $BCl_3/H_2 \leq 3 \times 10^{-4}$ causes the formation of a continuous layer of Fe_2B. If the correlation value of gas components is higher, FeB phase appears after 1, 0–1, 5 h of processing.

A large number of possibilities for controlling the phase composition of boronized layers, reducing toxicity and explosibility is provided by the device which exploits the use of dissociate ammonia and trichloride boron (Fig. 8.6) [19].

The dissociator is filled by catalyst ЦЕЧМ-1 and heated up to 800–900 °C. The obtained mixture of 75% H_2 and 25% N_2 is dried at −60 °C and filtered from ammonia residues in a tube filled with zeolite (NaX or CaX). The use of ammonia reduces the process costs as one container of liquid ammonia substitutes 10–12 containers of gaseous hydrogen.

After the working space (converter) is blown by dissociate ammonia and tested for the backfire, the gas is burnt at the exit and combustion products are exhausted

[1]Patent № 225454 Germany, publ. 31.07.85.

by the extraction system. Then boron trichloride is fed to the mixer made out of glass in the form of T-joint. This allows to control the level of zeolite exhaustion. The generation of white coating on the mixer's walls signals about zeolite regeneration. The regeneration is done by heating the tube with zeolite up to 300–400 °C followed by its blowing by a neutral gas (argon or nitrate) during 3–4 h until the complete stop of water steam and ammonia secretion. After BCl_3 is fed, in 20–30 min details go through a flame initiated in the moment of the lid's opening. The upper part of the converter and lid are positioned in the zone of temperatures that guarantee the inflammation of hydrogen at the moment of loading and unloading of the details. The short-term exposure to the hydrogen flame doesn't impact the quality of boriding. Large-sized details under boriding are put into a cold furnace before its blowing to prevent the unrational consumption of dissociate ammonia.

The proposed type of loading and unloading for small and medium-sized details allows to combine boriding with heating under any variant of thermal treatment (normalizing, quenching, ageing, high-temperature tempering).

The saturating process normally goes at temperatures 600–950 °C during 2–10 h. The consumption of gases for the converter d = 100 mm is: BCl_3—up to 4–6 l/h, dissociate ammonia—up to 160 l/h. If the size of the converter is different, the consumption of gases must be changed in correlation with the sectional area of the reaction zone, but the correlation of $BCl_3/\Sigma\ (N_2+H_2)$ is recommended to be in the range 1/20–1/50.

The reduction of saturating medium toxicity in comparison with other mentioned mediums is reached by using the mixture based on boron trimethyl $B(CH_3)_3$ and helium (He) with their content of 5 and 95% accordingly [20]. The process of saturation is highly technological in this case.

Two-phase boronized layers with different correlation of phases are formed by a glow current in gaseous mediums which contain either B_2H_6 or BCl_3 [21–24] or boron trymethyl and hydrogen with the pressure 3–8 GPa [25]. The decrease in the amount of high-boron phase FeB ensures the plasticity growth in obtained boronized layers. Its minimal amount is acquired when boron trimethyl is used (<20%).

Boriding in the plasma of gas currents may involve the use of pure boron trimethyl or boron trimethyl diluted by argon and hydrogen [25–27]; a mixture on the base of BCl_3-H_2-Ar may also be used [28]. The obtained layers' properties don't concede to the properties of the layers obtained in powder mixtures. Moreover their phase composition and the speed of formation are easily and efficiently controlled; the conditions for a complete automated process are provided.

Two-phase layers are also formed in the luminescent current in the gas mixture containing 0.1–20 vol% BCl_3 (with additions CH_4, TiCl etc.) and the mixture of transport gases Ar (45–90%) and H_3 (10–50%) at the pressure 300–1000 Pa and saturating temperature 750–1000 °C.[2] If the pressure is dropped to 200–250 Pa, the generation of one-phase boride layers with high plasticity is expected.

[2]Patent № 87616 SRR, 30.09,85.

A significant improvement of plasticity can be achieved by boriding methods which are aimed at obtaining the structure on the base of boride inclusions positioned in a solid solution. In particular, the use of plasma in gaseous mediums of low pressure containing BCl_3 and H_2 causes the formation of similar layers. The continuous layers of borides are formed only at the high temperature of the gas flow [29].

The boriding in the cold high-frequency plasma in the mixture of B_2H_6 and Ar involving the use of a high frequency generator (27.12 MHz) lead to the formation of an alloy boronized layer in steel AISI 4340; the layer is free from inclusions of FeB, Fe_2B, $Fe_{4.5}Ni_{18.5}B_6$ in a solid solution. The microhardness of the layer with high plasticity is 1000 HV [30].

The electrophysical methods which are capable of controlling the structure of boronized layers during saturation in gaseous mediums include ionic implantation of boron itself or in combination with other elements (Ti, Cr, S, N etc.). The most effective is the multielement implantation N^+, Cr^+ (dose 3×10^{17} cm^{-2}), Mo + (dose 10^{17} cm^{-2}), B (dose 5×10^{17} cm^{-2}). The layer's phase composition is represented by the compositions of oversaturated solid solutions with the compounds of non-stochiometric composition. The layer's width is 5 mkm; the plasticity of the layer is efficient.

The phase composition and the plasticity of a boronized layer can be controlled by the preliminary nickel plating or titanium-nitriding of the processed surfaces [31, 32]. The preliminary nickel plating and consequent gas boriding in the medium, containing 1–5 vol% BCl_3 and the mixture of H_2 and N_2, help to obtain a layer on the base of $(Ni, Fe)_2B$ and decrease the surface roughness caused by the impact of BCl_3. The saturation is carried out at 650–1050 °C.

The preliminary titamium-nitriding carried out in the medium $TiCl_4$–H_2–N_2 at the pressure 90–120 kPa and 1000–1100 °C, and the subsequent boron-titanizing at 850–950 °C in the medium containing (vol%): 2–25% $TiCl_4$, 0.02–0.2% BCl_3, 25–35% N_2, 60–75% H_2 and 0.005–0.004% H_2O—provides the formation of a nitro-boron-titanized layer with improved properties.

The preliminary treatment methods aimed at the quality improvement of boronized layers and the phase composition regulation include preliminary chemical precipitation (CVD-process) of iron oxide or other elements with catalytic properties during boriding. It leads to obtaining layers with improved bonds with the substrate and corrosion resistance.[3] The boriding of similar residues in gaseous mediums, containing hydrogen, is accompanied by the reactions of hydrogen reduction of these oxides. The obtained porous residues have an increased reactive ability and take part in the formation of borides from the external side of the detail's surface. An increase in size is also observed.

[3]Patent № 2001358, Germany, publ. 23.03.84.

8.3.2 The Connection Between Non-contact Processing Conditions in Gaseous Mediums and the Plasticity of Boronized Layers

Non-contact boriding in gaseous mediums is carried out in special sealed equipment which ensures the division of working space into gas-producing and saturating parts. Gaseous mediums are normally represented by powder, compact, grain or liquid substances and compositions which meet the following criteria:

- they are characterized by a sufficient durability and chemical capability during the circular change of temperature mode (they don't break and destruct when the working space is opened for loading and unloading of details without cooling and endure a multiple number of heating and cooling cycles);
- they provide the uninterrupted formation of saturating gas mediums during heating and isothermal soaking;
- they are highly porous, i.e. the gas-formation ability of their surfaces is maximum.

Gas-forming powder and compact mediums must contain following ingredients:

(1) a source of the saturating element: boron, boron carbide, borides of different metals, boron anhydride, alkiborides etc.;
(2) a salt filler (the activator of the process) which guarantees the uninterrupted formation of a gas saturating medium during heating and isothermal soaking; it interacts with the source of the saturating element—sodium fluoride, aluminum fluoride, cryolite, ammonium chloride, potassium or sodium fluoroborate etc.;
(3) an inert filler; it forms a solid carcass and improves the hardness of grain mixtures or compact material after drying; it also functions as a diluter of a saturating substance and determines the level of saturating capability. The widely used substances are: aluminum oxides, silicon oxides, magnesium oxides, calcium oxides and other technical materials—alumina, carborundum, fire clay, fine quartz powder etc.;
(4) a bonding substance; the following materials may be used: hydrolized ethyl silicate, liquid glass, BF_2 glue solution in acetone, spent sulfite liquor, Zaponlack, dextrin glue and other known bonding substances, glues and polishes.

Other parameters, apart from the temperature and time, used to control the technological process and its constituents—the speed of boride layer formation, its phase composition and the correlation between phases, the stress state and mechanical properties of boronized layers and complete details—are:

- the level of overpressure in a saturating medium;
- the gas-forming ability of the mixture which is determined by the composition, the area of the gas-forming surface, the size of the particles and the amount of boron-containing substance;
- the intensity of gas medium circulation in the saturation space.

The parameters of electrical-physical impact play a special role in controlling the process (the position of details in the electrical current chain, the electromagnetic field etc.).

Blowing a powder saturating mixture by a gas carrier is one of the methods for obtaining a gas medium for boriding. A saturating gas compound is formed as a result of the interaction between powder mixture components. For instance, blowing a powder mix by dissociate ammonia ensures some stability of the saturation process and the diffusion layer formation on the base of Fe_2B, given that the powder mixture contains NH_4Cl or CCl_4 or amorphous boron [33]. The same mechanism was used to generate a saturation medium after blowing the mixture containing B_4C or ferroboron with activators $NaBF_4$ or KBF_4, hydrogen, a mixture of hydrogen with inert gases (Ar, N_2, He) or a mixture of dissociate ammonia with CO or CH_4. The details, positioned below the powder mixture or in a separate container with a saturating gas mixture, obtained plastic one-phase boronized layers [34].

Non-contact boriding methods in gaseous mediums include, among all, EKabor technological processes (Table 8.1) [35]. They presuppose the limited contact and saturation in grain mixtures. The existing contact between the working surface and the saturating mixture in these cases is not a primary process which controls the boron mass transfer. The control over the phase composition of boronized layers at the first stage is determined by the content of mixture components and the grain size (Fig. 8.7) [36].

The EKabor compositions are primarily intended for one-phase boriding (Fe_2B) with an insignificant amount of FeB phase. EKabor 1.2 powders may be used multiple times (up to 6 times) with 20–50% refreshing after each use. The refreshing percentage depends on the saturation conditions.

Normally the treatment is carried out at temperatures 800–1050 °C. The exclusion is high-alloy steels and solid smelts as the generation of a one-phase boride layer requires the temperature higher than 900 °C which is the point when boriding influences the matrix metal and encourages W, Ta, Cr, Mo, Ti and other carbides to be replaced by corresponding borides.

The optimal temperature for the boriding of solid smelts is 1000 °C; it leads to the obtaining of a layer up to 10mkm with hardness >1500 HV [37].

The processing temperature for most steels lies in the range of 850–880 °C. The processing temperatures higher than 950 °C reduce the quality of the surface because of the melting of phosphide eutectics. The details, partly protected from boriding by a layer of copper, require the saturation temperature lower 950 °C. A higher temperature makes copper react with EKabor components and cause the melting of the working surface.

Table 8.1 Characteristics and purpose of EKabor technology [35]

No.	The name of the compound	Grain size, mkm	Density, t/m^3	Quality of the layer / Usage
1	EKabor 1	⩽150	~1.8	High quality and good interaction with the substrate. High reproducibility of results. Details' processing with minimal tolerance
2	EKabor 2	⩽850	~1.5	The surface layer is of exceptionally high quality. High reproducibility of results. Details' processing with minimal tolerance. Easy to unpack after processing
3	EKabor 3	⩽1400	~0.95	The surface layer of high quality. Good fliability of powder. Preferable for detauls with small width of a layer
4	EKabor HM	⩽150	~0.95–1.5	Good layer surface. Preferable for details of small profile and size and thin boride layer. Appropriate for all kinds of processed materials
5	EKabor–Paste	–	~1.9	Economically efficient for large details' local hardening. Paste is applied by dipping or spraying. Protective atmosphere (NH_3, Ar, N_2) is preferable for processing. Endothermic atmosphere and gases containing CO are not recommended
6	EKrit	⩽	~1.55	Covering protective filling, which protects boriding mix from air oxygen
7	EKabor WB	250–350	~0.95	Used for processing in fluidizated mediums in non-oxide atmospheres

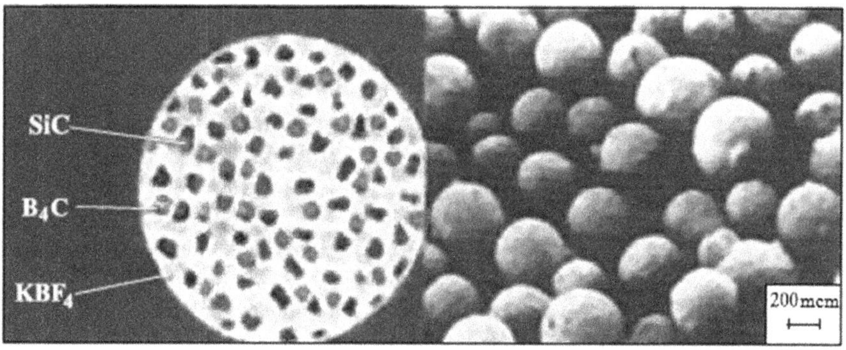

Figure 8.7 The composition, structure and type of the mixture EKabor WB [36]

The duration of boriding in EKabor compounds is between 15 min for the small details, processed in the spinning converter of the furnace, to 30 h. Normally the duration is 1–8 h. The most economically, technologically and quality efficient variant of processing is boriding at the maximum temperature and shorter duration.

The disadvantage of the processes, carried out in containers, is the tendency for the formation of brittle FeB phases during the heating stage; it isn't eliminated later due to the high activity of the saturating medium. At comparatively slow heating in the range of 700–800 °C in the typical container, a change of the saturating medium's activity is not possible. Blowing the container with a neutral gas in this temperature interval allows to reduce the activity of the saturating medium. The negative influence of FeB phase actively manifests itself in the layers of a significant width ($\sim$ 100 mkm).

Treatments in fluidizated mediums (EKabor WB) eliminate the mentioned disadvantages and are characterized by the following benefits:

- the wide temperature range of heating and cooling procedures in short operation cycles;
- the exceptional temperature stability of the process due to the high speed of heat transfer between suspended particles and boronized details;
- the possibility of uninterrupted processing and automatization of operations;
- the possibility of direct soaking after boriding on the same equipment;
- it decreases the processing time and energy costs during the thermal processing of mass-produced details.

The layer's width in all cases is chosen according to the processing conditions and depends on the type of the substrate material and its thermal processing. In particular, a high level of alloying corresponds with a slow speed of boronized layer formation (Fig. 8.8). The obtaining of layers of maximum width is not necessary.

Heating with a purpose of hardening must be carried out in an inert atmosphere, a medium of protective gases, a vacuum or in a neutral salt bath. The tempering of boronized details with the layer >120 mkm is undesirable as it leads to the formation of cracks.

The combined process which offers a reliable control over the activity of a saturating medium is the Borudif method. This method involves the usage of BF_3 gas for blowing a saturating mix on EKabor or the mixture of B_4C and SiC powders. The gas preliminary goes through SiO_2 washed by H_2SiO_3 at 450 °C. Silica (SiO_2, quartz sand) reduces the activity of BF_3gas. The result of the interaction is the formation of an intermediate unstable compound $(BOF)_3$ which is fed to the saturating medium. It helps to obtain plastic one-phase boronized layers of acicular structure in low- and medium-carbon steels under saturation [3, 38].

The scheme for the Borudif techonological process is given in Fig. 8.9. The hardened dctails may be also positioned in the volume behind the powder saturating mixture on the way of the gaseous medium movement.

The technology involves the following order of procedures:

- degreasing details' surfaces;
- positioning a layer of the mixture of B_4C (20%) and SiC on the reactor's bottom;
- putting details into a reactor;

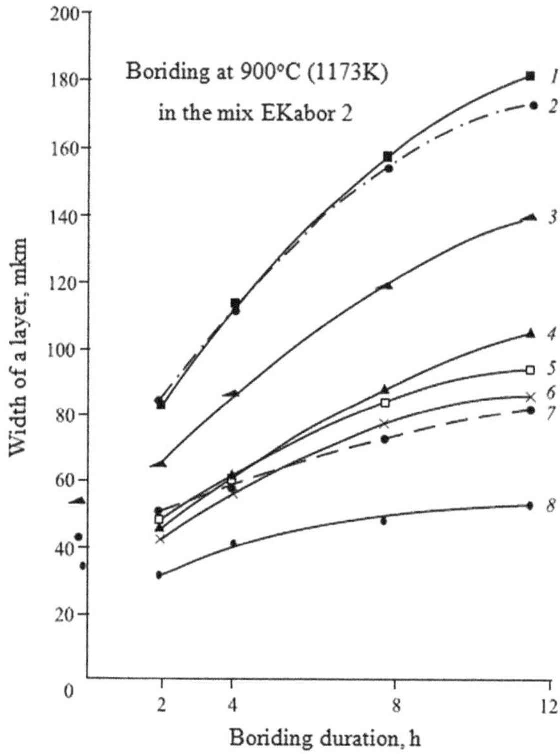

Figure 8.8 The dependence of boronized layer width from the steel brand [35]: *1*—steel 0401, C15; *2*—0503, C45; *3*—X, ШХ15 (2067, 100 Cr 6); *4*—100ХМФ (2363, 100 CrMoV 5 1); *5*—30ХМФ (2365, 32 CrMoV 3 3); *6*—Х12МФ (2379, 155 CrVMo 12 1); *7*—Х12 (2080, 210 Cr 12); *8*—Х12В (2436, 210 CrW 12)

- covering the details by the saturating mixture of B_4C (20%) and SiC (in case if details are put into the mixture);
- positioning a gas-permeable convector with a reaction decelerator SiO_2;
- sealing the reactor;
- blowing the reactor with nitrate (the volume of a blowing gas must be 5 times larger than the reactor's volume);
- positioning the reactor in a heated furnace;
- feeding BF_3 into the reactor (consumption 1.0–10.0 l/h, depends on the reactor's volume and desired level of gas activity reduction; during heating, the gas feeding might be reduced);
- the treatment of the details (the starting point is setting the temperature in the reactor in the range 800–1000 °C; the treatment duration is 3–8 h);
- blowing the reactor with nitrogen during the whole cooling period.

The reduction of BF_3 activity and the regulation of the feeding speed in this case allows to change the saturation ability of a gaseous medium in wide range and to ensure the obtaining of boronized layers from α-solid solution to two-phase layers (FeB + Fe_2B).

The circulation of the gaseous medium in some cases is substituted by the reverse-oscillating movement near the working surface. The reciprocal movement

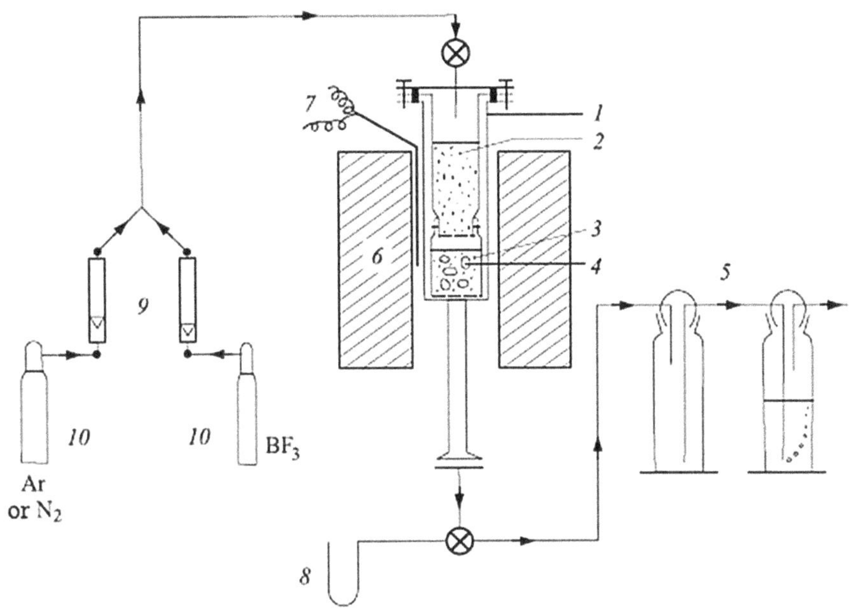

Figure 8.9 The experimental setting for boriding in the flowing gaseous medium [3, 38]: *1*—convector (reactor); *2*—reaction decelerator; *3*—saturating mix; *4*-processed details; *5*—neutralizer; *6*—resistance furnace; *7*-thermocouple; *8*—manometer; *9*—rotameters; *10*—containers of initial gases

of the gaseous saturating medium in a closed volume as a part of the gas contact or non-contact methods is created by bellows-sealed or cylinder piston systems (Fig. 8.10) [39]. The oscillation frequency of the gas medium under the non-contact method must provide the flow's turbulence; it is the laminarity that works under the contact method. If the pulsation frequency is up to 5 c^{-1}, the acceleration of the diffusion layers' formation would be double.

The use of boiling fluidizated and liquefied mixes, which generally realize the non-contact mechanism of the mass transfer thorough a gaseous medium, opens wide possibilities for the generation of boronized layers with different levels of plasticity. The regulation of boron subions' amount in gaseous mediums, i.e. the power of diffusion source, is carried out by choosing an optimal correlation between a boron-containing powder additive and a heat-carrier (diluter) material. The existing powder mixes for processing in containers for one-phase and two-phase boriding as well as specially developed substances may be used for saturation.

For instance, two-phase layers are obtained with the argon fluidization of the mixture containing (mass%): 72–75% FeB, 21–23% B (amorphous), 7–12% B_4C [40]. Processing at temperatures 950–1100 °C for 5–7 h on steel 45 leads to the formation of the layer which is 80–120 mkm wide.

One-phase layers are obtained when technological mixtures on the base of B_4C and Al_2O_3 are used; the content of B_4C must be less than 15 mass%. The role of the

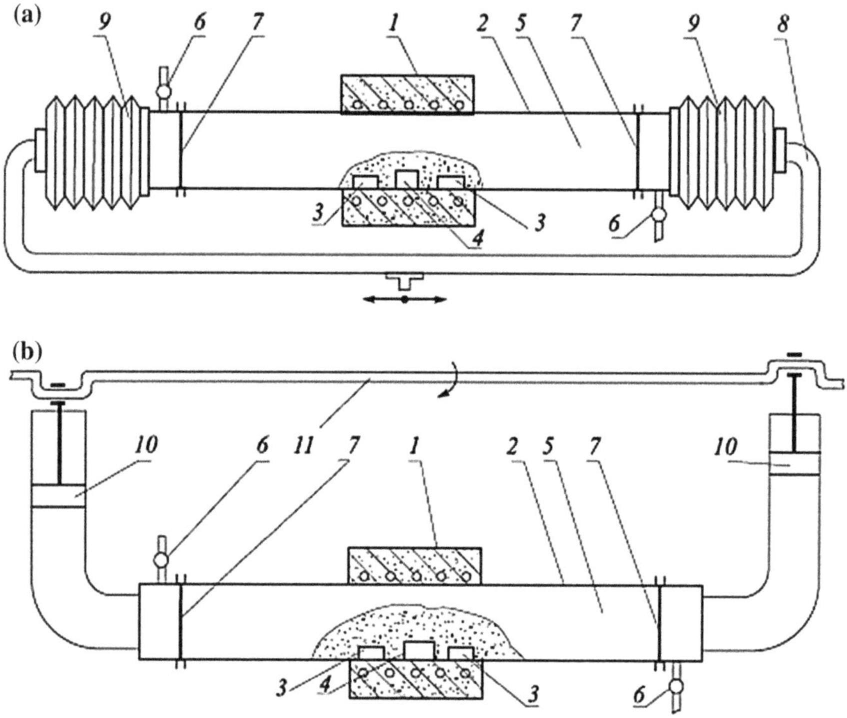

Figure 8.10 The scheme for setting a reverse-pulsating boriding process with bellows-sealed (**a**) or cylinder piston (**b**) systems [39]: *1*—electrical furnace; *2*—muffle; *3*—activator; *4*—processed detail; *5*—reaction space; *6*—faucets; *7*—reflecting nets; *8*—carrier; *9*—bellows; *10*—pistons; *11*—crankshaft

gaseous liquefying medium may be played by the following mixtures: 50% N2 + 50% NH_3; 90% N_2 + 10% NH_3; 95% N_2 + 5% C_3H_8; 85% N_2 + 10% NH_3 + 5% C_3H_8 [41].

As it was previously mentioned, the plasticity of boronized layers significantly depends on the level of their alloyness. In particular, certain compounds for silicon-boriding, nitrate-boriding, nickel-boriding, copper-boriding and chrome-boriding were developed for saturation in fluidized mediums [42–46]. For instance, the compound for silicon-boriding have the following components (in mass%): 5–25% B_4C; 2–10% Si; 0.05–0.25% CuO; 0.1–0.5% $NaBF_4$; 0.1–0.25 $NaBF_4$; 0.1–0.25% NH_4Cl; 64–92.75% Al_2O_3.

The process of saturation by boron and silicon is carried out at 600–1000 °C with some soaking for 10–20 min at the beginning of the processing at 350–500 °C. The total time of the treatment is 5–25 h. All the processes of consecutive and simultaneous saturation guarantee the increase of diffusion layers' plasticity, the acceleration of their growth, the increase of the wear resistance accompanied by a slight fall in hardness, the improvement of the working surface quality in comparison with the surfaces boronized differently. The processes are recommended for hardening cutting and die instruments and for processing precision details.

One of the disadvantages of fluidizated mediums is the loss of high-disperse materials in the working space, especially at the high consumption of gas in accordance with the conditions of technological processes; another one is the dependence of the process from aerodynamic properties of the flow. These disadvantages are less characteristic for grain mixtures with a definite specific grain weight (EKabor WB).

The total elimination of these disadvantages is provided by the use of boiling liquefied layers where the turbulence of the mixture is created by mechanical or ultrasound vibrations. The liquefaction is carried out by vibrating a pod of the heating setting or the processed detail itself.

The same mixtures can be used for liquefaction and fluidization. For instance, obtaining plastic one-phase boronized layers, alloyed by copper, requires the composition containing (in mass%): 5–35% B_4C; 0.05–0.25% CuO; 0.05–0.25% NH_4Cl; 64–94.9% Al_2O_3.

The heating goes up to 450 °C in a still saturating medium for 10–20 min. The saturation is carried out at the range of temperatures 600–950 °C while liquefying a heated mix. The layer's width at 950 °C in steel 45 reaches 0.11 mm, its hardness —1400 HV [47–50].

If certain detail's surfaces need protection from boriding in liquefied mediums, a special covering of sodium silicate may be used [51]; it can be replaced by a special paste which contains liquid glass (30–60%) and talc (40–70%) [52].

The mass transfer of boron in grained mixtures occurs generally in gaseous mediums. However this technique leads to the appearance of another factor for controlling the properties of obtained layers. It is the size of grains which in combination with a gas-forming surface determine the level of contact role in the general mass transfer. The increase of the grain size leads to the reduction of the gas-forming surface and increases the role of boron mass transfer through the gaseous medium. It is accompanied by the area expansion in the place of interaction and a sharp increase in the boron concentration. This leads to the occurrence of variable boron concentrations on the surface and the formation of separate boride zones which are alternated with unhardened surface sections. This layer's structure on the surface prevents the formation of high residue tensions and greatly improves the plasticity of the boronized layer. In some conditions of alternating-sign loading this structure is preferable—it increases the wear resistance of the hardened detail more effectively than the continuous boronized layer.

The use of grained mixtures simultaneously improves the working conditions for boriding in containers. The mixes are grained by producing pellets in special barrels or making a dough-like mass with the bonding material (e.g. wood resin) which is later molded and heated at 450 °C [16, 53–57]; the process is also carried out by making a thick mixture based on a melted salt followed by cooling, crushing and riddling [58]. Grained mixtures may be produced from any known powder compounds for one-phase and two-phase boriding. The saturating ability of grained mixtures is practically the same as in powder mixtures.

Boron saturation normally goes in sealed containers with fusible locks or protective fillings over the grained mixture made from silicon carbide and carbon black. The isolating filling is more technological while fusible lock provides a better sealing for containers and protects the saturating mixture from the influence of air oxygen. The level of sealing determines the speed of boronized layer formation and its phase composition. The more sealed a container is, the faster layers are formed and thus the possibility of two-phase boronized layers formation is more real.

The introduction of crushed charcoal or used carburizer in the amount of up to 10% may increase the gas-forming surface of the granulated material.

Boriding in grained mixes on the basis of B_4C leads to obtaining two-phase boride layers. For instance, the saturation of a grained mixture containing 68% B_4C, 16% $Na_2B_4O_7$, 10% KF and 6% C at 900 °C during 5 h with fusible locks in steel 45 leads to obtaining a 0.24 mm layer; the width of the layer changes to 0.19 mm after 4 uses. The decrease in the saturation activity of the medium due to its exhaustion leads to the decrease of the high-boron phase in boronized layers [16, 53]. The use of high frequency current heating allowed to get composite boronized layers with disperse boride inclusions [55–57] which are characterized by their high plasticity.

Two-phase boronized layers with 100% FeB phase on steel surfaces are formed in the grained mixtures containing 40–65% B_4C and 60–35% NaF (a fusible lock is used). The reduction of the brittle FeB phase in the layer down to 20–30% and the plasticity improvement is reached by introducing 50% of Al_2O_3 or 30% of carbon black into the composition; the other way is the use of the isolating medium of SiC filling. The latter method makes the container's packing and unpacking easier while limiting the maximum temperature of this granulated material's use to 850 °C due to the possibility of the saturation mixture fritting in hotter conditions.

One-phase boriding requires the granulated mixture containing 8–10% B_4C, 25–30% NaF, 60–67% ferromanganese (Mn_5). The simultaneous boron and manganese saturation guarantees the formation of the boronized layer on the base of phase $(Fe,Mn)_2B$.

A granulated material is produced the following way. A necessary amount of sodium fluoride is melted in a separate crucible at 1050 °C. Then powders of boron carbide and ferromanganese are added to the liquid salt. The obtained mixture is cooled, crushed and sifted to get grains of 2.5–3.5 mm. A smaller fraction of the mixture is used in the secondary liquid production.

The processed details are put into a container; the bottom is filled with a part of the granulated material and a process activator NH_4Cl (1–2% of granulated material's weight, necessary for saturation). The details are completely covered with the granulated material. Silicon carbide is used as a filling to protect the details and the granulated material from corrosion; it is positioned as a solid layer over the saturation mixture, 30 mm wide (minimum).

The process of one phase boriding is carried out at 750–950 °C for 1–10 h. The saturation of iron and steels leads to the formation of plastic faultless layers with

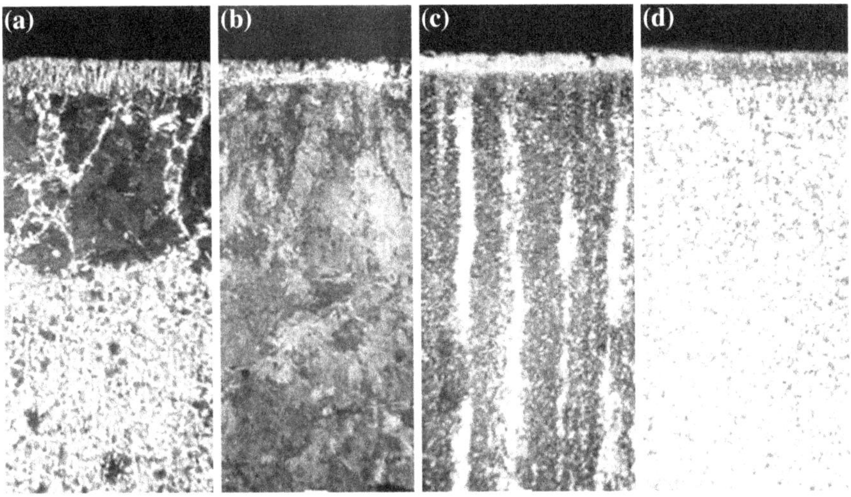

Figure 8.11 The microstructures of the boronized layers in steels after saturation in granulated mixtures containing 40% B_4C and 60% NaF at 850 °C for 4 h (×200): **a**—steel 35; **b**—steel У8; **c**—7X3; **d**—X12

1650–1800 HV microhardness (Fig. 8.11). Boriding of steel 45 at 950 °C during 4 h led to obtaining a 100 mkm layer [59].

The detailed study of the composites for one- and two-phase boriding with silicon carbide isolating filling showed that 40% is the optimal amount of B_4C. The grain size of 0.25–1.0 mm provides an optimal saturating ability due to the favorable combination of a gas-forming grain surface and gas permeability of a saturating volume.

To maintain equal saturation abilities of granulated mixtures at a maximum level, 30% of newly-produced mix should be added every time the mixture is used. If charcoal grains are present, the amount of the mixture may be reduced to 10–15%. When the fusible lock is used, the exhaustion of mixtures is slower and the granulated material can be used up to 5 times without refreshing.

Currently, the granulated mixtures used for obtaining one- and two-phase boronized layers are mass produced and delivered worldwide. They are DURBORID, Ekabor (Germany) [35, 36, 60], BKB1, BKB2, BTC [61, 62] and Borkorund (Russia) [63].

Non-contact saturation in gaseous mediums is characteristic for boriding processes in refractory-lined containers. A gaseous saturating medium is formed as a result of the interaction between the components of the solid compact fettling which cover the container's walls or sections in the injector [58]. The mass transfer of boron going from the saturation fettling to the processed surface is maintained in a self-organizing mode. The regulation of phase composition is carried out by selecting a saturating fettling composition and setting a certain pressure in the container.

The technology of producing refractory-lined containers for boriding in gaseous mediums and processing includes the following operations:

- producing the powder saturating mixture;
- producing the fettling by adding bonding materials into the powder mixture;
- covering the container's walls and lid with the fettling (an appropriate width of covering is 5–10% of the container's profile);
- drying (in order to obtain a gas-forming compact material);
- putting the activator on the bottom of the container ($\sim$1% of the fettling covering mass);
- positioning the processed details;
- sealing the container after closing the lid to provide a certain level of pressure in the container (boron anhydride (B_2O_3) or a sodium silicate rock can be used as fusible locks; silicon carbide is used as an isolating filling);
- the process of boriding;
- opening the lid and getting the processed details (the lid can be opened without cooling if the hardening is necessary).

Sealing is needed due to the slow gas medium formation and fast exhaustion of the fettling. The use of a fusible lock in combination with reliable sealing and higher pressure in the gas medium provides an equal saturation ability along the height of the refractory-lined container. The use of the isolating filling causes the reduction of saturating ability in the top part of the container. The level of reduction depends on the lid tightness and can reach 10%. The sign of exhaustion of the gas forming t container's fettling is the occurrence of a glass-like coating on the working surface. In this case it should be replaced.

This technology of boriding in gaseous mediums is recommended for processing medium and small details in large scale production. It reduces the risks of toxicity and explosibility in comparison with gas boriding in gaseous mixtures, creates the conditions for further thermal treatment (direct hardening) and loading of a new portion of details without cooling the container.

One of the peculiarities of the given technology is that internal walls of a fire-resistant crucible may melt during gaseous boriding at temperatures >1000 °C; this happens as a result of the boriding and formation of fusible eutectics. It can be prevented by using an intermediate (positioned between the crucible and the saturating fettling) ceramic inductor or by using an inductor made from non-alloy low-carbon steel at 1100 °C.

The technological factors have a significant impact on the formation speed of boronized layers. The study of the effects of boron carbide content in the saturating fettling showed that its optimal amount is 45%. The composition has following ingredients (mass%): 45% B_4C + 30% KBF_4 + 20% Al_2O_3 + 5% C (charcoal).

If the amount of boron carbide is higher, the boronized layer's width grows insignificantly. The fivefold use of the composition without refreshing causes the layer to reduce by 20–25%. It can be corrected by putting a portion of activator

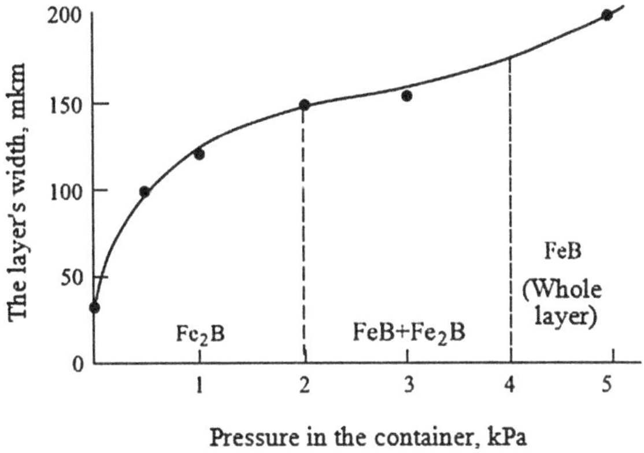

Figure 8.12 The influence of the extra pressure on the width and phase composition of boride layers in the gaseous medium of the refractory-lined container

KBF_4 at the bottom of the container in quantity of 2% to the weight of the saturation fettling before loading the details.

One of the most important technological characteristics is a coefficient of the refractory-lined container's loading; it is determined as a correlation between the detail's surface area to the area of the fettling surface ($K = S_{det}/S_{det}$). It must be around 0.8.

The gas-forming ability of the fettling is determined by the reaction surface of B_4C components and the activator and the porosity of the fettling itself. A small size of B_4C particles means a larger reaction surface but a smaller gas permeability of the fettling's compact material. The maintenance of the necessary porousness is provided by the introduction of charcoal in its composition; it not only creates the reductive atmosphere but provokes the emergence of the necessary porousness by burning. The optimal size of charcoal particles is 2–3 mm. If the container's loading coefficient is low, the influence of B_4C content is less significant; especially if its particles are not large.

The study of the boride layers' phase composition in steel 45 showed that it also depends on the gas-forming ability of the fettling and the level of sealing (Fig. 8.12). The increase of the pressure in the container leads to the transformation of the boride layer: from one-phase composition on the base of Fe_2B to a solid layer on the processed surface on the base of phase FeB.

The layer's structure obtained in the self-organizing gas medium doesn't differ from the traditional needle structure in carbon steels (Fig. 8.13). The layer's acicularity is less apparent in alloy steels.

Thus, the control over the structure formation of boronized layers during saturation in gaseous mediums of refractory-lined containers is possible through regulating the pressure, the alloyment level of used materials and the composition of the saturation mixture. For instance, a pseudoeutectic structure is obtained at appropriate temperature conditions (1050 °C) when nickel-based alloys are processed.

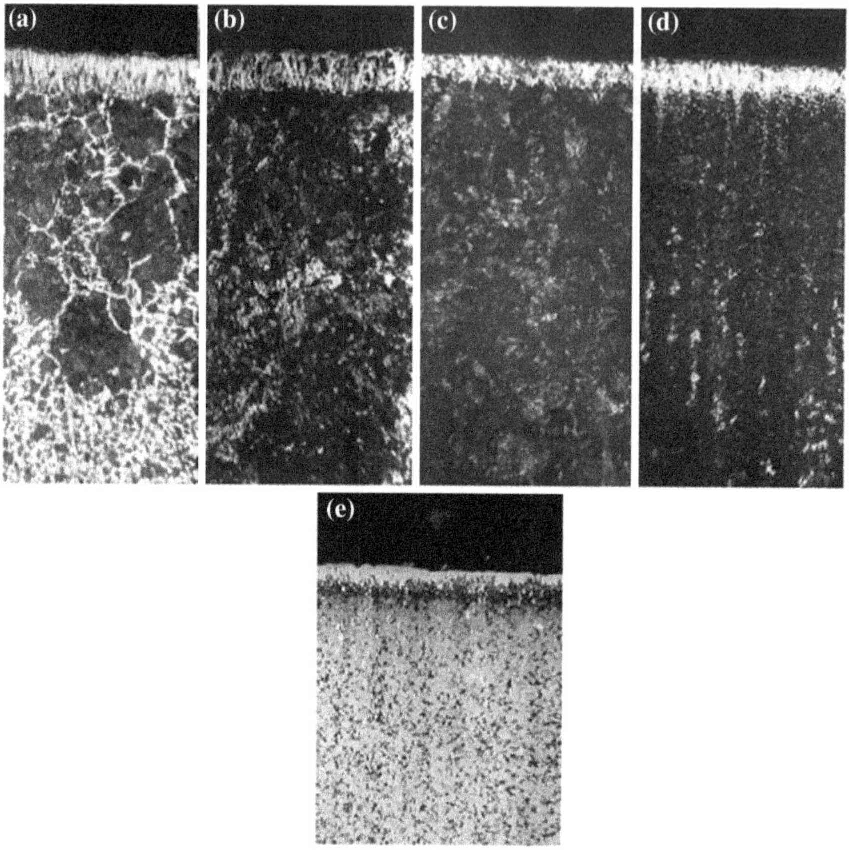

Figure 8.13 The microstructures of boronized layers in steels after saturation in gaseous medium in the refractory-lined container at 900 °C for 4 h (×200): **a**—steel 20; **b**—steel 45; **c**—steel У8; **d**—steel ХВГ; **e**—steel Х12М

Alloy boride coatings in carbon steels can't be obtained by the simultaneous deposition of boron and alloy elements (Cr, Ti, V, W, Mn etc.) in the wide range of their content in the saturating mixture as they bond with boron in the saturating fettling and form highly stable compounds. If there is a necessity, these plastic layers can be obtained in the conditions of gas contact method. The consecutive saturation requires the preliminary galvanic or electrochemical precipitation of alloy elements or the diffusion saturation with gas boriding that follows.

The elements which don't form borides in the saturation fettling (C, Si, Al, S, Zn, Sb, P and others) may be simultaneously deposited on the working surface from the gaseous medium leading to the formation of alloy plastic boronized layers [64].

The compositions for low-temperature boriding in the gaseous medium, formed in the refractory-lined container (Table 8.2), are characterized by the absence of charcoal.

Table 8.2 The recommended compositions of fettling for low-temperature boriding in a gasesous medium [64, 243–245]

The fettling's composition	Steel brand	Layer's width, mkm	Layer's phase composition
30–50% B_4C + 30–40% Al_2O_3 + 18.5–29.5% KBF_4 + 0.5–1.5% PTFE	Steel 45	10	FeB + Fe_2B
38–42% B_2O_3 + 24–31% Al (powder) + 0.5–1.5% NH_4F + 28–33% KBF_4 + 0.5–1.5% PTFE	Steel 45	10	Fe_2B
38–42% B_2O_3 + 24–31% Al (powder) + 0.5–1.5% NH_4F + 28–33% KBF_4 + 0.5–1.5% S (handle)	Steel У8	10	Fe_2B (layer alloyed by sulfur)

Notes
1. The low-temperature boriding was carried out at 600 °C for 2 h
2. Metal-thermal mixtures 2 and 4 were reduced at 950 °C in the containers with fusible locks before the coating production (without adding KBF_4 and PTFE). The reduced powder mixture was additionally enriched with a certain amount of KBF_4 and PTFE
3. PTFE—Polytetrafluoroethylene

Carbon, which isn't well soluted in the processed material at low temperatures, decelerates the boriding process in iron and steels. It is recommended to use boron anhydride (B_2O_3, softening temperature—450 °C) or other fusible salts with an appropriate viscosity as a fusible lock. For instance, one of the variants is to use a fusible lock which contains PbO and B_2O_3 in proportion 1:1 and provides the corrosion protection upon reaching 380 °C.[4]

The growth kinetics of boronized layers in low-temperature saturation processes in gaseous mediums of refractory-lined containers shows that soaking over 8 h leads to the formation of technologically appropriate layer widths. Simultaneously the increase of the processing temperature causes the layer to grow and decrease its compactness (Fig. 8.14). Plastic boronized layers are normally formed; in some cases—with a small quantity of FeB phase.

The saturating fettling used for low-temperature boriding can be made from any powder mixtures for one-phase and two-phase boriding with introducing an increased amount of KB_4 and PTFE according to the recommended compositions.

The technology of producing saturating refractory-lined containers for saturation in gaseous mediums has a range of disadvantages. To eliminate them, special containers are developed; their working volume is divided into gas-forming and saturating mediums by porous and perforated gas-permeable barriers. Instead of a coating-fettling, the use of granulated or powder mixtures is allowed. The container's structure provides the circulation of the gaseous saturating medium which is formed as a result of evaporation products going trough the powder mixture or activators dissociating there. The same level of sealing in same compounds but in

[4]N.G. Kuhareva, M.G. Krukovich, G.M. Levchenko. The author's certificate № 889738 USSR C23C. The composition of fusible lock. Publ. B.I. № 46, 15.12.81.

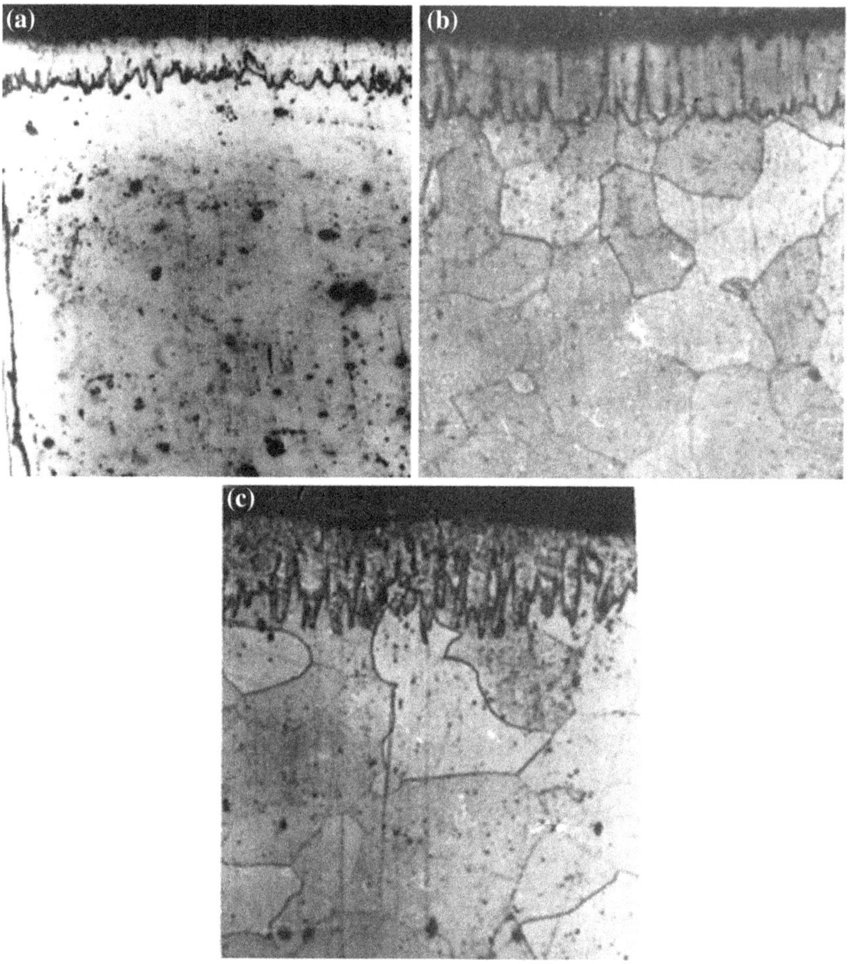

Figure 8.14 The microstructures of boride low-temperature layers in ARMCO-iron: **a**—electrolysis saturation at 550 °C for 16 h (×520); **b**—electrolysis saturation at 600 °C for 10 h (×300); **c**—saturation in the gaseous medium in the refractory-lined container at 700 °C for 8 h (×300)

different containers (one—refractory-lined; another—with divided saturating medium) has practically the same saturating ability. The second type of container is better to use for one-type details [58].

Multicomponent saturation in the gaseous medium, obtained as a result of dissociation of liquid solutions, is carried out at low temperatures. In particular, the low-temperature O, S, B, C and N saturation of high-speed steel P18 in the gaseous medium, obtained as a result of formamide and ethanol solution with diluted S, B etc.—containing components, increased the instruments' durability twice.

One of the perspective methods aimed at simplifying the apparate boriding in containers in gaseous mediums is to feed capsulated gas-forming ingredients into the saturating medium [65].

8.3.3 The Influence of the Conditions of Contact Saturation in Gaseous Mediums on the Plasticity of Boronized Layers

The contact saturation in gaseous mediums covers the boriding processes in gaseous powder mixtures, compact materials (coatings), saturating ribbons and models. Normally, the boron mass transfer goes through a gaseous phase [8, 66] with the insignificant participation of the mass transfer in contact points; in some cases—in combination with the mass transfer through gaseous and liquid mediums [9].

Powder medium saturation is carried out in containers. Details are put into a container, covered with a saturating mixture; the lid is then closed. Then the mixture and the details are isolated from the oxide environment.

The control over the structure and properties of boronized layers is done by following these technological methods:

- the composition of a saturating mix is selected;
- the optimal temperature and time of processing is set (taking into account the changing conditions of saturation in the container);
- the level of sealing for a saturating medium is set by using fusible locks of different width and viscosity, isolating fillings or coatings (SiC, cast-iron chips, mix of quartz sand and carbon black, used carburizer for cementing) or by carrying out special processing without isolation;
- the details are introduced into an electrical current chain; electromagnetic fields are used;
- the controllable atmospheres are applied;
- the local heating sources are used;
- the boronized layers are alloyed while boron and alloy elements are simultaneously precipitated; the alloying may be followed by the precipitation or by the alloying of the subtrate's material.

The most widely used phase in boriding in powder solid mediums is B_4C. The compounds most often include (mass%): 80–50% B_4C, 5–30% $Na_2B_4O_7$, 1–5% of one of the activators (NH_4Cl; NH_4F, NaF, $NaHF_2$, KF, AlF_3, Na_3AlF_6, CH_3COONa, CH_3COOK, KBF_4, $NaBF_4$, Na_2BH, $BaCl_2$, NaCl etc.), 10–15% of the inert additive (A_2O_3, MgO, CaO, crushed fire clay etc.) [8, 9, 58, 66–69]. Some recommended compounds may contain KBF_4 or other borates instead of $Na_2B_4O_7$. The saturation of steels and cast iron at the range of temperatures 900–1000 °C in containers with fusible locks causes obtaining two-phase boride layers. At the same time the use of isolating SiC filling creates the conditions for obtaining one-phase Fe_2B-based boride layers of increased plasticity.

The necessary remark must be made about a large number of B_4C based compounds (a combinatory type) mentioned in the literature: they don't affect the properties of the process itself and layers in particular. The multiple effects of the increased saturating ability and improved technology are connected with the choice of appropriate comparison object of literature data or the comparison with the results obtained in non-optimal conditions.

The improvement of quality and plasticity of two-phase boride layers is reached by introducing different elements into the saturating mixture: copper powder [70], BaF_2 [71], iron scales [72], abrasive metal mixture [73, 74], $(NH_4)_2Ni(SO_4)_2$ [75, 76], $CuCl_2$, NiO [75], zircon powder and talc [77–81], charcoal carburizer [78, 80, 82, 83], salicylic acid [84], aluminum-nickel catalyzer [85], Na_2SiF_6 [86] etc. Some of the mentioned additives are industrial waste products or don't have wide use, that's why their involvement is quite appropriate.

The stabilization of saturation from powder mixtures and obtaining reproductible results are reached by the use of the containers put into a heating furnace with a controlled atmosphere [87–89]. The products of pyrolysis of saturated and non-saturated carbohydrates, nitrogen, argon, hydrogen, products of triethanolamine decomposition are used as these atmospheres. To intensify the boriding in powders with an isolating SiC filling in protective atmosphere, it is recommended to add 2% (or less) of oxygen; it helps the formation process for a fritted oxide crust on the surface of SiC and maintains the necessary boron potential in the saturating mixture. Finally, it allows to lower the saturation temperature to 550 °C during boriding in powder B_4C and H_3BO_3.

The use of conducting saturating mixtures led to the possibility of boriding without heating furnaces. The most preferable conducting mixture components are graphite and boron carbide. The heating of the mixture and processed details is carried out by a stable or alternating current of industrial frequency 10–50 V with the use of two electrodes. The electrodes made of low-carbon steel are put from different sides of a muffle (Fig. 8.15). The speed of heating is determined by the electrode current density, the level of conducting mixture compression and its electrical conductivity, the level of the container's sealing and its heat-isolation.

Energy-saving requires three stages of the process:

- the first stage provides an un-even heating of the conducting mixture for 40 min up to ~40–50 °C (which is lower that the set temperature; the current density ~0.5 A/cm^2);
- the second stage involves the alignment of the temperature along the saturating medium section (the current density 0.2–0.3 A/cm^2; 30–40 min);
- the third saturating stage is carried out when the current density is 0.1–0.2 A/cm^2 during the set time.

The correlation between the masses of saturating mixture and details must be in proportion (2–2.5):1 [90]. The most appropriate composition for boriding is: 40%

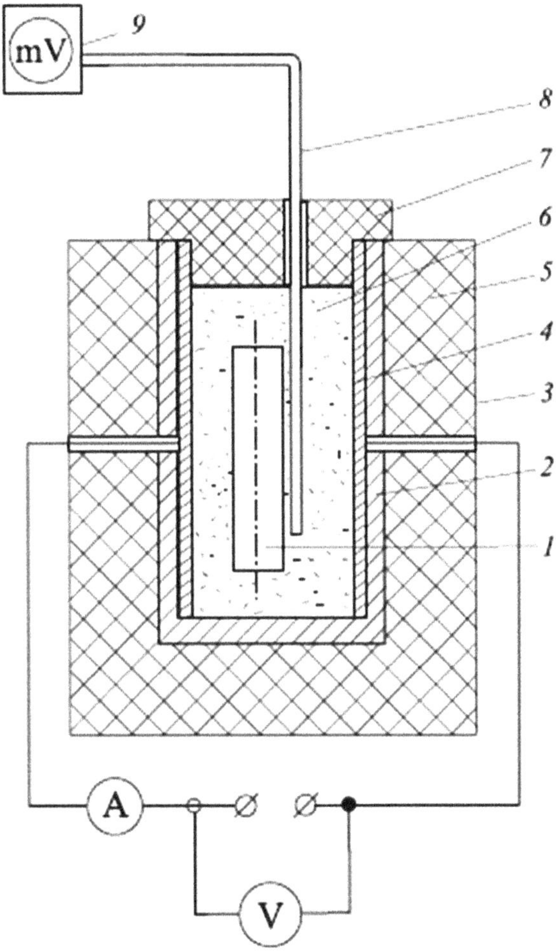

Figure 8.15 The scheme of the boriding setting in conducting powder mixtures: *1*—processed detail; *2*—muffle; *3*—corpus; *4*—electrodes; *5*—heat-isolating material; *6*—conducting saturating mix; *7*—lid; *8*—thermocouple; *9*—millivoltmeter

B_4C, 5% KBF_4, 25% SiO_2, 30% graphite. Boronized layers, obtained at 1000 °C soaking for 1 h at the third period, consist mostly of Fe_2B boride 100–120 mkm wide (in steel Ст3). Two hour soaking causes the layer to compose of 1/3 FeB and 2/3 Fe_2B.

The activization of the surface consisting of B_4C particles is done by the preliminary processing of B_4C with an overheated steam; the purpose of the activization is to provide the formation of the increased number of compounds in the saturating medium [91]. The activization of the processed surface, i.e. creating additional number of defects of the crystal structure and increasing its reaction capacity, is done by putting details into the smelt of sodium cyanide [92] or 10% solution of sulfate-salicylic acid [93].

The reduction of the brittle high-boron phase in the layer or formation of Fe_2B-based one-phase layer is caused by the partial replacement of B_4C by Al_2O_3, SiC, amorphous carbon (graphitized), carbon black, charcoal or its total replacement by the mixture of ferroboron and SiC [58, 82, 83, 94–96]. Charcoal creates a reducing atmosphere in the container and protects the mixture from fusing. A one-phase layer is obtained by maintaining B_4C content in the mixture below 10%. This method significantly reduces the cost of the used mixture.

A comparatively effective way of controlling the structure-formation is the microalloying treatment of boronized layers by S, Mn, Si, Ti, al, Ze, Cr and rare-earth metals [58, 97–101]. These elements are fed into the saturating mixture in powder state in a pure form or in the form of alloys, oxides and salts.

In particular, the saturation of steels in a compound containing (in mass) 74–80% B_4C, 2–4% NaF, 1.5–2.5% Al, 1.0–1.5% Zr and ferrochrome, allowed to increase the wear resistance by 1.5 (if the layer's structure is homogenous) [101]. 10% of rare-earth metals in the content of the saturating mixture also leads to obtaining a Fe_2B-based layer in iron and alloys which is characterized by reduced brittleness and sufficient wear resistance [99]. The double stage of boron-nitration of stainless steels in the boron carbide based mixtures (at 580–600 °C for 2–3 h; then at 860–920 °C for 4–10 h) helps to obtain alloyed 0.75–0.1 mm wide one-phase layers [99]. The simultaneous saturation of steel 45 with boron, silicon and manganese at 900 °C in the mixture of 91–95% B_4C, 3–5% silicomanganese (Simn 17) and 3–6% KBF_4 for 4 h led to obtaining a one-phase layer of high plasticity and 0.21 mm wide [58].

One of the methods of accelerating the growth of boronized layers and increase their plasticity is boron-sulfurizing. Microalloying of borides with sulfur encourages the generation of $Fe_2(B,S)$-based one-phase layers. Sulfur is fed directly into the boriding mix in the amount of 1%; the other way is to have saturation in the sulfur-free mixture but in the sulfurized container.

The introduction of sulfur (0.5–1.0%) into the mixture containing 40–60% B_4C, 1.5–2.5%, 36–38%, increases the saturation capacity by 20–30% [97].

For instance, the use of a preliminary sulfurized container with sealing by a fusible lock in the mixture, containing 94% B_4C and 4% AlF_3, led to the formation of a 145 mkm boronized layer in steel У8 at 900 °C for 4 h. The containers were sulfurized from alloy X25H35 in the mixture of FeS (75%), Al_2O_3 (20%) and NH_4Cl (5%) at 950 °C for 4 h.

Saturation in non-sulferized containers in the same conditions leads to the formation of a boronized layer of only 80 mkm [98].

The analysis of mixtures based on amorphous boron showed 4 main possible variants:

- saturation in the mixtures containing 94–98% boron; other components are activators (carbon salts of alkaline metals, AlF_3, NaF etc.) [102];
- saturation in the mixtures containing 4–6% B, 87–89% Al_2O_3, 4–6% MgO, 1–3% NaF [103];
- hydrogen medium annealing of the details covered with a layer of amorphous boron (0.1–0.5 mm wide) [104, 105];
- vacuum saturation in containers [106, 107].

The generation of two-phase boride layers is guaranteed in all cases; in the second case the amount of a brittle high-boron phase is minimal. One-phase layers are obtained if SiC-based isolating filling is used.

The last variant presupposes the multiple use of amorphous boron; it is preliminary processed in vacuum (10^{-4}–10^{-5} torr) for 3 h and then in argon medium, then it is cooled in this medium. This type of treatment allows to cleanse boron from sorbent oxygen and prevent further adsorption of oxygen from air under other technological operations. Boriding is carried out in containers; the optimal pressure is 10^{-3} torr. Argon is fed to the furnace at the final stages of the process; the argon medium is used for the further cooling of the processed details. When boron is used multiple times (up to 50), it doesn't have a tendency to fuse; details after the treatment are easy to clean [107].

Boriding and the use of metal borides Saturation in gaseous mediums, acquired from powder mixtures, often involves the use of borides Mg, (MgB_{12}), Ca (CaB_6) and Ba (BaB_6) (apart from the usual borides of Fe, Ti, Cr) [108–112]. The phase composition and properties are regulated in the same way as in the mixtures with B_4C—by reducing the content of borides in the mixture by replacing them with Al_2O_3 and SiC, controlling the level of sealing or adding the substances which have a stabilizing effect on FeB phase on the working surface. In particular, rod sulfur in the amount of 0.5–1.5 mass% stabilizes Fe_2B phase, increases the width of plastic boride layers and improves their tribology properties (Table 8.3).

Boriding in metal-thermal mixtures One of the directions in developing resource-saving technologies is the search for the energy efficient hardening treatment. The process of boriding in energy producing powders and compact materials (coatings) is one of these energy efficient processes. It is especially appropriate for a local treatment of large-scale details and instruments. The

Table 8.3 The influence of sulfur on the width of boride layers obtained in powder mediums on the base of borides of Mg, Ca and Ba [58, 108–111]

The composition of saturating mixture, mass%	Width of a layer, mkm	The intensity of a wear process, $I_h \times 10^{12}$
45–50% CaB_6 + 40–50% CaO + 1–3% KBF_4 + 3–7% NaF	110	1.1
45–50% MgB_{12} + 40–50% MgO + 1–3% KBF_4 + 3–7% NaF	140	1.15
45–50% CaB_6 + 40–50% CaO + 1–2% KBF_4 +3–7% NaF + 0.5–1.0% S (handle)	160	0.65
45–50% MgB_{12} + 40–50% MgO + 1—2% KBF_4 + 3–7% NaF + 0.5–1.0% S (handle)	180	0.60

Notes

1. Boriding of steel У8 was carried out in the containers with a fusible lock at 900 °C for 4 h

2. The intensity of the wear process was determined under sliding friction in the presence of border greasing (v = 0.57 m/s; p = 20 kgs/mm^2)

materials (coatings) used in the process are materials which provide the process of self-spreading high-temperature synthesis (SHS).

SHS is carried out in vacuum, open air, in inert or reacting gases under the pressure. The main conditions for selecting SHS-system components are:

- the exothermicity of reagents' interaction;
- the generation of effective solid products;
- the technical and economical expediency.

SHS processes are used in order to obtain alloy powders which can be used as one of the components of the boriding mixture or as a reducing agent for saturation in liquid mediums. In the first case details are put directly into the powder or the compact saturating furnace charge; one of the charge's components is a material obtained during SHS process. In the second case the obtained powder is added to a salt smelt where it functions as an electrochemical reducing agent.

Another application of SHS is the technology directed at the obtaining of saturating mixtures while combining the burning and boriding processes or during their separate progress. Boriding is carried out either in powder or compact mixtures.

The mixture ignition, or the initiation of local reaction on the mixture surface, is done by supplying a short-time heat impulse (electric spiral, electric-spark current, laser beam, fuse mix etc.) which leads to the formation of a burning wave and its distribution along the heated charge's material and the detail's surface. The duration of the initiation is normally much shorter than the time of the charge's burning. In some cases (e.g., for low-exothermal reactions) the process is initiated by heating of the whole surface of the charge in the furnace and carried out in a slow mode of burning.

The burning wave includes different chemical reactions, physical and physical-chemical processes connected with elements' atoms' diffusion, evaporation and precipitation etc., which provide a necessary heat emission. The wave has a certain length and consists of several zones:

- a zone of warming and a pre-flame zone where the heat transmission and heating of the charge take place;
- zones of the reaction with main reactions of burning which provide a necessary heat emission;
- zones of final combustion, where chemical reactions continue to emerge but they no longer affect the speed of heat emission;
- zones of secondary physical-chemical processes which determine the composition and structure of final products of composition and the structure of the surface of diffusion layers.

The main stage of the SHS process is the occurrence of the burning wave.

The main characteristics of this treatment are: the burning speed 0.1–20 cm/c; the burning temperature 2300–3800 K; the speed of charge heating in the wave

10^3–10^6 gr/c; the ignition capacity 10–200 kal/(cm^{2}c); the ignition delay 0.2–1.2 s; the ignition temperature 800–1200 K.

The economic factor of boriding, which often turns out to be the most important, led to the development of cheaper compounds for boriding; these compounds are based on metal-thermal mixtures. They are 7–15 times cheaper than traditional ones [8].

The carrier of boron atoms is B_2O_3. Boron is reduced from the oxide under heating by the elements which resemble oxygen in many ways (Li, Na, K, Ca, Ba, Mg, Sr, Ak, Zr, Ti, Si etc.). The process goes according to two technological variants:

- a combined process where boron reduction is followed by boriding in one container during one heating; a layer is formed during the further delay;
- a separate process where boron reduction and boriding go separately—in terms of place and time: first the mixture is reduced, then after crushing and riddling it is used for boriding.

The combined process is preferable for the mixtures where reduction is accompanied by insignificant heat emission. The optimal correlation between B_2O_3 and a reducing agent is calculated by the equation of thermal balance of the redox reaction. For instance, the correlation of B_2O_3 and Al is 65:35 (taking into account the level of reducibility). To prevent the fusion during reduction and heat absorption, oxides Al_2O_3, MgO, CaO etc. are added to the boriding mixtures. The function of activators is given to haloid compounds of alkaline and alkaline earth metals which reduce the temperature of the redox reaction's start and help the formation of boron subcompounds in a gaseous saturating atmosphere.

The aluminum-thermal mixture is recommended for practical use. It contains (mass%): 69–79% Al_2O_3, 13–22% B_2O_3, 5–10% Al and 0.5–1.0% NaF. Boriding is carried out at 850–1000 °C for 4–8 h. Saturation of steel У8 (950 °C, for 4 h, in the mixture of 70% Al_2O_3, 22% B_2O_3, 7.5% Al and 0.5% NaF) led to the formation of a two-phase 105 mkm layer. The separate process was carried out; a sodium-silicate rock was the base for the fusible lock.

The role of reducing agents is given to different ligatures and alloys: calcium silicon, mischmetal silicon, barium silicon, calcium-manganese ligature, $CaAl_2$, SiC etc. The best combination of technological effectiveness and saturation capacity is provided by the use of the mixture: 79.5% Al_2O_3, 10% B_2O_3, 10% calcium silicon (CK25) and 0.5% NaF. The saturation of steel У8 in this mixture at 950 °C for 4 h led to the formation of a two-phase 95 mkm layer.

The reduction of FeB high-boron phase content is witnessed when Al content is less than 8%; the use of SiC, manganese silicon, mischmetal silicon etc. as a reducing agent causes the formation of one-phase layers. For instance, one-phase layers in steels are formed in the mixture of 30–70% SiC, 20–45% B_2O_3, <8% KBF_4, <6% NH_4Cl.

In metal-thermal mixtures, a crushed smelted borax ($Na_2B_4O_7$) is used as a boron transmitter. The necessary technological effectiveness of boriding is provided by the use of the mixture which contains: 57% $Na_2B_4O_7$, 15% Al (powder), 25% C

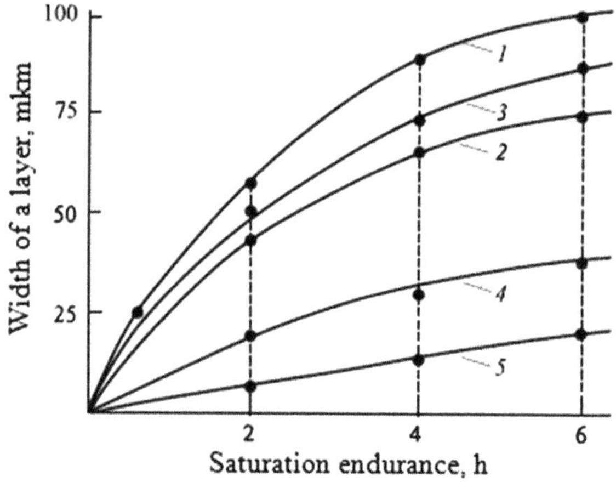

Figure 8.16 The influence of saturation endurance at 900 °C on the width of boride layers in steels: *1*—65Г, *2*—ХВГ, *3*—У8, *4*—4Х3В3МФС, *5*—Х12М

(carbon black), 2% $NaBF_4$. One-phase layers are formed when SiC is used as a corrosion-isolating medium. The amount of carbon black, as well as the level of sealing (a type of the lock), influences the speed of the boride layer formation. To prevent the formation of the aluminum solid solution in the layer, saturation with the carborundum isolating filling should involve a lesser amount of aluminum powder than the calculated amount. The saturating capacity of this compound is at the technologically appropriate level (Fig. 8.16) [58].

Boriding in metal-thermal mixtures can be repeated several times. To maintain a high saturation capacity it is necessary to add a new portion of activator and its mingling. After the triple use, the mixture demands the addition of 30% of a newly-produced mixture or the mixture of 3–5% Al and 0.5% NaF.

Alloying of boronized layers. The simultaneous increase in the layer's width and the reduction of FeB-phase content is reached by the addition of FeO, Fe_3O_4, NiO, MnO [113], TiO_2, Cr_2O_3 [114, 115] into the metal-thermal mixture. In particular, the boron-titanizing of steel У8 (950 °C, 4 h, the compound of 23.5% B_2O_3, 4% TiO_2, 48% Al_2O_3, 22.5% Al, 1.0% NaF and 1.0% S) leads to obtaining a 135 mkm wide layer. The processing is carried out in the container with the fusable lock. Boronized layers of increased plasticity in these fast-growing compounds are obtained if Cr_2O_3 is added instead of TiO_2. The boron-chrome plating on steel 45 (950 °C, 4 h, compound of 20.5% B_2O_3, 7% Cr_2O_3, 48% Al2O3, 22.5% Al, 1.0% NaF and 1% S) leads to the formation of a (Fe, $Cr)_2B$-based layer (width 220 mkm); on steel У8–175 mkm. The microhardness of chrome and titanium-alloyed boronized layers exceeds the microhardness of non-alloy borides by 100–150 HV.

One of the effective methods of controlling the phase composition and plasticity of boronized layers is their alloying by introducing metals and ferroalloys into the powder mixture. The process involves the use of mixtures which contain B, B_4C and metal-thermal compounds. The saturation causes the formation of alloy borides (Fe, Me)B and (Fe, $Me)_2B$ which provide pressure tension on the details' surface [116–119].

The optimal additives to introduce into technical B_4C are clear elements in a powder state (mass%): 1.5–2.0% Cr; 6–9% Mo; 18–20% W; 10–15% Cu, 12–17% Ti; 15–17% V; 3–7% Ta; 5–7% Zr; 3–7% Mn; 13–18% Nb; 7–15% Co [8, 68, 69, 120–129]. Saturation causes the increase in the layer's width by 1.2–1.5 times; Ti, Nb, V, Mo, Ni and Co increase the content of FeB in the phase layer; Cu-Fe_2B, Zr increase the dispersability of the structure and the layer's porousness. The microhardness may increase up to 2600 HV.

Simultanesous boron and chrome saturations may be carried out when the chrome content is even higher. For instance, processing in the powder composition, which includes 20–25% Cr, 45–67% B4C, 10–15% FeCr, 1–5% $Li_2B_4O_7$, 1–5% $Na_2B_4O_7$, 1–5% KBF_4, accelerates the boride layer growth 1.7–2 times [128].

The optimal content of Mo and W powder additives in the metal-thermal mixture is 10–45 at. % from the calculated boron content. If this amount goes from 30 to 15%, the amount of FeB phase slowly decreases; this creates the possibility for controlling the layer's properties [129].

The wide opportunities of controlling the phase composition and plasticity are created for simultaneous saturation. Different combinations may be used. In particular, preliminary boriding and further chrome plating, vanadium processing, titanizing, sulfurizing lead to the formation of a layer which is characterized by the 1.5–2 times increased wear resistance in the conditions of grease-free friction and 10–15 times improved heat-resistance [130]. Plastic $(FeCr)_2B$-based boronized layers are formed as a result of contact boriding from gaseous mediums in the powder mixture containing B_4C and NaF and further vacuum chrome plating. The layer's microhardness is estimated as 2150 HV [131]. The further vanadium processing reduces the defectiveness of the boronized layer and its brittleness. The steel surface (0.18–0.23% C, 0.2–0.35% Si, 0.7–0.9% Mn, 0.4–0.6% Cr, 0.4–0.7% Ni) acquires a layer of V_4C_3 and later Fe_2B [132]. The formation of vanadium carbide, apparently, can be explained by the presence of carbon on the surface after the preliminary boriding in the mixtures which contain boron carbide.

One of the variants of the consecutive saturation is preliminary diffusion chrome plating at 1200 °C (high frequency current heating) from the gaseous medium obtained from a compact material (coating) which contain 75% Cr and 25% Na_3AlF_6. The further boriding from the powder mixture (81% B_4C, 14% $Na_2B_4O_7$ and 5% NH_4Cl), in the container with the fusable lock at 1000 °C led to obtaining the layers with 2500 HV microhardness in steels 15Х2М2ФБС, 15ХМФ, 15ХМФ, 15Х12ВМФ and 15ХМФК [133].

Alloyed boronized layers with advanced exploitation properties are formed as a result of saturation of high-alloy heat-resistant, stainless and maraging steels and alloys. The structure and properties of these layers are determined by:

- the substrate material's composition;
- the saturation medium's composition;
- the process technology.

The formation of boronized layers goes according to the diffusion or diffusion-crystallization mechanisms. The first mechanism slows down the layer's growth and helps to even the borderline between the boride layer and the base. The second mechanism significantly accelerates the growth of layers. In all the cases the formation of quality layers has certain limits of the substrate's composition, medium and process temperature.

In particular, boriding of the chrome-nickel steel in EKabor-2 mixture (containing Si) causes the formation of Ni_2Si and Ni_3Si compounds in the layer which reduces the layer's quality [134]. To prevent defects, it is recommended to reduce the content of Ni in boronized steels to 20%. Boriding of maraging steels in compounds containing B_4C or amorphous boron is not dictated by these limitations [135]. The increase of molybdenum content in alloys is accompanied by the increase of the borderline zone and reduction of the boride width (to 5% Mo); at the interval 5–8% Mo—shortening of the borderline zone and a slight increase of the boride layer. All the cases are characterized by the reduction of the layer's width.

Boriding of Fe-Ni-Cr-based heat-resistant alloys according to the polythermal sections (Fig. 5.15) of the four-component equilibrium diagrams (<1000 °C) causes the formation of MeB, Me_2B and α-solid solution based layers (where Me is a component of Fe, Ni, Cr alloys). The correlation between these elements may change scarcely, that's why the formed boride phases are solid solutions based on one element which has the most resemblance to boron. Saturation in mixtures containing 20–76% B_4C, 76–20% Al_2O_3 and 4% KBF_4 leads to the generation of alloy CrB and Cr_2B-based boride layers with 2100–2300 HV microhardness (the condition is 10–25% chrome content in alloys). Saturation is carried out in the containers with the fusible lock. To obtain quality plastic boronized layers it is necessary [136, 137]:

- to reduce the amount of B_4C in the saturating mixture;
- to use the fusable lock and the saturating mixture which would exclude the use of silicon-containing materials;
- to reduce the soaking time period to 2 h;
- to carry out boriding at 950–1000 °C.

Boriding of corrosion-resistant steels (13% Cr) in the mixture which contains 20% B_4C, 2.5% KBF_4 and SiC, is accompanied by the deceleration of the borides' growth and smoothing of the borderline. The saturation process was carried out at 850 °C for 15 h [138, 139]. It is accompanied by the porousness and splits formation if the chrome and nickel content exceeds 15–17%. The formation of splits is explained by the redistribution of alloy elements in the sublayer and occurrence of stretching residual stresses on the surface [140, 141]. Normally, the layers' width doesn't reach 0.15 mm which is sufficient for precision details which go under thermal and mechanical processing after boriding; it is also appropriate for mechanical relieving of the porous zone due to the redistribution of stresses which provoke the formation of new splits.

Boriding from the gaseous medium, obtained from powder materials in containers (1000 °C, 4 h for stainless steels containing 6–22% chrome) proved the tendency of boronized layers to acquire porousness [142]. The interlayer of small lamellar boride formations appears when the chrome content in steel is 13–22%. This interlayer is formed after borides come to the surface; its development speed and width depend on the temperature and the process duration. Its appearance is caused by the chrome exhaustion of a particular zone of the boronized layer. Chrome which is characterized by its strong resemblance to boron diffuses to the surface and forms its own borides.

Layers of increased width may be obtained by preliminary vacuum annealing of details (e.g., for steel 12X18H10T—1200 °C) or protective atmosphere annealing which causes chrome to evaporate from a certain amount of metal and makes the surface less alloyed. This leads to the acceleration of the boronized layer's growth [143–146]. The reduction of the alloy level of surface layers is also reached by preliminary ironing [146].

Preliminary electrolyte ironing of depreciated surfaces and further boriding allow to restore and harden the details which have a wear indicator below 0.6 mm. Boriding of electrolyte residues leads to the 2–10X increase of the diffusion layer's width in comparison with boriding steels. The maximum acceleration is caused by putting "soft" residues which have relatively small and balanced grains. The boride layer is better to be formed at the boundaries of electrolyte iron coating.

The result of diffusion boriding of the electrolyte residues is obtaining one-phase boride layers which are characterized by 1500 HV hardness. This technology is recommended for restoring and hardening the details of agricultural mechanisms [147].

Layers of increased width are also formed in steels when boriding is preceded by applying galvanic nickel, nickel-phosphorous and cobalt coatings [148, 149]. This leads to the formation of plastic composite boronized layers with increased porousness and the porousness acquired according to the Sharpie principle (as a result of combining solid and soft composites). The increased porousness under exploitation helps to keep grease; the structure with alternating solid and soft inclusions is favorable for friction points which work under alternating loading in grease conditions.

The intensification of boriding from gaseous mediums in powder mixtures can be achieved by letting changing or stable electrical current go through it (10–55 V; current density during heating—0.2–0.5 A/cm^2; during saturation—0.1–0.2 A/cm^2). Conducting mixture fillings are graphite, electrode middlings, coke, graphitized anthracite, charcoal etc. The content of conducting fillings should be 30–32%, B_4C—40–45%, inert filling—30–35%, activator—1–3%. The mixture and details' heating is done by the electric current going through the mixture. The composition of this mixture (1000 °C, steel 30) leads to the formation of a two-phase boride layer with 30% FeB. Temperatures in the range of 1050–1100 °C cause the layer to consist of round Fe_2B boride inclusions which are evenly spread in eutectics; the temperature 1100 °C leads to the eutectic structure of the layer. The layer's width under 2 h saturation was 0.3 mm [90]. Electric currents create a directed flow of boron subcompounds in the saturating gaseous medium and activates the saturated surface in the presence of electric microcurrents; this determines the final result.

A more advanced boriding technology in electrical conducting mixtures is saturation in a spinning injector. Graphite or coke are recommended to be used as conducting elements; the best boron-containing materials are B, B_4C, B_2O_3, $Na_2B_4O_7$, the dilutes—Al_2O_3, SiC, the activators—NH_4Cl, KBF_4. As in the previous case, saturation is carried out by the passing electrical current. The steel surface ensures the generation of plastic one-phase Fe2B-based boronized layers [150].

If the saturation mixture is surrounded by the high voltage electrical field (U ~2.0 kB, I = 1–2 mA) after some sufficient furnace heating, the accelerated layer's growth occurs on the surface which is positioned perpendicularly to the field's acting direction. The difference in the layer's width after 4 h-soaking is 0.1 mm [151]. The layer's plasticity is improved by the reduction of the boron carbide content to 25% of the mixture. Apart from the directed subions flow in the saturating gaseous medium, the applied electrical field causes the structure of the material and the sublayer to reduce which has a positive effect on the mechanical properties of the boronized detail.

The inclusion of details into a stable electrical current chain and their functioning as a cathode during saturation in gaseous mediums obtained from the powder mixture containing 20–50% B_4C and Al_2O_3 also accelerate the formation of boronized layers by 1.5–2.0 times [152] due to the easier reduction of boron subcompounds. The efficiency of the mass transfer increases when a layer of solid electrolytes (alloy of $Na_2B_4O_7$ and Li or eutectic alloy $Na_2B_4O_7$ and NaF) is preliminary applied on the processed surface.

The fast growth of boronized plastic layers is acquired when they are formed according to the diffusion-crystallization mechanism where the processed surface is in LCD state. This state is more often acquired while boronizing high-alloy steels and smelts. The temperature and concentration conditions for processing of these materials are set according to the multi-component equilibrium diagrams. In particular, for smelts of Fe-Cr-Ni and Fe-W-V-Cr systems, boriding conditions were determined by the four- and five-component equilibrium diagrams: Fe-Cr-Ni-B and Fe-W-V-Cr-B [58, 136, 137, 153].

For heat-resistant smelts of Fe-Cr-Ni system, the boriding temperature, according to the diffusion-crystallization mechanism of the layer's formation, is 1022 °C; the gaseous medium is formed during the contact saturation from powders. The maximum of the process temperature is 1050 °C (if the nickel content is ~40%); 1150 °C—when the nickel content is ~20%. Higher saturation temperatures, in accordance with the smooth reduction of boron concentration into the alloy's deeper levels, lead to the formation of a continuous zone of liquid solution. This causes the local volume boost, tearing of the boride coating and partial displacement of liquid solution. Finally, this leads to the detail's deformation and the increase of the surface's roughness.

The most appropriate boron concentration on the surface which ensures the controllable formation of a boronized layer according to the diffusion-crystallization mechanism is 4–8.5%. Higher boron concentrations cause the occurrence of an excessive amount of liquid solution in the sublayer (due to the polythermal sections of the equilibrium diagram schemes—Chap. 5) and its displacement through the ruptures in the boride layer. Lower boron concentrations decelerate layers' growth or cause the formation of α-solid solution—based layers on the processed surface accompanied by the occurrence of a big amount of liquid phase on grain borders. This also leads to the increase of roughness.

The quality structure-formation pattern during isothermal soaking and cooling can be seen in the equilibrium diagram schemes and their sections. The number and size of borides which are formed from liquid solutions are determined by the crystallization conditions: the level of overcooling, the number of crystallization centres, the speed of their growth. The form of borides is determined by the diffusion anisotropy of crystallites and the presence of extrinsic atoms which can block the growth of certain sides.

Due to the fact that borides are compounds with a low crystal lattice symmetry and high anisotropy of the diffusion coefficient in different crystallographic directions and as a result of alloy elements blocking certain planes, growing crystals of primary borides obtain an elongated form. Their spontaneous occurrence provides their spontaneous orientation in the layer. Taking into account that studied alloys mostly consist of boride-forming elements and the process of the layers' formation is accompanied by a gradual melting (dilution) of matrix solid solution and crystallization of borides, the surface layers' matrix foundation is gradually replaced by a high-dispersible pseudo-eutectic structure. The replacement process may stop if the boron concentration on the surface decreases or stops during processed details' cooling at temperatures which are lower than the temperature for the formation of eutectics for Me_2B + α-solid solution.

The excessive eutectics content in the layer causes the inclusion of the cooling speed factor into the structure's dispersibility control, so the use of accelerated cooling after boriding is necessary to provide the disintegration of the eutectic composite.

The generation of pseudo-eutectic layers demands the optimal composition of the powder mixture which contains 20% B_4C, 76% Al_2O_3, 4% KBF_4. A higher amount of boron carbide in the mixture sets >9% boron concentration on the processed surface which leads to the surface melting and the deformation of the detail at 1050 °C. If the boron carbide content is <10%, the width of obtained layers is significantly reduced (<0.15 mm) [137].

The formation speed of plastic boronized layers of pseudo-eutectic structure (Fig. 8.17) is 6–10 times faster than for the diffusion mechanism formation. The width of pseudoeutectic layers is 0.5–1.0 mm. The obtained layers have a heterogeneous structure consisting of alloy borides on the base of CrB, Cr_2B, Ni_2B, Ni_3B which are distributed in α-solid solutions. The microhardness of structural composites varies from 445 to 1620 HV. The general manner of microhardness change

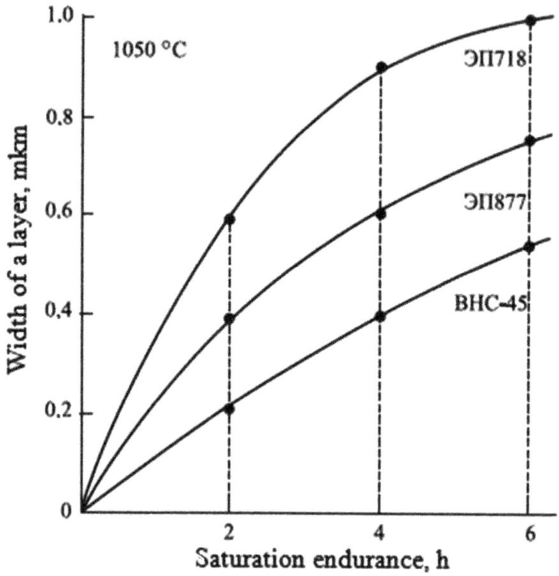

Figure 8.17 The growth kinetics of boronized layers in heat-resistant alloys in accordance with the diffusion-crystalizing mechanism of formation

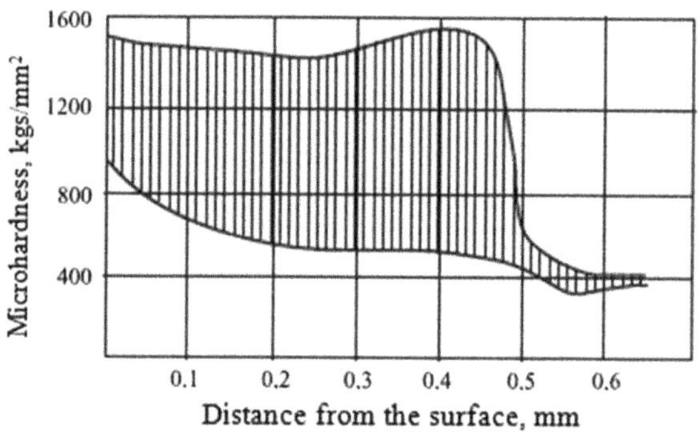

Figure 8.18 The microhardness distribution in the pseudo-eutectic layer

in the layer's width depends on the form, size, amount and mutual position of phase composites and can be shown as an interval of microhardness change (Fig. 8.18).

The best quality of obtained layers can be acquired if the Fe-Cr-Ni system alloys are saturated at temperatures <1050 °C and for no longer than 8 h. The violation of these limitations would cause the defects in the structure and the layer's shelling.

The lasting use of powder mixtures for maintaining a saturation capacity and technological effectiveness demands the addition of 35–45% of a fresh mixture with every new cycle. Stable results can be obtained if the distance between the container's walls and the processed details' surface is respected as well as the distance between different surfaces of the processed details. In the first case the distance should be 40–50 mm, in the second—25–30 mm.

High-speed steels are included in the system Fe-W-V-Cr as carbon in steels is located in carbides and doesn't take part in the formation of boronized layers for some time. The conditions for forming pseudo-eutectic structures in this system are: the temperature interval of 1160–1200 °C, the boron concentration on the surface 4.0–7.0%. The presence of non-melting and hardly dissoluble carbides at saturation temperature reduces the risk of the surface's deformation. However it works only at short-time soaking. Soakings over 3 h may cause the dissolution of carbides and the increase of the liquid phase on the surface. The control over the layer's width, size, form and the amount of boride inclusions is done by selecting a saturating mixture composition, setting temperature and temporal parameters and cooling speed after processing.

Borides in the pseudo-eutectic layer's structure in this system of alloys have mainly a globular structure. The hardness of alloyed borides $(Fe, W, V, Cr)_2B$ is 2200 HV. The general character of the microhardness change in the layer can be represented by the microhardness change interval.

Boriding in compact materials, obtained due to the application of pastes or coatings on hardened surfaces and their consecutive drying, is a contact method of saturation in gaseous mediums. This method is used when there is the need for local hardening of large-scale details, construction elements and stamping tools.

The structure of boronized layers under saturation in solid compact materials (coatings) is determined by the following factors:

- the temperature and temporal parameters;
- the technological methods;
- the composition and type of saturating materials and their aggregate states.

The change in the process temperature allows to carry out saturation by the diffusion or diffusion-crystallization mechanism, transform the mechanism of boron mass transfer through a gaseous medium into the compact boron mass transfer or the mass transfer in liquid melted mediums or choose variants of their combinations.

The conditions of the boronized layers' formation according to the diffusion-crystallization mechanism depend not only on the use of high alloy steels and smelts but also the high speed of surface heating and fast growth of a boronized layer which is accompanied by the accelerated accumulation of boron, carbon and alloy elements in the sublayer that provokes some zones of the sublayer to melt. It is also important that the durability of effective use of a compact material is limited so it is necessary to take into account the changing character of its saturation capacity.

The technological methods are connected to the choice of heating methods, the width of compact materials and the use of multilayer coatings which include saturating, energy-releasing and protective coatings.

The composition of saturating compact materials determines the initial state of boron, the formation mechanism of saturating atoms and the level of boride alloy level. The developed compounds are divided into the compounds of short-term and long-term effect. The first type of compounds provides the high speed formation of boronized layers and aimed at hardening small details under the heating for tempering. The second type provides a slower growth of layers and they are used together with heating for tempering and other types of thermal processing of large-scale details and instruments (stamps) during long soaking.

Several boron sources may be used: amorphous boron, boron carbide and ferroboron, metal-thermal mixes of combined and separate boron saturation from its oxides. Different types of heating are used for saturation: high frequency current, electro-contact, furnace, laser, electronic beam, glowing current heating, heating in melted mediums and their combinations. Many compositions aimed at powder boriding are used for producing pastes and coatings. The recommendations for accelerating boronized layers' growth, achieving higher exploitation characteristics, obtaining plastic one-phase layers and composite structures coincide with the methods mentioned above.

Saturating B_4C-based compositions normally contain (mass%): 20–75% B_4C, 30–50% (Na_3AlF_6 or NaA or AlF_3); 8–22% ($Na_2B_4O_7$ or B_2O_3 or G_3BO_3); 2–10% (NH_4F or KBF_4 or $NaBF_4$); 5–40% (Al_2O_3 or bentonite or marshalite). Liquid glass, hydrolyzed ethyl silicate, sulfite-spirit grain, dextrin solution, BF_2 glue solution in acetone, dibutyl phthalate, polyvinyl acetate, epoxy resins, modified with polyvinyl butyral, heat-resistant enamel etc. may be used as bonding materials [9, 154–162].

The effective exploitation period of compact materials is one of the most important technological characteristics which helps to obtain high-quality boronized layers with a certain phase composition and high-quality hardened surfaces. Ignoring the time limit leads to the complete stop of boride layer growth and the reduction of its width which is determined by the deboriding process itself and corrosion of detail's surfaces. Deboriding is accompanied with the formation of an iron layer on the boronized surface; it is easily oxidized and attacked by salt composites of the saturating mix (NaF, Na_3AlF_6, $Na_2B_4O_7$ etc.) [58].

The thickness of applied coating also influences the technological characteristics and saturation capacity of the compact material. If the mixture has a solid carcass which holds the mixture on the processed surface, it is covered by a glass-like crust during saturation; this crust prevents the free exit of the gaseous saturating medium. At the same time the over-excessive thickness decelerates its heating speed and prolongs the technological process. The most appropriate width of the coating is 6–10 mm.

The heating speed for the processed surfaces (when the compact material is 4–6 mm wide) has the following indexes:

- for the electrical resistor furnace—15–17 °C/s;
- for the electrical resistor furnace with the additionally applied energy-releasing paste—25–28 °C/s;
- for the salt chloride-barium bath—30 °C/s.

The combination of these heating speeds (if directions of heat and diffusion flows with a high saturating capacity are the same) provides the replacement of

Table 8.4 The technological characteristics of compact saturating materials

No.	Coating composition, mass%	Phase composition of a layer	Effective period of exploitation for a compact material, X hours at temperature, °C	
			800	1000
1	50% B_4C + 50% NaF	FeB, Fe_2B	2	0.5
2	37.5% B_4C + 37.5% NaF + 25% Al_2O_3	–	8	6
3	25% B_4C + 25% NaF + 50% Al_2O_3	–	16	8
4	65% B_4C + 35% Na_3AlF_6	–	6	4
5	40% B_4C + 60% $Na_2B_4O_7$ (contact heating)	Fe_2B (t61000 °C) FeB (t > 1050 °C)	10	3
6	84% B_4C + 16% $Na_2B_4O_7$	FeB, Fe_2B	16	10
7	50% B_4C + 3% NaF + 2% $NaBF_4$ + 43% Al_2O_3	–	18	12
8	60% Na_2B_4O + 40% Каси (silicocalcium)	–	3	1
9	60% Na_2B_4O + 40% Mg	Fe_2B	1	0.5
10	20% B_4C + 48% B_2O_3 + 32% Mg	–	6	2
11	97% B + 2% Al + 1% MgF_2	FeB, Fe_2B	10	6
12	**95% Al_2O_3 + 4% B + 1% NH_4F***	Fe_2B	10	8

Note *The effective period of exploitation for a 4 mm compact material was determined during the furnace heating. **The composition was obtained after the preliminary reduction of the mixture: 60% Al_2O_3 + 24% B_2O_3 + 15% Al + 1% NH_4F. ***NH_4F was additionally fed during reduction

carbon and alloy elements into the sublayer by growing borides. The sufficient sublayer's temperature creates a liquid-crystal state zone. Thus, the conditions for the growth of a boronized layer are created: it can grow according both to the diffusion mechanism on the surface and the diffusion-crystallization mechanism in the sublayer.

In particular, the heating of samples, made from steels 45 and У8 and covered with a coating based on 50% B_4C and 50% NaF, in the electric resistor furnace heated up to 1000 °C led to obtaining a layer consisting of a boride zone and a heterogeneous sublayer zone (Table 8.4).

The soaking for 10 min causes the formation of a 0.10–0.12 mm boride zone and a 0.2 mm sublayer. The thickness of the coating in this case is 2 mm. Slower heating would increase the volume of the coating but eliminate this effect in carbon steels in the same conditions. This effect is not witnessed while heating melted salts (as a result of the free exit of gaseous saturating mediums in the smelt and continuous decrease of the compact material's width due to the dilution). The

Table 8.5 The compositions of protective and energy-releasing coatings

The coating's composition	Bonding material	Critical point, °C
65–99% marshalite + 1–15% fireproof clay	Processing with 35% $AlCl_3$ water solution, drying	1100
70% SiO_2 + 30% B_2O_3	Hydrolized ethyl silicate	–
100% marshalite	Liquid glass, dextrin glue	–
70% marshalite + 30% B_2O_3	Liquid glass, hydrolized ethyl silicate	1000
22% Al (powder) + 58% iron mill scale + 10% SiO_2 + 10% sodium nitrate	Liquid glass	1100

processing of high alloy steels according to these methods leads to obtaining a pseudo-eutectic structure of increased plasticity.

The obtained structure of the boronized layer provides a smooth transition from the boride layer (1800 HV) to the substrate's material (650 HV after hardening); this creates a more favorable distribution of residual stresses and increases the reserved plasticity of the layer.

If a more durable exploitation of saturating compact materials is needed, their composition may be amplified by carcass-forming substances (refractory oxides—Al_2O_3, CaO, MgO etc.) or an additional protective coating (Table 8.5.)

The compact energy-releasing coating works in the mode of thermal spontateous combustion of self-propagating high-temperature synthesis. The use of the saturating (energy-releasing) coating which ensures SHS progress is also possible.

A short-term use of the compact saturating material for boriding (high frequency current heating, electrical contact heating, laser heating) involves applying one layer of paste and coating. Long-term heating combined with heating for tempering requires the use of pastes with FeO, Fe_2O_3, B_2O_3, abrasive-metal mixture, pyrite cinder, iron powder, ferromanganese, silicon oxide, silicon carbide etc. The amount of one of the substances is 10–50%. The mentioned additives in the composition of the saturating material increase the durability under high-temperature soaking, take part in the formation of the boronized layer and protect the detail from corrosion. The width of the obtained boride layer is 0.05–01 mm and depends on the duration of saturation and the temperature. In case when the coating contains >30% of iron oxides, ferromanganese, silicomanganese, silicon oxide or pyrite cinder, iron and steels generate plastic one-phase boride layers (Fe_2B). In other cases two-phase layers (FeB and Fe_2B) are formed [9, 154, 161].

The simultaneous saturation with boron or other elements from compact materials (coatings) is carried out in the following compounds [9]:

- boron-nitriding—60–65% B_4C; 5–6% NaF; 5–15% Potassium ferrocyanide; 20–30% mill scale or abrasive-metal mixture;
- boron-cuprum treatment—50–60% B_4C; 30–35 cuprum oxide; 5–6% NaF;
- boron-chrome plating—55–60% B_4C; 30–35% chrome oxide; 5–6% NaF;

- boron-silicizing—60–65% B_4C; 33–35% quartz powder; 5–6% NaF;
- boron-molybdenum treatment: 55–60% B_4C; 5–6% NaF; 25–30% mill scale; 60–70% B_4C, 5–6% NaF; 25–30% molybdenum oxides;
- boron-tungsten treatment—55–60% B_4C; 5–6% NaF; 10–12% tungsten carbide; 60–70% B_4C; 5–6% BaF; 25–30% tungsten oxides;
- boron-zirconium treatment—55–60% B_4C; 5–6% NaF; 8–10% zirconium diboride; 25–30% mill scale;
- boron-titanizing—55–60% B_4C; 5–6% NaF; 10–12% titan; 25–30% mill scale.

The saturating compact materials produced from these compounds preserve their initial form quite well under high-temperature soaking, protect the processed surface from corrosion and effectively work more than 8 h at 1050 °C.

The prolonged working time of a compact material can be also achieved by applying a layer of protective coating, e.g. containing marshalite, cryolite and bonding material.

The saturation of alloy steels in described compounds at 900 °C demonstrate an appropriate speed of plastic boronized layers' growth (Table 8.6).

The metallographic and X-ray diffraction analysis show that the boride layers consist primarily of Fe_2B phase (its amount is 97–75% depending on the steel brand). In particular, boron isn't found after boron-silicizing of steel 5XHB (according to the X-ray diffraction analysis). After all kinds of processing, steel 5X3B2MΦC generates layers which contain 40–55% of FeB phase as a result of the decelerating growth of boride phases and increased saturation medium activity. At

Table 8.6 The influence of the working period on the growth of diffusion layers [9]

The type of processing	Steel	The layer's width, mkm, at different working periods			
		1	2	4	6
Boriding	5XHM	55	80	125	160
	7X3	50	75	120	145
	5X3B3MΦC	7	17	28	36
Boron-nitriding	5XHM	75	120	160	185
	7X3	70	90	125	155
	5X3B3MΦC	18	33	47	60
Boron-silicizing	5XHM	55	75	105	130
	7X3	40	60	85	100
	5X3B3MΦC	8	20	32	42
Boron-cuprum treatment	5XHM	50	65	100	120
	7X3	30	55	75	90
	5X3B3MΦC	5	12	27	33
Boron-chrome plating	5XHM	40	60	80	110
	7X3	25	47	65	80
	5X3B3MΦC	14	25	37	50

the same time the reduction of the boride layer's thickness causes the initiation of critical layer's width factor for this particular steel brand; it is done to obtain non-defected layers and prevent spontaneous chipping. Higher saturation temperatures lead to the formation of wider layers with high defectiveness.

Amorphous boron may be also used for obtaining a compact saturating material; the content of amorphous boron is 6–90%. In this case the following activating additives can be used: sodium fluoride (5–50%) if furnace heating or high frequency current heating is carried out; carbon black (15–40%); borax (10–45%); sodium chloride (1–10%); triethanolamine (3–5%) etc. [154, 163–166]. Two-phase boride layers are obtained in all cases. Wider layers are obtained when manganese powder is used, in particular during high frequency current heating [167]. If boriding is carried out without melting the processed surface and with a minimal boron content in the compact material, steels generate one-phase boride layers.

The process of producing and applying pastes can be facilitated by spraying borax on the preliminary heated (850 °C) surface; then B_4C is sprayed over and protective coating is applied [168]. The mass transfer in saturating medium is provided by spontaneous redox-reactions which are usual for ionic melted mediums.

The technology of coating application is also facilitated by covering the processed surface with a corrosion-erosion layer which contains $Li_2B_4O_7$ in methanol; it is done by splattering the solution onto the heated (100 °C) surface with further boron carbide (B_4C) application. Saturation is carried out at 850 °C in argon or any other protective atmosphere, vacuum or furnace for 2–48 h. This technology is aimed at obtaining a Fe_2B layer characterized by reduced brittleness [169, 170].

In metal-thermal mixtures boron reduces from B_2O_3 or from $Na_2B_4O_7$ by aluminum, silicocalcium, manganese etc. The coatings additionally contain Al_2O_3, NaF, KBF_4, FeO etc. This ensures their stable workability during long soakings. Saturation can be done according to the combined or separate variants. These mediums normally generate plastic one-phase boride layers [154].

Plastic boronized layers are also formed when they are alloyed with chrome, manganese or cuprum [167, 171–173]. Chrome is added to the coating in the form of metal powder or oxide Cr_2O_3. For instance, saturation in the following composite —70 B_4C, 25% Cr_2O_3 and 5% NaF—for 2.5 h at 850 °C in steel 45 led to the formation of a one-phase layer (0.05 mm wide). The obtained layer is characterized by the increased corrosion resistance.

Saturation in molds during metal casting is one of the ways for obtaining thick layers with composite structures in casts [9, 174, 175]. The acquired width of boronized layers reaches minimum 1.0 mm. The composition of the coating saturation mixture, which is preliminary applied onto the surface of the clay form and stays unmelted during casting, includes 30–40% B_4C, 2–5% NaF and 40–45% of one of the following materials: zirconium sand, zirconium powder, mill scale. The latter component functions as a fire-resistant material and stabilizer of redox-reaction processes. The bonding function is given to hydrolyzed ethyl-silicate or liquid glass (15–16%). The width of the coating mixture is less than 5 mm. The obtained layers are boride eutectics containing iron borides and perlite. The boron mass transfer goes through a gaseous medium which is formed as a result of interaction between the coating mixture components—boron carbide and sodium fluoride—and moistening of boron carbide particles with metal.

The temperature of casting and the speed of form cooling after metal casting are the additional factors of the control over the structure and the width of boronized layers. In particular, accelerated cooling (the duration of the contact between a liquid metal and saturation coating is less than 3 s) prevents the generation of a boronized layer wider than 0.5 mm (in steel 45). Mold and form cooling in air significantly increases the diffusion period which leads to the thickening of a boronized layer up to 1.0–2.0 mm. The use of decelerated cooling of the form in the furnace, heated to 1100 °C, or by isothermal soaking of the form in the furnace for 2 h (950 °C) leads to the increase in the layer's width up to 1.6 mm. The duration of the soaking effects the amount of perlite found in the layer's structure.

A high quality of casting surfaces is reached by the use of ceramic forms. The ceramic coating of molds consists of a fire-resistant filling based on zirconium oxide (zirconium powder and zirconium sand 1:1). Hydrolyzed ethyl-silicate is used as a bonding material; ethyl-silicate is amplified by 40% water solution of tri-ethanolamine to ensure its calcification. To saturate mold's surfaces, a boride component is added to the ceramic suspension instead of a part of the fire-resistant filling (boron carbide and sodium fluoride in correlation 10:1). The width of a boronized layer depends on the amount of boronizing component in the coating layer and casting conditions (temperature and cooling) (Table 8.7).

In case when a ceramic material contains up to 50% of a boronized component, the structure of a boronized composite layer is pre-eutectic and contains boride eutectic and perlite elements. If there is more of the boronized component, the perlite elements decrease their number. The casting temperature increases to 1650 °C and the increase of the boronized component up to 70% or isothermal soaking allow to obtain eutectic layers of enhanced hardness up to 1.00 mm wide with the total boronized layer's width of 2 mm (steel 45).

The boronized layer of steel 5XHM contains Cr_2B and Mo_2B borides and carbides of these alloy elements. The microhardness of boride eutectics in steel 45 and 5XHM is 700 and 900 HV accordingly; after quenching, these numbers rise to 1100 and 1300 HV due to the formation of martensite.

Table 8.7 The influence of boriding conditions on the layer's width under the treatment of different steels [9]

Steel brand	Casting temperature	Layer's width, mm, depending on the boronized component content in the ceramic material, %				
		20	40	60	70	80
Steel 45	1650	0.65	1.15	1.6	1.8	2
	1550	0.5	0.8	1	1.1	1.1
5XHM	1650	0.3	0.65	0.80	0.85	0.9
	1550	0.25	0.4	0.5	0.55	0.55

The optimal composition of boronized ceramics provides the formation of developed plastic composite boronized layers combined with a high-quality hardened surface: 50–60% B_4C, 4–6% NaF, 35–45% zirconium oxide-based filling (the mix of zirconium sand and zirconium powder in proportion 1:1) [9].

Boriding in gaseous mediums obtained upon the interaction between the components of compact materials is carried out by laser heating while the processed surface is protected from melting; it results in the formation of narrow boride layers. The process involves using compact materials containing boron, B_4C, FeB and transition metal borides, a process activator, carbon black for increasing the heat absorbance, and a bonding material. Due to the shortness of the process, the obtained layers are narrower than 20 mkm. Wider surfaces can be generated after multiple surface scanning. A change of the compact material is necessary after 3rd scanning. The phase compositions on the processed surface may largely vary from FeB to Fe_2B.

Boriding in compact materials in a glow current is also quite productive for obtaining plastic boronized layers. The potential of 40–50 V with a 5 mm distance between the processed surface and electrode increases the mass transfer because the processes of gas formation and electrolysis are activated during salt component melting. Finally it results in the boronized layer's growth acceleration [176, 177]. The paste contains B_4C, KBF_4, Na_3AlF, $Na_2B_4O_7$; alloy elements oxides (e.g. ZrO) or metal powders may be also used if it's necessary to get alloyed layers. The paste width is 0.5–4 mm. The saturation goes at 500–900 °C on the surface in the air atmosphere or nitrogen or argon atmosphere at consumption 50–100 l/h [178]. The optimal period for the compact material's layer exploitation should not exceed 3 h. Wider layers are obtained by applying a new paste layer on the detail's surface after cleaning and repeating the process.

Low-temperature boriding ($T_{sat} < Ac_1$) needs the inclusion of KBF_4 ($\sim$40%) into the paste composition. Boriding at <600 °C results in obtaining Fe_2B-based layers; at higher temperatures—FeB and Fe_2B layers. Medium-temperature boriding (<800 °C) causes the formation of Fe_2B-based layers from the composition containing 60%B and 40% $Na_2B_4O_7$; higher temperatures (>850 °C) lead to the formation of FeB and Fe_2B layers. The layer's width at low-temperature boriding is 0.05 mm, at medium-temperature—0.1–0.2 mm. Wider layers may be obtained if the preliminary iron cladding is done [179].

Boronized layers are also acquired after the use of concentrated heating sources. In particular, chrome-alloyed layers are generated if the surface, covered with a compact saturation material, is impacted by a relativistic electron beam. The compact material was obtained after applying a coating containing B_4C and chrome powder [180].

Boriding from ribbons, forms and models is another variant of boriding. This method is used for producing small details and instruments of saturation "forms" for boriding. These details and instruments have a mould relief; their surface is covered with a saturation mixture. Boriding of large-scale details and constructions involves the use of ribbons and technological models produced in the same fashion [58].

The production of a "model" is done by adding a thermo-reactive powder polymer (10% of the total mass) to the saturation mixture. The obtained mixture fills the stamp's engraving with a sinkhead which exceeds the engraving depth by 50%. A layer of the isolating mixture with 5% combination of phenol formaldehyde resin and urotropine ("pulverbakelite") is put over the initial mixture; it helps to retain a gas medium better. The isolating mixture contains 47.5% SiO_2, 47.5% B_2O_3, 5% pulverbakelite or 91% SiO_2, 4% carbon black, 5% "pulverbakelite". The thermo-reactive polymer assists in obtaining a compact saturation material during its polymerization and fusing at 250 °C. The engraving is preliminary covered by paint, graphite coating or aluminum foil to ensure an easier extraction of the model from the stamp. The obtained saturation "model" can be used multiple times which turns out to be an advantage.

When the "model" is set into the stamp's engraving, the borders of the isolating sinkhead should be better covered with carbon black or coated with fire-resistant clay. Comparison characteristics of the layers obtained upon boriding from a saturating compact material and a saturating "model" of the same composition, containing 70% B_4C and 30% Na_3AlF_6 with the same type of the isolating mixture (50% B_2O_3 and 50% SiO_2) are demonstrated in Table 8.8. The difference lies in the use of different bonding materials at their production: for compact materials it is the solution of BF_2 glue in acetone; for "models"—"pulverbakelite".

Reusable technological saturation "models" are made mechanically from popular cheap materials (fire clay, fire-resistant clay etc.). A saturation mixture is applied on the porous surface of the model in the form of water suspension, paste or coating. The model is used after a drying stage according to the described technology or in combination with another type of sealing.

Compact elements (applications and side plates) are also produced by pressing a saturation mixture. For example, pressing of the mixture—60–80% amorphous boron, 30–40% cryolite, 3–6% sodium fluoride, 3–6% boron trioxide—led to the

Table 8.8 The comparative results of boriding involving the saturation "model" and the saturation compact material

The number of uses	The roughness of the stamp's working surface, R_a, mkm		The width of boride layer, mkm
	Before boriding	After boriding	
Compact material			
1	0.75	1.3	75
Reusable model			
2	0.79	0.85	105
3	0.75	0.90	98
4	0.75	0.80	92
5	0.79	0.85	85

Note Saturation conditions: steel 45, 870 °C, 4 h. Each use of the reusable saturation model required positioning of 2% KBF_4 (from the initial mass of saturation mixture) on the bottom of the engraving formation of side plates for boronizing necks of large-scale shafts (the press form was preliminary produced on the common equipment). Side plates are produced in the form of separate parts which comprise a muff divided into 2–3 parts. These elements are fastened to a processed shaft neck by a narrow elastic material (e.g. polytetrafluoroethylene). The processed zone is put into a ring inductor to be heated by a high frequency current. Processing details of other shapes requires the use of an induction loop. The heating goes up to 1050–1150 °C for 1.5–2 h. After 0.2–1.0 min when the lining softens, the detail is repositioned by 0.5–1.0 mm relative to the compact material to tighten the contact by squeezing the lining by the elastic material. When the heating is done and the inductor is turned off, the details are soaked for 1.5–2 min. Then the process of heating (1050–1150 °C) is repeated followed by cooling without repositioning the detail. The width of a diffusion layer normally reaches 50–200 mkm. The required width of a boronized layer is obtained by changing the heating time, saturation temperature and the number of heating and cooling cycles. Thermal circulation causes a more even distribution of residual stresses in the layer and the improvement of the boronized layers' plasticity [39, 181].

A better protection of the linings and the processed surface from corrosion is provided by adding a calcite powder into the upper layer of the contact side plates during their production.

The main reason for such an effective development of diffusion layers during thermal circulation is the formation of the fields of significant stresses in saturated metals; they are defined by:

- reaching and preserving (during the whole period of processing) a small-grain structure with a developed inter-grain and inter-phase surface;
- oversaturation of the metal with crystal-structure defects (borders between grains and subgrains, dislocations and vacancies);
- the homogenous temperature field along the section due to the alternating heating and cooling modes;
- the homogenous field of stresses occurring during the diffusion of saturating element and phase transformations [182].

The properties of boronized layers during their saturation from compact materials can be controlled by:

- selecting the saturation mixtures which ensure the formation of plastic boronized layers;
- setting the temperature which provides the layer's formation according the diffusion or diffusion-crystallization mechanism;
- thermocircular processing;
- setting a level of saturation medium sealing.

8.3.4 *The Influence of Saturation Conditions in Liquid Mediums on the Plasticity of Boronized Layers*

According to the general classification, boriding in liquid ionic mediums is divided into electrolysis and non-electrolysis boriding. The process is carried out in smelts of boron-containing and neutral salts, metal smelts and water solutions or suspensions.

Electrolysis boriding in salt smelts The phase composition of salt smelts, their phase correlation and dispersability, the alloyness level of acquired layers and their structure (under the processing of alloys and steels) can be controlled by the following procedures:

- Setting the density of cathode currents. Saturation in $Ba_2B_4O_7$-based smelts, which is normally used for electrolysis boriding, causes the generation of one-phase layers in 0.025–0.06 A/cm^2 current range in iron and steels; the layers consist of Fe_2B phase or $(Fe, Me)_2B$ in alloy steels.

When the current density is >0.12 A/cm^2, two-phase boride layers FeB + Fe_2B with a continuous FeB zone are formed on the surface. The interval of the current density 0.06–0.12 A/cm^2 and 2 h soaking cause the formation of two-phase layers with different phase correlations (Fig. 8.19) [58].

- The introduction of manganese oxide (MnO) into the composition of the borax-based smelt. The optimal amount of MnO, introduced to the smelt to obtain one-phase $(Fe, Mn)_2B$ boride layers, is 7–15 mass% (Fig. 8.20a). A smaller percentage of manganese oxide causes the structure formation to be determined by the current density; a larger amount of MnO leads to the formation of the layers which consist of α-solid solution of Fe (Mn) with boride inclusions. These layers are characterized by high plasticity and wear resistance if disperse boride inclusions exceed 15%. They are used for details of contact

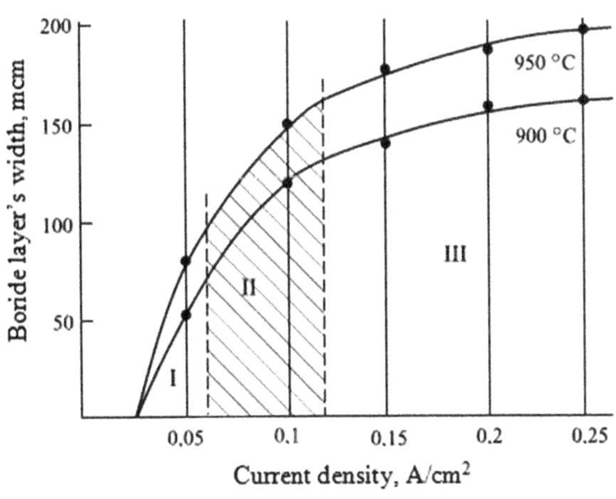

Figure 8.19 The dependence of the boride layer's width and phase composition on the surface from the density of the cathode current (steel 35, 2 h). *I*—Fe_2B, *II*—Fe_2B + FeB, *III*–FeB

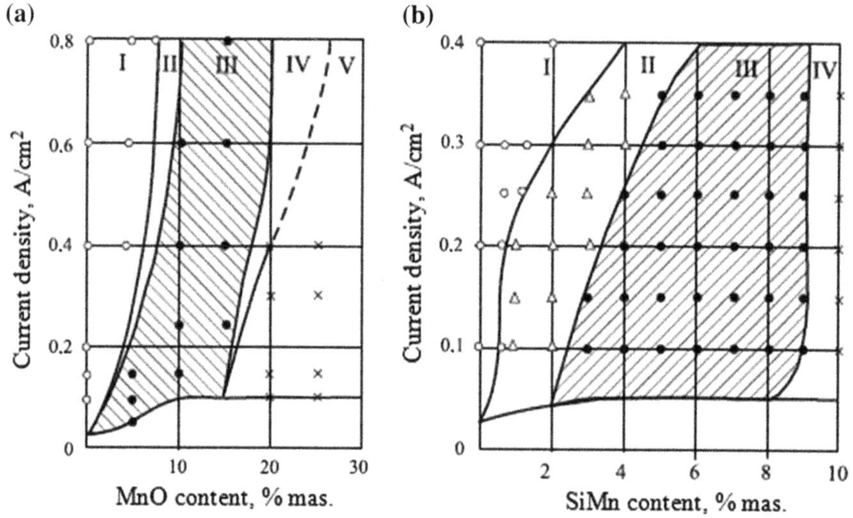

Figure 8.20 The structural diagrams of electrolysis boriding in the borax smelt containing MnO (**a**) and silicomanganese SiMn17 (**b**). The structure of the layers by zones: *I*—FeB + Fe_2B; *II*—Fe_2B with separate needles FeB; *III*—(Fe, Mn)$_2$B; *IV*—(Fe, Mn)$_2$B + α-solid solution of Fe(Mn); *V*—the boride layers are absent [α-solid solution of Fe(Mn)]

points at alternating loads. The composition, which provides the stable generation of plastic one-phase boride layers in the range of current density 0.08–0.3 A/cm^2, contains 80–94% $Na_2B_4O_7$, 3–5% NaF, 3–15% MnO.

- Introducing powder silicomanganese (grain size—0.1–0.25 mm) into the borax smelt. (Fe,Mn)$_2$B-based one-phase layers are constantly formed at the appropriate combination of current density (0.05–0.4 A/cm^2) and silicomanganese content (6–8%) (Fig. 8.20b). If the amount of silicomanganese is smaller (area II), two-phase boride layers (Fe, Mn)B and (Fe, Mn)$_2$B are formed with additional inclusions of the high-boron phase. These layers are also characterized by a high reserve of plasticity [58, 183, 184]. The formation speed of one-phase layers is preserved at a sufficiently high level (Table 8.9) due to the elimination of the only factor which limits the boron mass transfer on the processed material—a reduced boron diffusion speed in the high-boron phase.

One-phase boride layers are also formed when borax solution Fe_2O_3 or NiO (5–15%) is added [58, 183]. The saturation is carried out at the current density 0.05–0.3 A/cm^2 and 850–950 °C.

- Implementing stage modes for the current density and temperature. The first temperature stage is 750–800 °C at the current density 0.15 A/cm^2; the duration is 5–15% of the whole process duration. Soaking at this stage causes the generation of multiple boride crystals (more than at 950 °C) which are perpendicularly oriented towards the diffusion front in the crystal-graphic orientation

Table 8.9 The formation kinetics of one-phase $(Fe, Mn)_2B$ boride layers

Steel brand	Layers' width (mkm) in different processing conditions (temperature, time)								
	800			900			950		
	2	4	6	2	4	6	2	4	6
20	75	115	155	105	155	200	150	200	240
45	65	90	120	80	130	140	120	160	190
У8	55	75	90	55	90	105	95	130	150
ХВГ	45	65	75	50	75	95	75	95	110

Note The density of the cathode current is 0.15 A/cm^2. The smelt's composition: 88% $Na_2B_4O_7$, 4% NaF, 8% MnO; 91% $Na_2B_4O_7$, 4% NaF, 5% silicomangamese. Both compositions are characterized by the same saturation capacity

[001]. The growth speed of these crystals is faster than those which are differently oriented at the same saturation conditions.

The second stage—950–1000 °C—provides a further growth of a big number of positively oriented boride crystals. This creates the conditions for accelerated formation of boride layers even at the cathode current density reduced to 0.08 A/cm^2. The duration of the transition between the first and second stages should not exceed 20% of the total process time.

Electrolysis boriding of steel 20 in the borax solution at the two-stage mode (1 stage—800 °C, 0.25 h, j = 0.2 a/cm^2; 2 stage—950 °C, 2 h, j = 0.08 a/cm^2) leads to the formation of a two-phase boride layer with the total width of 340 mkm with a solid layer of 250 mkm. The duration of heating from 800 to 950 °C was 0.75 h. The electrolysis saturation of this steel at 950 °C for 3 h at j = 0.2 A/cm^2 allows to obtain a 300 mkm boride layer with a solid 190 mkm zone.

The layers obtained at the stage mode are characterized by the increased plasticity and acicularity, they have a favorable distribution of residue compression stresses and reduced defectiveness [58].

The stage mode also includes the change of saturating capacity which leads to the reduction of the boronized layer's brittleness: in particular, by preliminary non-electrolysis saturation at a switched-off current [185] and multiple alternation of soakings in the smelt at electrolysis (40–60 min) and without electrolysis at a switched-on current (annealing for 20–30 min) [186]. The same results are achieved when the boriding process takes two stages: 1—constant current electrolysis for 10–15 min; 2—alternating current electrolysis for 80–130 min. In both cases the second stage is characterized by the resorption of the high-boron phase FeB and Fe_2B phase's growth [187].

Alternating soaking during the process of electrolysis with impulse currents of 10^{-5}–10^{-6} duration and 10^{-1}–10^{-2} fill ratio reduces the process duration and electrical power 1.7 times and accelerates layers' growth speed 1.8 times [188].

- Using a reverse current. The optimal total period of reversion is 1.2–2.2 s if the cathode period duration is 0.8–1.8 s and the anode period duration is 0.40–0445 s (the necessary ratio τ_a–τ_k = 0.3–0.5). The intervals of cathode and anode current densities on the processed surface to sustain the maximum boriding speed in borax solution are $j_k = 0.2–0.4\ A/cm^2$ and $j_a = 0.2\ A/cm^2$ [39, 189]. Initially the electrolysis boriding process is carried out by a constant current for 5–10 min to provide a stable cathode potential for the processed surface; it will prevent its dissolution in an anode period.

Boriding with the reverse current leads to the formation of a Fe_2B-based layer, thickening of boride needles, reduction of voltage and voltage swing in the layer's width that lead to the increase of the plasticity.

- Introducing metal oxides or their chlorides, smelts and powders into the electrolyte (borax smelt). The metals—Cr, V, W, Ti, Ta, Zr, Nb, Mo and others—can be used [39, 189–191]. The number of introduced oxides is 10–20% in mass. The formation speed of boride layers is reduced by 20–40%. The change in the percentage ratio of oxides and cathode current density leads to the change in the layer's structure in iron and steels—from $(Fe, Me)_2B$ to (Fe, Me)B. The hardness of Fe_2B boride under alloying increases from 1600 HV to 2000 HV; FeB boride—from 2000 HV to 2400 HV.

In particular, the use of reverse currents in simultaneous saturation of steels with boron and zirconium or chrome and tantalum ensures the formation of layers with layer-needle structure. The surface generates layers containing intermetallic compounds of alloy elements (e.g., $FeZr_2$ and Fe_2Ta) and borides (ZeB_2 and TaB correspondingly). Boron-chrome plating causes the formation of layers which contain (Fe, Cr)B, $(Fe, Cr)_2B$ and Cr_5B_3. Borides with the reduced content of chrome are found in the distance from the surface. The obtained layers are characterized by increased plasticity [39].

- Preliminary application of galvanic or electrochemical Ni, Cr, Ta etc. layers on the processed surface [192]. This technology allows to drastically change the layer's structure and plasticity from pseudo-eutectic high-dispersable plastic boronized layers to normal boronized layers with reduced ductility. The layers' structure in this cases is determined by the multiple-component equilibrium diagrams in accordance with temperature-concentration conditions, current density and processing duration.
- Saturation of alloy steels, i.e. the alloying of borides from the substrate's material. It mostly effects the substrate's structure in low-alloy steels during the diffusion formation of the layers and the diffusion-crystallization formation of the layer's structure.

In particular, the pseudo-eutectic structure of the boronized layer in steel ЭП 718 (0.6 mm wide) is obtained as the result of the electrolysis saturation in a borax smelt with cathode current density $0.04–0.06\ A/cm^2$ at 1050 °C for 2 h.

Low-temperature electrolysis boriding is carried out at the range of temperatures (550–700 °C) in the smelts based on borax or the eutectic mixture of tetraborates of alkali metals. To lower the melting point of borax ($Na_2B_4O_7$), the eutectic mixtures of the following elements are used: carbonates, chlorides and fluorides of alkali metals, lead and bismuth (30–35%) oxides, sulfates (30%), phosphates (15%) and alkali (10%), selenium amino acid (10–15%) etc. Practically all the cases result in the formation of two-phase boride layers. The process is carried out at the cathode current density 0.1–0.15 A/cm^2 for 5–20 h. Large current densities in the cathode region lead to the heating of the smelt and processed surface. The thickness of obtained layers reaches 0.01–0.04 mm [193–197]. The layers are characterized by the increased plasticity due to the low level of residue stresses as the processing is accompanied by the absence of phase transformation in the metal of the substrate and decreased thermal stresses.

The main disadvantage of the mentioned compounds is the medium castability of the smelt which is accompanied by several difficulties at loading the details and the loss of large amounts of salt from the bath (it goes away with details) after the end of treatment.

The bath compositions for electrolysis saturation with halogenide boron compounds are characterized by a higher technical effectiveness. For instance:

- the eutectic mixture KBF_4 + LiF + NaF + KF for the details, boronized at the temperature range 600–900 °C [198];
- the mixture 20KF + 30NaF + 49LiF + (0.7–1.0)BF_2 (%) for boriding at 800–900 °C in atmospheres containing 90% N_2 and 10% H_2;
- the mixture with correlation 72KF + 8LiF + 20KBF_4 (mass%) in the nitrate atmosphere [199];
- the mixture 27LiF + 63KF + 10KBF_4 (mass%) for boriding at 700–850 °C;
- the mixture 80$Na_2B_4O_7$ + 20NaCl (mass%) for boriding at 800–900 °C [200].

The last composition has a tendency for delayering at medium-temperature boriding.

Non-electrolysis boriding in melted salt mediums has an electrochemical nature of the boron mass transfer which allows to effectively control the whole technological process. Normally the saturation medium contains a salt component (which acquires a melted state at the saturation temperature) and a solid powder component (electrochemical reducing agent). The saturation is carried out by putting details into the smelt without including them into the external electrical current source chain. The process requires a wide range of temperatures 550–1100 °C (low-temperature, medium-temperature and high-temperature boriding) for 4–24 h.

All mediums are divided into three types according to their relation to boron:

(1) saturation element (boron) is a part of salt and powder components of the bath;
(2) boron is a part of the salt component;
(3) boron is a part of the powder component.

In all cases the boron mass transfer in saturating mediums goes spontaneously in a self-regulating mode.

In *the first type of mediums* the mass transfer is supported by boron subions which are easily reduced on the processed surface during the work of short-circuit microgalvanic elements. The salt component's function is carried out by the tetraborates or fluoroborates of alloy metals or their mixtures in amount 65–90%; powder additives are represented by amorphous boron, boron carbide, Mg, Ca, Ba polyborides, ferroboron or chrome, nickel, titanium, silicon and other borides (10–35%) [8, 67, 69, 201–204]. For instance, fast formation of boride layers is established by using salt baths containing 55% borax, 40–50% ferroboron and 4–5% ferro aluminum [205].

Two-phase boriding is mostly used for compounds which contain borates of alkali metals and boron carbide. The regulation of the phase composition is done by changing the smelt composition or setting the optimal temperature and saturation time conditions. In particular, to reduce costs (1.2–2.4 times) and layer's brittleness by reducing the content of the high-boron phase, sodium octaborate or potassium tetraborate should be used as a salt bath component [201]. Baths with B_4C at the saturation temperature <800 °C and 5-h-soaking in iron and steels generate sufficiently plastic boride layers with a medium amount of FeB boride [8]. For instance, boriding of steel 45 for 4 h at 800 °C in the compound, containing 30% B_4C and 70% $Na_2B_4O_7$, leads to the formation of a 80-mkm one-phase Fe_2B layer.

The use of amorphous boron as a powder additive also causes the generation of two-phase layers. The details of complex forms are recommended to have the composition of high castability and technological efficiency; they are supposed to contain (mass%): 51–53% $Na_2B_4O_7$, 13–15% NaCl, 17–20% KCl, the rest is amorphous boron [203]. Saturation of carbon and instrumental steels is carried out at 800–1000 °C. A longer period of bath exploitation can be achieved by the use of a composition with 51–53% $Na_2B_4O_7$, 34–36% $NaBF_4$ and the rest—B (amorphous) [204]. Boriding is carried out in furnace baths with electrical or inductor heating [203, 206].

The improvement of boriding technological effectiveness in melted salts is achieved by the use of electrode baths. Asymmetrical impact on the composition of electromagnetic field by a three-phase alternating current (3–6 V) ensures a constant bath stirring and effective withdrawal of cathode process reaction products. This finally leads to the increase of saturation capacity [206]. The decrease of FeB high-boron phase leads to the reduction of the amount of boron-containing powder additiv.

The boride layers, obtained in baths containing chrome, nickel and silicon borides, are characterized by the increased plasticity due to the chrome, nickel and silicon alloying [207, 208]. The compositions also cause the acceleration of boride layers' formation (Table 8.10).

Table 8.10 Bath compositions used for generating alloy boronized layers

Bath composition, mass%	Layer's width, mkm	Phase composition of a layer
55–65% $Na_2B_4O_7$ + 10–15% NaF + 25–30% CrB_2	170	(Fe,Cr)B, $(Fe,Cr)_2B$
60–65% $Na_2B_4O_7$ + 10–15% NaF + 20–30% SiB_6	180	FeB,Fe_2B, (3% Si on a surface)

Note Boriding of steel 45 was carried out at 900 °C for 4 h

The change in the correlation of phases and the consecutive improvement of plasticity is also achieved by introducing manganese, chrome and nickel oxides into the salt component composition. The regulation of the phase correlation in the layer at 800–1000 °C saturation temperatures is done by using the mixture of B_4C and SiC or other electrochemical reducing agents as powder additives [8, 209].

Alloyed layers can be also obtained by adding special elements into the salt smelt. These elements are: oxides V_2O_5, Cr_2O_3, ZrO etc., rare earth metal chlorides [8, 210] or specially developed synthesized electrochemical reducing agents [58]. The alloy elements solute in borides and improve the hardness (W, V, Cr), corrosion resistance (Mn, Al, Zr, Cr) and heat resistance (Mn, Ni, Cr) characteristics. The alloyed layers obtained by preliminary precipitation of nickel galvanic or electrochemical coatings with further boron saturation are also characterized by an increased plasticity. Boriding at 650–950 °C for 2–8 h in the smelt containing KCl, $BaCl_2$, NaF, B_2O_3 and ferroboron leads to the generation of nickel borides Ni_2B, NiB_2 alloyed by iron [8, 69, 211].

The layers, obtained as a result of boronizing high-alloy steels and smelts, are determined by the nature of alloy elements and their content in the smelt. In particular, nickel- and iron-based boride layers are formed while boronizing the alloy 55НХВКТБЮ in the smelt which contains $Na_2B_4O_7$ and MgB_{12}. The most appropriate combination of properties is obtained after saturation at 950 °C for 2.5–3 h and direct cooling (from boriding temperature) in oil and further double ageing (800 °C, 6 h and 700 °C, 4 h). The increase of hardness, wear resistance and contact durability is observed [212]. If the saturation takes more than 3 h, boronized layers wider than 80 mkm are obtained but the surface of details can generate spontaneous splits. The reason for splits is the big difference of coefficients of thermal expansion for the boronized layers and the matrix.

The optimal compound contains 79% $Na_2B_4O_7$, 15% NaCl, 6% MgB_{12}. Simultaneously, saturation at 900 °C after a 12-h bath work leads to the reduction of saturation capacity of the bath and the formation of layers of different widths. A stable bath work up to 300–350 h is reached by adding 0.3–0.4% MgB_{12} into the bath every 5 h. After 350 h the smelt is better to be changed and a bath 50% refreshing by the initial compound must be done.

The pseudo-eutectic structures of increased plasticity and significant width, obtained according to the diffusion-crystallization mechanism, are formed in baths with a smaller MgB_{12} content (to 4%). In particular, boriding of alloy 55НХВКТБЮ at 1050 °C for 3 h led to the formation of a 0.65 mm-wide layer; boriding of steel Р6М5—0.55 mm-wide layer. In the first case layers had a high-disperse structure, in the second—a low-disperse structure.

The first-type mediums for low-temperature boriding are eutectic mixtures of alkali metals tetraborates (Na, K, Li) and a boron-containing powder additive (B, B_4C, CaB_6, BaB_6, MgB_{12} etc.) [194, 213]. Some compositions are demonstrated in Table 8.11.

The saturation process is carried out at 550–700 °C for 8–24 h. It is recommended to use it in combination with high tempering and ageing to ensure non-deformative processing. Normally, the width of layers doesn't exceed

Table 8.11 Bath compositions for low-temperature non-electrolysis boriding

Bath composition, mass%	Layer's width	Phase composition of a layer
19% LiF, 7% NaF, 34% KF, 20% B_2O_3, 20% B_4C	0.035	FeB, Fe_2B
11% Li_2CO_3, 25% K_2CO_3, 20% Na_2CO_3, 12% $Li_2B_4O_7$, 5% $K_2B_4O_7$, 7% $Na_2B_4O_7$, 20% B4C	0.020	Fe_2B
22% LiCl, 27% KCl, 2% NaCl, 13% $Li_2B_4O_7$, 6% K2B4O7, 10% $Na_2B_4O_7$, 20% B_4C	0.035	FeB, Fe_2B
15% LiF, 30% KF, 6% NaF, 15% $Li_2B_4O_7$, 6% $K_2B_4O_7$, 8% $Na_2B_4O_7$, 20% B_4C	0.040	FeB, Fe_2B

Note Saturation of steel 45 was carried out at 450 °C for 8 h. Compositions are characterized by high viscosity

0.050 mm, which in many cases seems insufficient. The thickening of layers is reached by preliminary nitrating. The obtained layers are characterized by an increased plasticity [214].

The *second type of mediums* includes the tetraborates of fluoride borates of alkali metals or their mixtures as a salt component; powder additives in this case are elements and compounds (electrochemical reducing agents) which don't contain boron and have high electronegativity in a chosen salt smelt in correlation with the processed metal and formed boride phases. The amount of a salt component normally doesn't exceed 65–75%, a powder additive—25–35%. The boron mass transfer in these compounds is provided by the work of short-circuit microgalvanic elements (Chap. 7).

The saturation capacity of baths, phase compositions of formed layers and the level of their alloyness are determined by the type of the powder additive and the level of its stable electrode potential. Consequently the second-type compounds provide wide possibilities for regulating boronized layers' plasticity.

Solid components of baths may be represented by metal powders (Ca, Be, Ce, Li, Mg, Al, Ti) which are added to the alkali metals' tetraborates smelts in inert atmosphere [215]. The amount of added metal per 100 g of the bath should be <0.2 of its gram-equivalent. The most appropriate amount is 0.2–3.2 g-eq. For instance, 2.5–40 g Mg per a 100 g bath. Boriding is carried out at 650–1175 °C. The most effective temperature is >815 °C (Table 8.12).

It is more technologically efficient to use corrosion (elimination)-resistant alloys and compound powders which are exploited in different industries. They are silicon carbide, silicocalcium, silicomanganese, ferrosilicon, calcium carbide, calcium-aluminum alloy, nickel-aluminum ligature etc (Table 8.13) [8, 58, 69, 216–220]. The effectiveness of electrochemical reducing agents is determined by the difference between stable electrode potentials of a reducing agent and the processed metal. The electrochemical reducing agent must be more electronegative than the processed metal and a predicted phase or solid solution on the surface.

Table 8.12 The compositions for high- and medium-temperature boriding in melted salts [207, 218–220, 246–250]

No.	Smelts compositions	Electrode potential of a reducing agent	Electrode potential of Fe during saturation, mV	Optimal size of reducing agent's grain, mm	Layer's width, mm	Phase composition of the salt
1	75% $Na_2B_4O_7$ + 25% CaC_2	−680	−550	0.5	125	FeB, Fe_2B
2	60% $Na_2B_4O_7$ + 30% KM + 10%B_2O_3	−910	−630	0.25	145	FeB, Fe_2B
3	60%$Na_2B_4O_7$ + 10% B_2O_3 + 30%Симн	−710	−590	0.10–0.15	160	Fe_2B
4	65% $Na_2B_4O_7$ + 10% NaCl + 25% Сицр	−1040	−750	0.10–0.1	165	FeB, Fe_2B
5	60% $Na_2B_4O_7$ + 10% NaCl + 30% P70	−1020	−720	0.10	140	FeB, Fe_2B
6	70% $Na_2B_4O_7$ + 30% SiC	−350	−300	0.05	70	Fe_2B
7	70% $Na_2B_4O_7$ + 20% SiC + 10% B_4C	–	−480	0.05	130	FeB, Fe_2B
8	65% $Na_2B_4O_7$ + 10% NaF + 25%FeAl (50%Al)	−500	−420	0.05–0.1	130	FeB, Fe_2B
9	63%$Na_2B_4O_7$ + 12%NaCl + 25% NiAl + (50%Al)	−450	−380	0.05–0.1	210	(FeNi)B (FeNi)$_2$B,
10	55% $Na_2B_4O_7$ + 17%NaCl + 28% CCT	−1000	−720	0.10–0.15	145	FeB, Fe_2B

Note The layers were obtained in steel 45 at 950 °C and 4 h-soaking. The explanations for approximate names of powder electrochemical reducing agents are given in Annex 8.1

Table 8.13 The compositions of electrochemical reducing agents for boriding in melted boron-containing salts

No.	Material	Symbol	Chemical composition											
			Ca	Ba	Sr	Si	Mg	Ti	Al	Mn	P3M	C	Zr	Fe
1	Magnum-calcium ligature	KM	19–28	–	–	51–54	15–25	1.5–2.5	1.5–2.5	–	–	–	–	–
2	Silicomanganese	SiMn-17	–	–	–	17–20	–	–	–	65	–	1.75	–	Res.
3	Silicozirconium	SiZr-–50–1	–	–	–	40	–	–	–	–	–	0.09	40	Res.
4	Silicomishmetal	SMM	1.8–1.3	–	–	47–49	–	–	4.5–5.5	–	30–33	–	–	Res.
5	Silicobarium	SiBa	–	20–25	–	60–70	–	–	–	–	–	–	–	Res.
6	Silicocalcium	CaSi-1	28	–	–	62	–	–	2.5	–	–	–	–	Res.
7	Silicostrontium	SCT	–	–	20–25	60–70	–	–	–	–	–	–	–	Res.
8	Experimental ligature	P-70	–	–	–	30–40	–	–	–	–	60–70	–	–	Res.
9	E.l.	PK-30	10	–	–	50	–	–	–	–	30		–	Res.
10	E.l.	P-30	–	–	–	40–50	–	–	–	–	30–35	–	–	Res.
11	Calcium carbide	CaC_2	98(CaC_2)	–	–	–	–	–	–	–	–	2	–	–
12	Silicon carbide (green)	SiC	–	–	–	0.59	–	–	0.06	–	–	0.03	–	0.11
13	Ferro aluminum	Fe–Al	–	–	–	2	–	–	0.49	–	–	–	–	19
14	Nickel-aluminum ligature	Ni–A	–	–	–	4	–	–	48	–	–	–	–	48% Ni

Several compositions for one-phase and two-phase boriding are developed in accordance with the theoretical basis and recommended for use (Table 8.12). The growth of one-phase boride layers is faster than of two-phase layers' despite the significant difference in stable electrode potentials (different speeds of redox-reaction processes and boron concentration on the surface). It is explained by a bigger boron diffusion coefficient in the crystal lattice of Fe_2B boride than in FeB. Two-phase layers, formed in compounds, containing electrochemical reducing agents with stable electrode potentials from 400 too 900 mV, are characterized by the absence of continuous FeB phase layer and increased ductility.

Some contradiction between the speed of boronized layers' formation and stable electrode potentials is explained by alloying borides with powder additive elements, which change the smelt composition due to the presence of powder additive elements' ions and the shift of indexes for stable electrode potentials.

Low-temperature boriding in second-type mediums (Table 8.10) is carried out by replacing boron carbide with any of the following electrochemical reducing agents: silicocalcium, silicomanganese, silicon carbide, silicomishmetal. The speed of boronized layers' formation slightly reduces in comparison with boriding by the 1st technological variant; it happens because of different smelt ions are reduced on the surface. The fact that smelts have different viscosity in the range of temperatures 550–600 °C is worth special mentioning. It significantly complicates the process of low-temperature boriding and makes these compositions technologically inefficient in this temperature interval.

The third-type compositions use halogenides of alkali and alkaline earth metals or other boron-free salts as their salt components; solid powder components may be presented by following boron-containing compounds: boron, boron carbide, ferroboron, magnum, calcium, barium polyborides etc. The main advantage of these compounds is their high castability; the main disadvantage is the tendency for delayering.

The delivery of boron to the processed surface is provided by reducing boron ions of low valency accompanying the work of short circuit microgalvanic elements. Boron subions occur in the smelt some time after the bath starts to work because a salt component of the bath becomes a boron-containing salt.

The growing speed of a boronized layer and the correlation of phases in the layer may be controlled by changing the amount of a boron-containing additive in the bath. The most technologically efficient bath has chloride salts and amorphous boron (6–8%) foundation [221]; the alternative is the bath containing 99–95% $CaCl_2$ and 1 55% MgB_{12} [222]. The latter bath composition is characterized by waste-free technological cleanliness.

The bath for one-phase boriding, containing 85–90% $CaCl_2$ and 10–15% CaB_5, has good technological characteristics for medium-temperature boriding (750–900 °C) (Table 8.14). If the calcium carbide content is >13% and the soaking is longer than 3 h, the surface at 750 °C begins to generate a FeB boride layer (5–7 mkm wide). Lower process parameters in the medium-temperature interval causes the formation of plastic one-phase layers. The width of the general layer reaches

Table 8.14 The influence of temperature and time on the width of one-phase boronized layers in steel 45

Temperature, °C	The width of boride layer, mkm, obtained at different process duration, h		
	1	3	5
750	17	27	35
850	22	40	60
950	35	65	90

Table 8.15 The compositions for low-temperature one-phase boriding [251, 252]

The composition of a salt component, mass%	Boron-containing additive	Layer's width, mm
22–25% LiF, 9–10% NaF, 44–55% KF	15–25% (B или B_4C, CaB_6, MgB_{12}, BaB_6, ферробор, и др.)	0.030
32–36% LiCl, 2–3% NaCl, 41–46% KCl	15–25% (B или B_4C, CaB_6, MgB_{12}, ферробор, ферроборал и др.)	0.033

Note Saturation of steel 45 was carried out at 550 °C for 8 h

35 mkm. The presence of a small amount of FeB phase doesn't significantly effect the high plasticity of obtained layers.

Due to the delayering of the bath composition, it is better to avoid positioning processed details in the upper part of the composition (the reduced speed of layer formation) and the lower part (increased speed of layer formation). Each of these parts comprises 15% of the total smelt height. A periodic stirring helps to eliminate this disadvantage.

Special bath compositions have been developed for low-temperature non-electrolyze boriding; they are aimed at obtaining plastic one-phase boride layers (Fe_2B). These baths include the eutectic mixtures of halogenides of alkali metals as salt components and solid powder boron-containing substances as boron sources (Table 8.15).

The second composition based on alkali metal chlorides has a lower melting point which causes its increased castability in the recommended temperature range (550–700 °C), reduces the smelt withdrawal when the details are taken out from the bath and simplifies the process procedure. The compositions have practically the same saturation capacity. The obtained plastic boride layers have approximately 1700 HV hardness and are characterized by the increased corrosion and heat resistance.

Low-temperature boriding encourages the generation of boride layers wider than 0.080 mkm at 700 °C that in many cases seems insufficient. The thickening of the layers can be reached by preliminary nitrating [214]. The obtained layers have increased plasticity. The process of low-temperature boriding is recommended to use for non-deformative processing in combination with operations of high-temperature tempering of steels and ageing of alloys.

The protector method of boriding presupposes the production of a metal couple with the processed detail which acts like an electrochemical protector at their connection; they form a galvanic couple in the melted boron-containing salt. In accordance with the electrochemical nature of non-electrolysis processes carried out in melted salts, the protector must be more electronegative than the processed detail's material and the expected phase or solid solution of the saturation element on the surface. The protector is chosen from the electrochemical range of currents (where stable electrode potentials of different materials are put) built for the used salt solution.

The method involves the use of electrochemical protectors which can be found in solid or liquid state during boriding.

A solid protector may be compact (in the form of a metal plate or ribbon) or powder. It is mechanically fastened to the place where a boride layer needs to be obtained. A plastic protector may be connected directly to the processed detail or by a special conductor. In the first case the distribution of the occurring galvanic current is even along the border between the protector and metal and consequently the boride layer would be evenly distributed along this border. The second case presupposes that the maximum current density is found in the contact point of the conductor which determines the maximum density to be in this place. The density decreases with distancing from the contact point; this leads to the thinning of the layer. The defining effectiveness characteristics for the protector's activity is the correlation of the anode surface (protector) and cathode surface (detail) which determines the final current density and voltage.

When boride layers are obtained in a big area of the surface, the calculation of the necessary amount of protectors is done [223]; it should take into account the type of fastening and set current density.

The electrical content between the bath and the detail which are separated by a barrier is supported by the conductor when the protector is powder.

The liquid protector is positioned on the bottom of the bath; it is done by selecting unit weights for the salt smelt and protector's material. The electrical contact between the details and the protector is established by the conductor or by the walls of the metal crucible and curtain equipment for details.

Boriding of steel 35 in borax solution ($Na_2B_4O_7$) at 900 °C according to the protector method leads to the formation of plastic one-phase boride layers of adequate width (Fig. 8.21). Titan alloy BT1 plates were used as an electrochemical protector; it met all electrochemical and technological requirements. The estimated density of the spontaneously generated galvanic current was 0.4 A/cm^2 (with the protector efficiency ratio—0.5). The width of the boronized zone along the protector was 20 mm. The electrolysis from the external source of the stable current with the same cathode current density was ensured by the even width of the boronized layer (35–40 mkm, at 900 °C for 2 h).

The deduced dependence of the plastic boronized layers' growth on the protector's efficiency ratio ($K = S_{prot}/S_{det}$, where S_{prot} is the square of the protector's

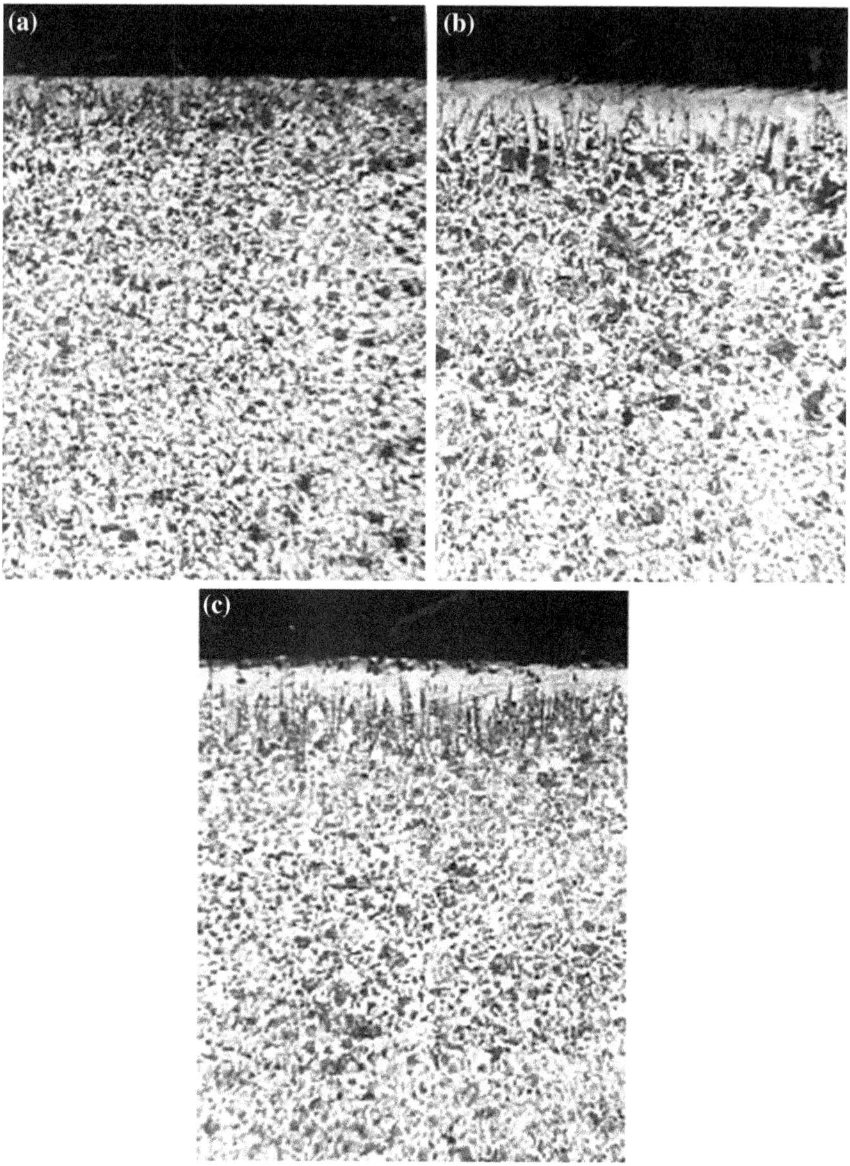

Figure 8.21 The microstructures of boride layers obtained with the protector method at 900 °C (steel 35, ×100): **a**—2 h, **b**—4 h, **c**—6 h

surface, S_{det}—square of the detail's surface), temperature and saturation duration during point connection of protector to the detail and along the surface makes it possible to recommend this method for the treatment of details and instruments. It has the following advantages:

- it creates the conditions for local non-corrosion boriding;
- it decreases the labor-output ratio in preparation of the saturation bath and cleaning details from smelts after processing [58].

Saturation in water solutions is attributed to the processing in liquid mediums by the type of used medium and to the saturation in gaseous ion mediums by the mechanism of boron mass transfer. The composition of the solution used for boriding is: 50–90% of the mixture (40–60% H_2O and 60–40% glycerine), 5–15% K_2CO_3, 5–35% B_4C. Boron carbide with grains № 5–6 is found in the solution in the form of suspension. A processed detail is included into the chain of electric current as a cathode. Nickel, lead or molybdenum plates are used as anodes. The correlation of anode and cathode surfaces must not exceed 10. Saturation is carried out at the current density 0.3–0.9 A/cm^2 and voltage 180–220 V. Steel 45 at 800 °C in 12 min generates a boride layer of 50–55 mkm. However the layer is distributed unevenly across the surface due to the composition's delayering. The replacement of boron carbide by amorphous boron eliminates this disadvantage and increases the processing effectiveness by 1.5 times [224]. In this case the composition contains 40–60% glycerine, 5–10% Na_2CO_3, 2–5% amorphous boron, the rest is water. The processing of details of complicated forms and the formation of evenly thick layers require the use of anodes reflecting the surface form and different kinds of screens. Normally saturation leads to the formation of two-phase layers. The decrease of a high-boron phase or preventing its occurrence is reached by reducing the content of boron in the composition or by manganese, nickel, chrome and other elements alloying of the formed layer.

Heating of the processed surface is done by a glow or spark current generated in a steam jacket which divides it from the water solution during treatment. The boron mass transfer goes through the gaseous medium with the help of the boron subions which are reduced on the metal surface in the plasma of the occurring current. The temperature of the metal surface during treatment reaches 650–1100 °C. The temperature of water solution should be maintained at 60–70 °C.

Thus, the plasticity of obtained layers may be controlled during non-electrolysis boriding in melted mediums (solution) with powder additives by:

- choosing a composition of a salt component;
- choosing a type of powder electrochemical reducing agent, its amount and grain size;
- using alloy steels as a substrate; alloy boride layers and alloy substrate are generated in this case;
- optimizing the temperature and duration of the process;
- choosing the temperature which ensures the formation of the layer by the diffusion or diffusion-crystallization mechanisms;
- thermal processing.

8.3.5 *The Influence of Processing Conditions on the Plasticity of Boronized Layers at the Atomic State of the Medium*

Atomic boron in a saturation medium is typical for boriding in melted metals, the formation of layers in molds (due to the fusing of metals and forms' coating components) and during additional treatment of different boron-containing layers or coatings with the use of concentrated energy sources. It is worth mentioning that fusing is accompanied by diffusion processes. These processes leading to the transformation of the chemical composition in the processed surface must be attributed to the processes of chemical-thermal treatment.

In particular, *boriding in melted lithium is recommended.* Melted lithium must contain 7–10% of amorphous boron; boriding must be carried out at 650–1000 °C for 3–10 h [225]. The process is accompanied by the formation of two-phase boride layers; it goes faster than in powder mixtures. Smelts of lead, cadmium bismuth and other metals may be used as melted saturation metal mediums. Plastic one-phase boride layers are obtained by reducing the boron content in the alloy.

One of the examples of contact boriding is *saturation in molds,* which is accompanied by the destruction of the saturation compact material. The generation of a boronized layer goes by fusing a boron-containing substance and a liquid metal. The result of the compact saturation material's destruction and its mixing with liquid metal is that obtaining of good-shaped processed surface is impossible. Boron distribution across the layer's width and correlation of its structural components depends on the compact saturation material width, the thickness of cast walls (cooling speed), the casting temperature and chemical composition of the metal cast [9, 226].

The formation of boronized layers accompanied by the use of concentrated energy sources can be carried out with solid or melting surfaces. The process is supported by the contact of compact material and the processed surface. In particular, boriding with laser emission used on the processed surface requires applying of saturation paste on the detail's processed surface; the paste must be based on boron, B4C, FeB, CrB, VB2. A better heat absorption is acquired by adding carbon black. Organic materials (e.g. ethanol) are usually used as a bonding material [227]. The width of the applied layer must be in the range of 0.01–0.08 mm [228].

The thickness of a boronized layer depends on the surface temperature and laser emission duration which are regulated by the repositioning speed of the detail or laser beam and its focusing. The treatment with impulse laser (E = 0.5–1.0 kJ; duration (1–6) × 10^{-3} c) of the steel 45 surface with B_4C later leads to the formation of a boronized layer containing a boride uncorroded zone 0.05–0.011 mm wide with 1950–2100 HV and a transition zone with 250–530 HV microhardness [229–232].

Laser treatment often leads to the formation of amorphous layers with high exploitation properties. If the energy is more densed (CO_2-laser) and the melting point is reached, an amorphous layer on iron is impossible to obtain due to the insufficient cooling speed at crystallization. Nd-laser treatment ensures the generation of an amorphous structure after melting. The Microhardness of the amorphous boronized layer depends on the carbon content in the substrate and the boron content in the layer. In all cases it is higher than Fe_2B microhardness and has an improved wear resistance [229]. Presumably it can be explained by a larger plasticity reserve of amorphous structure.

Laser treatment effectiveness may be improved by injecting B_4C in the processed zone. Obtained carbon boride structures in steel 45, У8, 4Х4М2ФВС and gray iron СЧ70 turn out to be flawless and can be compared to boride phases by hardness [233]. Laser boron and carbon saturation helps to obtain dispersed boride structures (FeB, Fe_2B) in solid solution. The best mechanical properties are ensured at 30–45% B_4C [234].

Alloyed boronized layers of enhanced plasticity can be generated during laser treatment by supplementing coating with carbon, boron, cobalt, vanadium. The resistance of metal-processing instruments increases in double [228]. Fusing during laser treatment of compact saturation material, containing Cr_2B and VB_2, also leads to obtaining plastic and wear-resitant composition structures on the processed surface. The processing is done at 2.5 kW voltage, scanning speed 0.2 m/min, lane width 2 mm and penetrating weld width 1 mm [235]. The saturation of steels У7, У12, Х12М from powders B_4C (75%) and Cr (25%) on the surface leads to the formation of an over-saturated solid solution of alloy elements in iron which is thermally stable up to 650 °C.

The obtained plastic composite layers ensure the improvement of exploitation resistance for cold and hot processing instruments 2.5–4 times [236, 237]. Non-defect boronized layers are formed if the layer has a preeutectic structure. In some cases eutectic and over-eutectic structures with cracks can also be used for increasing the wear resistance [229].

Electrolysis-beam chemical-thermal processing on the metal surface causes the occurrence of complicated physical-chemical processes; they are effected by the temperature, charged particles flux and diffusion elements. The changes in the crystal lattice simultaneously take place. The combination of high temperature and local point defects leads the stimulation of relaxing and diffusion processes and the formation of composite boronized layers according to the diffusion and diffusion-crystallization mechanisms [238–240].

It is stated that electron-beam boriding from clean boron carbide causes the surface layer's structure to be pre-eutectic and combine martensite formations obtained from primary austenite grains, and ledeburite, comprising of α-phase and cementite Fe_3C. Fe_3C lattice periods are larger than any estimated data which is the evidence for the replacement of some carbon atoms with boron atoms accompanied by the formation of carbon-borides $Fe_3(C,B)$.

The increase of the layer's plasticity during electron-beam boriding may be ensured by changing the morphology of borides. In particular, the boronized layer obtained during saturation from boron carbide-based coating, contains round inclusions located on the layer's surface and eutectics. The boronized layer, obtained during saturation from amorphous boron based coating, consists of differently shaped particles: rhombic, prism, dendrite. The surface is covered with a continuous light film with needles directed inwards the layer [240, 241]. Peculiarities of formation of composite structures are closely studied in Chap. 9.

8.3.6 The Influence of Additional Processing on the Plasticity of Boronized Layers

Additional processing during boriding is carried out before saturation, during treatment and after boriding.

Additional processing treatments before boriding, leading to the increase in the plasticity of boronized layers, involve a preliminary oxidation of the processed surface, applying and precipitation of different coatings (galvanic, CVD, PVD, diffusion etc.), putting details into smelts or spraying cyanic and other salts onto the smelts' surface, preliminary electron-beam heating etc. These procedures might prevent the formation of high-boron phases, activate the details' surfaces, increase the number of defects with a crystal structure in the surface volumes of the processed metals. They were described in the previous parts.

Preliminary electron-beam processing causes the generation of an increased number of vacancy and inter-joint defects of the crystal structure in near-surface metal layers. It depends on the intensity of irradiation. Further boriding causes 25–30% spurt in the boride layer growth (in comparison with boriding without irradiation); the boride layer would predominantly consist of phase Fe_2B [240, 242]. The increase of the layer's plasticity is also witnessed.

One of the types of preliminary processing is the use of deforming cutting (patent RF #2015202) which includes the formation of the micro-crack surface of the substrate in the form of narrow deep channels. Further boriding leads to the formation of a hardened layer up to 1.4 mm wide. The layer looks like alternating vertical and inclined zones of different hardness which positively effects the hardness, reduces the tendency to scoring, has a comparatively low friction coefficient. This technology allows to:

- obtain protective coatings for details produced from any metal materials;
- regulate the composition and properties of the composition structure of the coating;
- obtain self-greasing surfaces by adding liquid, consistent and solid (fluoroplast, molybdenum disulfide, graphite) greasing materials to the surface structure;
- introduce ceramic and hard metal powders into the micro-crack structure, combine processing with denotation, plasma or supersonic gas-flame spraying;
- obtain a composite structure able to work in conditions of dynamic and stress loadings.

Additional post-boriding treatment involves the use of certain technological methods. In particular, the use of salt and boron-containing solid elements for bath blowing leads to obtaining boronized layers with cell surfaces. The cells are round and are filled with oxides. This surface increases an oil-retaining effect during the work of hardened details in contact points and increases the ductility of layers.

The plasticity improvement for layers obtained from gaseous and powder mediums requires an impulse stable or an alternating current, magnetic fields, ultrasonic oscillations, reverse reciprocal movement of a gaseous medium etc.

Additional post-boriding processing is carried out in order to reduce the amount of the high-boron phase in the obtained fragmented layer's structure or create a special modified structure of the boride layer and sublayer or to create a compressive stress on the surface. These effects are achieved by a further thermal treatment involving different types of heating which bring their own peculiarities into changing the layer's structure. In particular, quenching heating of boronized steel details in melted salt baths on the boronized surface after 30 min soaking causes the formation of a continuous layer of clean iron 15–20 mkm wide. It is accompanied by the growth of a boronized layer by 10% and the increase of ductility; it happens due to the resorption of the high-boron phase. These changes may be attributed to the mechanical treatment allowance. The maximum effect of ductility improvement free of an iron layer formation on the surface is reached by heating boronized details in baths with a decreased boron content. For instance, if boron or magnum, calcium, barium polyborides powders are used, the sufficient number of these materials introduced to the smelt is 2–3%. The presence of B_4C increases this number to 4–6%.

Heating in the furnace oxidation atmosphere for 3–5 h leads to the compensation of boronized layer's thinning by the layer's widening due to the high-boron phase resorption. The increase of Fe_2B phase amount improves its plasticity. The increase of the detail's size is 10–15% from the boride layer's width which is attributed to the preliminary mechanical processing allowance.

Heating of boronized details in the redox controlled atmosphere after 3–5 h soaking also leads to the formation of iron layer on the surface. Thus, all these cases of additional heating of boronized details for preserving a boride layer on the surface require the creation of gaseous or liquid mediums with the boron potential which would maintain the boron concentration at 4–5% on the surface.

The use of high frequency currents or laser heating, aimed at obtaining a fragmented or composite structure of a boronized layer with high plasticity, also requires following the same recommendations taking into account the heating speed and temperature.

Scanning of the boronized surface by the beam of electrons with insufficient power density (2.5×10^{-3} W/cm^2) doesn't cause deboriding. At 2.0×10^4 W/cm^2 the process is accompanied by intensive resorption of the high boron phase with the formation of a composite layer based on Fe_2B and the formation of a porous zone on the surface.

References

1. Samsonov V.G., Stasovskaya V.V. // Poroshkovaya metallurgiya. 1966. №12. pp. 95–97.
2. Tarasov S.Yu. Struktura poverkhnostnykh sloev treniya i uprochneniye vysokonagruzhennykh tribotekhnicheskikh kontaktov diffuzionnym borirovaniem. Dissertaciya kandidata tekhnicheskikh nauk. Tomsk, 1994. 138 p. [The structure of surface layers of friction and hardening of high-loaded tribotechnical contacts with diffusion boriding. Candidate of Science dissertation].
3. Gæriot P., Thevenot F., Driver J.H., Laurent A. Borudif: Nouvelle technique industrielle de boruration des aciers // Traitement thermique. 152–81. pp. 21–28.
4. Dukarevich I.S., Mozharov M.V., Shigaev A.S. Pereraspredeleniye elementov v borirovannom sloe // MiTOM. 1973. №2. pp. 64–66 [The redistribution of elements in a boronized layer].
5. Lyakhovich L.S., Puchkov E.P., Voroshnin L.G. // Izv. AN BSSR. Physico-technical science. 1970. №2. pp. 110–114.
6. Gulyaev A.P. Metallovedeniye. Moscow, 1986. 544 p [Physical Metallurgy].
7. Lyakhovich L.S., Vasilyev L.A., Krukovich M.G., Turov Yu. V. Issledovaniye srednetemperaturnogo eletroliznogo borirovaniya stalei 38XC i 40X // Izv. AN BSSR. Physico-technical science. 1976. №3. pp. 48–51.
8. Voroshnin L.G., Lyakhovich L.S. Borirovaniye stali. Moscow, 1978. 240 p. [Steel boriding].
9. Belskyi E.I., Sitkevich M.V., Ponkratin E.I., Stefanovich V.A. Khimiko-termicheskaya obraborka instrumentalnykh materialov. Minsk, 1986. 247 p. [Chemical-thermal processing of tool materials].
10. Gopalakrishnan P., Shankar, Balamrugan S.S., Thiagi A.K. // J. Mater. Sci. and Technol. (Bulgaria). 2002. V. 10, № 4. pp. 45–51.
11. Gurevich B.G., Govyazina E.A. Elektroliznoye borirovaniye stalnykh detalei. Moscow, 1976. P. 70 [Electrolysis boriding of steel details].
12. Boiko V.A., Gabinskaya G.D., Epik A.P. Izmeneniye razmerov borirovannykh detalei v protsesse nagreva i okhlazhdeniya // MiTOM. 1979. №1. pp. 8–10 [The change in sizes of boronized details during the process of heating and cooling].
13. Kunst H., Schaaber O. // Harterei-Tech. Mitt. 1967. Bd. 22, №1. S. 1–27.
14. Kastner B., Przybylowiz K., Grygoruk I. // Harterei-Tech. Mitt. 1979. Bd. 34, №4. S. 173–179.
15. Lakhtin Yu. M., Pchelkina M.A. Gazovoe borirovaniye stali // Izv. vuzov. “Mashinostroenie”. 1961. №2. pp. 171–174 [Gas boriding of steel].
16. Koeling R. Boriren van stahl // Polytechn. Tijdchrift. 1970. V. 25, №22. S. 962–977.
17. Kuleshov Yu. S., Nefedov V.G. Neitralizator dlya ustanovki gazovogo borirovaniya // MiTOM. 1983. №10. P. 7 [A neutralizer of a setting for gas boriding].
18. Pchelkina M.A., Lakhtin Yu. M. Borirovaniye stali v gazovykh sredakh // Izv. vuzov. Chernaya metallurgiya. 1960. №7. pp. 163–170.
19. Kuleshov Yu. S., Nefedov V.G., Sergienko R.I. Ustanovka dlya gazovogo borirovaniya // Zaschitnye pokrytiya na metallah. Kiev. Iss. 15. pp. 31–33. [The setting for gas boriding].
20. Filding S.J., Hanks S.D., Johnson P.S., Kislyakov A., Wintee J. // G. Nucl. Mater. 1992. V. 186, №3. S. 217–226.
21. Prokoshkin D.A., Arzamasov B.N., Ryabchenko E.V., Mikhailov I.A. Polucheniye pokrytii na metallah v tleyuschem razryade. // Zaschitnye pokrytiya na metallah. Kiev, 1970. Iss. 3. pp. 7–10 [Obtaining of coatings on metals in glowing discharges].
22. Prokoshkin D.A., Arzamasov B.N., Ryabchenko E.V. Khimiko-termicheskaya obrabotka metallov v tleyuschem razryade // Khimiko-termicheskaya obrabotka stali i splavov. Moscow. 1969. pp. 107–109 [Chemical-thermal processing of metals in a glowing discharge].
23. Wierschon T., Michalski S., Karpinski T. // 2 nd Int. Congr. Heat Treat. Mater. IFHT. Milano, 1982, P. 375–383.

24. Casadesus P., Frantz C., Gantois M. // Rev. Metall. 1979. V. 10, №11. P. 1739–1743; // Mem. Sci. Rev. Metall. 1979. V. 76. P. 9–21.
25. Kuper A., Qiao X., Jams C., Stock H.-R. Plasmauntersturts Boriezen mit Trimethytilborat // HTM: Harter-techn. Mitt. 1988. Bd. 53, №6. P. 374–379.
26. Kuper A., Stock H.-R., Mayr P. Heue crkenntniss rum Plasma-boriezen mit Trimethytilborat // HTM: Harter-techn. Mitt. 2000. Bd. 55, №6. P. 353–360.
27. Wierschon T., Bogacki I., Karpinski T. // Proc. 18 Th. Int. Conf. Heat Mater. Detroit, 1980, «Mitals Park».
28. Rodriguer Cabeo E., Laudien G., Biemer S., Rie K.-T., Hoppe S., Frick M. Plasmaboriezen in einer BCI3-H2-Ar-Atmosphare // HTM: Harter-techn. Mitt. 1999. Bd. 54, №2. P. 110–115.
29. Farras S., Filier E., Kolzsvary Z. // Metalozn i obrob ciepl. 1981. №51–52. S. 66–68.
30. Grill A., Finberg J., Avini R., Spalvins I., Bukly D.H. Boriding of steel in a Could K.F. plasma of diboran in argon // Mater Sce. And Eng. 1986. V. 78, №2. P. 201–207.
31. Sposob polucheniya mnogofaznykh boridnykh sloev na detalyakh iz splavov zheleza / Pat. 224877 Germany, C232 C. Publ. 17.05.85. [A method for obtaining multi-phase boride layers on iron-smelt details].
32. Sposob polucheniya poverkhnostnykh sloev iz nitrida titana i bora / Pat. 224882 Germany, C23C. Publ. 17.07.85. [A method for obtaining surface layers from titanium nitride and boron].
33. Kuleshov Yu. S., Nefedov V.G., Sergienko R.I. Sposob gazovogo borirovaniya izdelii. / Cert. № 478892 C23C. Publ. 1979. [A method of gas boriding of tools].
34. Mazumoto K., Adzuma Ya., Mazuda S. Saturating medium for boriding metals and alloys / Pat. 52-63388 Japan. Publ. 24.12.80.
35. Fichtl W., Trauser N., Graf von Matuschka A. Boronizing with EKabor / German patent published: DE 3431044A1, DOS dd Mar. 6, 1986.
36. Graf von Matuschka A., Trauser N., Ziese J. Borieren im Wirbelbett // Harterei-Tech. Mitt. 1988. V. 43, №1. S. 21–26.
37. Genel K., Ozbek I., Bindal C. // Mater. Sci. and Eng. A. 2003. V. 347, №1–2. P. 311–314.
38. Guriot P., Thevenot F., Driver J.H. // Thin solid Films. 1981. 78, №1. P. 67–76.
39. Afanasyev A.A. Elektroliznoye borirovaniye reversirovannym tokom konstruktsionnykh detalei. Dissertatsiya doktora tekhnicheskikh nauk. Voronezh, 2001. 311 p [Electrolysis boriding with reverse currents of construction steels].
40. Gutman M.B., Mikhailov L.A., Kaufman V.G. Sposob borirovaniya. Cert. №223662 C23C. Publ. 1968.
41. Rojahn P. Ausbildung und Wachstum von Boridschichten nach Behandlung im Wirbelbett // Harterei-Tech. Mitt. 1988. V. 43, №1. S. 56–62.
42. Balandin Yu. A. Sposob borosilitcirovaniya stalnykh izdelii v psevdoozhizhennom sloye. Patent 2190689 Russia. МПК7 C23C 8/72, 12/02. Magnitogorsk. Publ. 10.10.02. [Boron-silicon treatment of steel tools in fluidizated layers].
43. Balandin Yu. A. Sposob boroazotirovaniya stalnykh izdelii v psevdoozhizhennom sloye. Patent 2194793 Russia. МПК7 C23C 8/26, 8/70, 12/02. Magnitogorsk. Publ. 20.12.02. [Boron-nitrate treatment of steel tools in fluidized layers].
44. Balandin Yu. A. Kompleksnoye nasyschcniye stalnoi poverkhnosti borom, azotom i medyu v psevdoozhizhennom sloye // Izv. vuzon. Mashinostroeniye. 2004. №9. C. 39–42. [Complex saturation of steel surfaces with boron, nitrate and cuprum in fluidized layers].
45. Balandin Yu. A. Izucheniye processov borirovaniya, borosilitsirovaniya i boronikelirovaniya instrumentalnykh stalei v psevdoozhizhennom sloye // Materialovedenie, 2005. №3. C. 47–51 [The study of boriding, boron-silicon and boron-nickel treatment of tool steels in fluidized layers].
46. Balandin Yu. A. Uprochneniye poverkhnosti shtampovykh stalei diffuzionnym borirovaniem, boromedneniem i borokhromirovaniem v psevdoozhizhennom sloe // MiTOM. 2005. №3. C. 27–30 [Surface hardening in stamp steels by diffusion boriding, boron-cuprum and boron-chrome treatment in fluidized layers].

47. Grachev S.V., Zavarov A.S., Kolpakov A.S., Balandin Yu. A. Sposob boromedneniya stalnykh izdelii v vibrokipyschem sloye. Patent 2004619 Russia C23C. Publ. 1993. №45–46. [The method of boron-cuprum treatment of steel tools in vibro-boiling layers].
48. Grachev S.V., Kolpakov A.S., Balandin Yu. A. Boromedneniye shtampovykh izdelii v vibrokipyaschem sloye // Ural SU. 18.06.93. №5966-4M93. 8 p.
49. Balandin Yu. A. Struktura i svoistva borirovannykh sloev, poluchennykh na instrumentalnykh stalyakh v vibrokipyaschem sloye. Magnitogorsk, 2003. 22 p. [The structure and properties of boronized layers obtained in tool steels in vibro-boiling layers].
50. Grachev S.V., Maltseva L.A., Maltseva T.A., Kolpakov A.S., Dmitriyev M. Yu. Borirovaniye i borokhromirovaniye v vibrokipyaschem sloye // MiTOM. 1999. №11. P. 3–6 [Boriding and boron-chromizing in vibro-boiling layers].
51. Hoasoda K., Komuro D. A method of partial protection from the formation of boronized layers // Patent proposal № 379752 Japan C23C. 1994. №24. Publ. 04.04.91.
52. Dvoretskaya G.F., Mishin V.N., Schadrichev E.V., Surov E.F. Pasta dlya zaschity poverkhnostei stalnykh detalei pri borirovanii / Cert. № 876779 C23C. Publ. 1981 [The paste used for protection of steel details under boriding].
53. Kunst H., Schaaber O. // Harterei-Tech. Mitt. 1967. 22, H 3, №4. S. 275–292.
54. Krzhminski G. Pat. № 2127096 Germany C23C. Method of metal boriding (with the focus on steels and iron). Publ. 06.11.80.
55. Rusin P.I., Stepanovich A.D., Nikolaev V.E., Stepanovich V.A. Sposob uprochneniya izdelii. Cert. № 9200780 C23C. Publ. 1982.
56. Nikolaev V.E. Struktura i svoistva instrumentalnykh shtampovykh stalei pri borirovanii v granulakh. Novocherkassk, 1984. 16 p. [The structure and properties of tool stamping steels under boriding in grains].
57. Nikolaev V.E. Borirovaniye shtampovykh stalei v boronasyschayuschikh granulakh // Povysheniye nadezhnosti i dolgovechnosti detalei selskokhozyaistvennykh mashin metodami termicheskoi i khimiko-termicheskoi obtabotki. Moscow, 1981. pp. 137–138 [Boriding of stamping steels in boron-saturating grains].
58. Krukovich M.G. Razrabotka teoreticheskikh i prikladnykh aspektov upravleniya strukturoi i svoistvami borirovannykh sloev i ikh ispolzovanie pri proizvodstve transportnoi tekhniki. Dis. dokt. tekh. nauk. Moscow, 1995. 416 p. [The development of theoretical and applied aspects of controlling the structure and properties of boronized layers and their use in producing transport machinery. Doctorate dissertation].
59. Turov Yu. V., Lyakhovich L.S., Krukovich M.G. Sreda dlya borirovaniya. Cert. № 549502 C23C. Publ. 1977 [A medium for boriding].
60. «Degussa»-Hart // Techn. Mitt. 1973. Bd. 28, №2. S. 113–117.
61. Kotlyarenko L.A., Kadyrov V. Kh., Pikalov S.N., Epik A.P. Borizator dlya konteinernogo borirovaniya stalei // Tekhnologiya i organizatciya proizvodstva. 1987. №4. pp. 48–50. [A boronizer for a container type of steel boriding].
62. Savintsev M.I. Uprochneniye stalei termodiffuzionnym borirovaniem // Tekhnologya metallov. 2004. №9. pp. 20–23. [Steel hardening with thermal diffusion boriding].
63. Trunin V.S., Petrova L.V., Abramova N.A., Savinykh Yu. G. Tekhnologiya proizvodstva alumoboridnoi kompozitcii dlya poverkhnostnogo uprochneniya metallicheskikh izdelii // Uralsk Scientific Chemistry Institute Bulletin. 1991. №68. pp. 45–48 [The technology of producing aluminum-boride composition for surface hardening of metal details].
64. Krukovich M.G., Kukharev B.S. Sostav dlya khimiko-termicheskoi obrabotki stalnykh izdelii. Cert. № 885343 C23C. Publ. 1981. №44. [A composition for chemical-thermal processing of steel details].
65. Naumov A.G., Latyshev V.N., Godlevskiy V.A., Korotkov Yu. A., Onashin N.M. Sposob khimiko-termicheskoi obraborki. Cert. № 4870051 Russia C23C. Publ. 1993. №17.
66. Samsonov G.V., Epik A.P. Tugoplavkiye pokrytiya. Moscow, 1973. 400 p. [High-melting point metals].
67. Minkevich A.N. Khimiko-termicheskaya obrabotka metallov i splavov. Moscow, 1965. 491 p. [Chemical-thermal processing of metals and alloys].

68. Glukhov V.P. Boridnye pokrytiya na zheleze i stalyakh. Kiev, 1970. 208 p. [Boride coatings on iron and steels].
69. Lyakhovich L.S., ed. Khimiko-termicheskaya obrabotka metallov i splavov. Spravochnik. Moscow, 1981. 424 p. [Chemical-thermal processing of metals and alloys. Reference book].
70. Gabinskaya G.D., Dzyuba V.a., Bukreeva N.A. et al. Povysheniye stoikosti shtampovogo instrumenta borirovaniyem v usloviyah PO "Novokramatorskiy mashinostroitelnyi zavod" // Tekhnologiya zagotovitelnogo i mekhanosborochnogo proizvodstva. Kramatorsk, 1982. pp. 109–111 [The improvement of resistance in stamping tools with boriding at Novokramatorsk machinery plant].
71. Zanevskiy S.S., Prosvirova S.K. Sostav dlya borirovaniya stalnykh izdelii. Cert. № 1002402 C23C. Publ 1983. №9. [A composition for boronizing steel tools].
72. Gabinskaya G.D., Kolesnik N.M., Tovpyga V.N. et al. Sostav dlya borirovaniya stalnykh izdelii. Cert. № 926069 C23C. Publ. 1982. №17. [A composition for boronizing steel details].
73. Sitkevich M.V., Belskiy E.I., Stefanovich V.A. et al. Sostav dlya diffuzionnogo nasyscheniya stalnykh izdelii. Cert. № 937533 C23C. Publ. 1982. №23. [A composition for diffusion saturation of steel details].
74. Golubev E.I., Kulpina V.A., Savelyev V.A. Sostav dlya borirovaniya stalnykh izdelii. Cert. № 1027281 C23C. Publ. 1983. №25. [A composition for boronizing steel dedtails].
75. Kukharev B.S., Borisenok G.V., Vaschev S.E., Levitan S.N. Sostav dlya borirovaniya stalnykh izdelii. Cert. № 996513 C23C. Publ. 1983. №6. [A composition for boronizing steel details].
76. Borisenok G.V., Kukharev B.S., Vaschev S.E., Levitan S.N. Sostav dlya borirovaniya stalnykh izdelii . Cert. № 954504 C23C. Publ. 1982. №32. [A composition for boronizing steel details].
77. Arkelyan V.S., Boldovskaya S.O., Sluttskii I.A. Sostav dlya borirovaniya stalnykh izdelii . Cert. № 986969 C23C. Publ. 1983. №1. [A composition for boronizing steel details].
78. Kukharev B.S., Terekhov V.K. et al. Sostav dlya borirovaniya stalnykh izdelii . Cert. № 954504 C23C. Publ. 1982. №32. [A composition for boronizing steel details].
79. Voroshnin L.G., Kukharev B.S., Nedvitskiy A.A. et al. Sostav dlya borirovaniya stalnykh izdelii . Cert. № 1010160 C23C. Publ. 1983. №13. [A composition for boronizing steel details].
80. Voroshnin L.G., Kukharev B.S., Terekhov V.K. et al. Sostav dlya borirovaniya stalnykh izdelii . Cert. № 986958 C23C. Publ. 1983. №1. [A composition for boronizing steel details].
81. Filippova N.V., Bugrov V.P. Sostav dlya borirovaniya stalnykh izdelii . Cert. № 1601195 C23C. Publ. 1990. № 39. [A composition for boronizing steel details].
82. Asadula Sharif. Issledovaniye zakonomernostei formirovaniya svoist odnofaznykh boridnykh sloev. Minsk, 1980. 19 p. [The study of the formation pattern for properties of one-phase boride layers].
83. Vaschev S.E. Uprochnyayuschaya obrabotka mashinostroitelnykh stalei odnofaznym borirovaniem v poroshkovykh sredakh. Minsk, 1986. 18 p. [Hardening processing of machinery steels with one-phase boriding in powder mediums].
84. Bartashov V.G., Dmitriev M.A., Kutsepalov A.N., Chernyshov V.G. Poroshkoobrazniy sostav dlya borirovaniya stalnykh izdelii. Cert. № 933801 C23C. Publ. 1981. №21. [Powder composition for boronizing steel details].
85. Krasner M.S., Metrik D. Ya. Sostav dlya borirovaniya. Cert. №1121318 C23C. Publ. 1985. №40. [A composition for boriding].
86. Utezava Kh., Khiyami S., Yamasita M. The composition for boriding medium. Pat. №55-32786 Japan C23C. Publ. 27.08.80.
87. Minkov A.N., Kolesnik M.N., Shulga T.I. Borirovaniye shtampov v poroshkovoobraznykh smesyakh s primeneniem zaschitnoi atmosfery // Razrabotka, proizvodstvo i primeneniye instrumentalnykh materialov. Kiev, 1990. pp. 39–40. [Boriding of stamps in powder mixtures with the use of protective atmosphere].

88. Zieger H., Rihle M., Marks G., Vagner V. The method of boronizing stamps on the base of iron. Pat. № 224973 Germany C23C. Publ. 24.02.80.
89. Krasner M.S., Puchkov E.P. Borirovaniye v poroshkovykh smesyakh na osnove karbida bora s primeneniyem zaschitnykh atmosfer // Metallurgiya. Minsk. Iss. 20. 1986. pp. 30–32. [Boriding in powder mixtures on the basis of boron carbide with the use of protective atmospheres].
90. Gerasimov L.V., Prosvirnin V.I. Borirovaniye v tokoprovodyaschikh poroshkoobraznykh smesyakh // Skorostnye protsessy khimiko-termicheskoi obraborki s primeneniyem past i suspenzii. Riga, 1972. Iss. 200. pp. 91–95 [Boriding in conducting powder mixtures].
91. Arques J.L. Introduction ala boruracio del acero // Temple. 1979. №13. P. 57–61.
92. Muta A., Toad G., Simanoki Kh. // 1967. Vol. 31, №3. pp. 296–300.
93. Surmit A.M., Gabinskaya G.D. Sposob borirovaniya. Cert. № 779436 C23C. Publ. 1980. №41.
94. Redi A.B. Boronising behavior of Ni-Si-Cr steel // Surface Eng. 1992. №2. P. 136–140.
95. Lu Ming - Giong. // Harter.-Techn. Mitt. 1983. V. 38, №3. C. 156–169.
96. Gabinskaya G.D., Kolesnik N.M., Tovpyga V.N. Sostav dlya borirovaniya stalnykh izdelii. Cert. № 908943 C23C. Publ. 1982. №8. [The composition for boronizing steel tools].
97. Kukharev B.S., Krukovich M.G., Levitan S.N., Kukhareva N.G. Sostav dlya diffuzionnogo nasyscheniya stalnykh izdelii. Cert. № 876776 C23C. Publ. 30.10.81, №40. [The composition for diffusion saturation of steel tools].
98. Kukharev B.S., Krukovich M.G., Kukhareva N.G., Stasevich G.V., Levitan S.N. Sposob borirovaniya stalnykh izdelii. Cert. №876775 C23C. Publ. 30.10.81, №40. [The method of boronizing steel details].
99. Wu Hulan, Ping Jeoung. // Jiechu rechuli. 1989. №6. pp. 49–53.
100. Lu Yaokun, Li Jiahguo, Yuan Guangde. Jiechu rechuli // Heat. Treat. Metals. 1994. №3. S. 26–28.
101. Putilin V.G., Rustamov V.Kh. Sostav dlya borirovaniya stalnykh izdelii. Cert. № 1014982 C23C. Publ. 1983. №16. [The composition for boronizing steel tools].
102. Sobolev Yu. A. Sreda dlya borirovaniya. Cert. № 1643680 C23C. Publ. 1991. №15. [The medium for boriding].
103. Kukharev B.S., Levitan S.N., Borisenok G.V. Poroshkoobraznyi sostav dlya borirovaniya stalnykh izdelii. Cert. № 908944 C23C. Publ. 1982. №8. [The powder composition for boronizing steel tools].
104. Isakov S.A., Deineko V.A., Pahadnya V.P. Borirovaniye stali iz pokrytiya v srede vodoroda // Metallurgy. Minsk, 1983. Iss. 17. pp. 132–135. [Boronizing steel in the surface in the medium of hydrogen].
105. Lyakhovich L.S., Dolmanov F.V., Isakov S.A. Borirovaniye stalei v gazovoi srede // MiTOM. 1982. №4. pp. 25–28 [Boriding of steels in gaseous mediums].
106. Zemskov G.V., Kaidash N.G., Pravenkaya L.L. // MiTOM. 1964. №3. C. 61–63.
107. Krzyminski H., Kunst H. // Hart.-Techn. Mitt. 1973. Bd 28, H. 2. S. 100–112.
108. Krukovich M.G., Kukharev B.S., Levitan S.N. Sostav dlya borirovaniya stalnykh izdelii. Cert. № 834241 C23C. Publ. 1981. №20. [The composition for boronizing steel tools].
109. Krukovich M.G., Kukharev B.S., Levitan S.N. Sostav dlya borirovaniya stalnykh izdelii. Cert. № 885342C23C. Publ. 1981. №44. [The composition for boronizing steel tools].
110. Krukovich M.G., Kukharev B.S., Levitan S.N., Kukhareva N.G. Sostav dlya borirovaniya stalnykh izdelii. Cert. № 872598 C23C. Publ. 1981. №38. [The composition for boronizing steel tools].
111. Krukovich M.G., Kukharev B.S., Levitan S.N., Kukhareva N.G. Sostav dlya borirovaniya stalnykh izdelii. Cert. № 962331 C23C. Publ. 1981. №36. [The composition for boronizing steel tools].
112. Zilbergleit B.I., Mamontov V.B., Plyshevskii Yu. S. et al. Sostav dlya termodiffuzionnnogo borirovaniya stalnykh izdelii. Cert. № 985140 C23C. Publ. 1982. № 48 [The composition for thermal diffusion boriding of steel details].

113. Kukharev B.S., Tamelo S.A. Intensifikatsiya protcessa poroshkovogo borirovaniya // Metallurgiya. Minsk, 1983. Iss. 17. pp. 147–149. [The intensification of powder boriding].
114. Kukharev B.S., Stasevich G.V., Levitan S.N., Kukhareva N.G., Krukovich M.G. Sostav dlya termodiffuzionnnogo borirovaniya stalnykh izdelii. Cert. № 831859 C23C. Publ. 1981. № 19 [The composition for thermal diffusion boriding of steel details].
115. Kukharev B.S., Levitan S.N., Stasevich G.V., Kukhareva N.G., Krukovich M.G. Sostav dlya borirokhromirovaniya stalnykh izdelii. Cert. № 865968 C23C. Publ. 1981. №35. [The composition for boron-chromizing of steel details].
116. Kuzma Yu. B., Chaban N.F. Dvoinye i troinye sistemy, soderzhaschie bor. Spravochnik. Moscow, 1990. 320 p. [Double and triple boron-containing systems].
117. Bannykh O.A., Budberg P.V., Alisova S.P. et al. Diagrammy sostoyaniya dvoinykh i mnogokomponentnykh sistem na osnove zheleza. Spravochnik. Moscow, 1986. 440 p. [The equilibrium diagrams of double and multicomponent systems on the basis of iron. Reference book].
118. Koskov V.D. Povysheniye dolgovechnosti detalei burovoi tekhniki putem formirovaniya na poverkhnosti kompleksno-legirovannogo boridnogo sloya. Sverdlovsk, 1986. 20 p. [The improvement of durability of oil-rig machinery by forming complex alloyed boride layer on the surface].
119. Chernega S.M., Loskutov V.F., Yakovchuk Yu. E. Ostatochnye napryazheniya v legirovannykh boridnykh pokrytiyakh // Zaschitnye pokrytiya na metallakh. Kiev, 1987. Iss. 21. pp. 87–89. [Residual stresses in alloyed boride coatings].
120. Krukovich M.G., Kukharev B.S., Levitan S.N. Sostav dlya kompleksnoi khimiko-termichekoi obrabotki stalnykh izdelii. Cert. № 962022 C23C. Publ. 1982. №37 [The composition for complex chemical-thermal processing of steel details].
121. Zemskov G.V. Mnogokomponentnoye diffuzionnoye nasyscheniye metallov i splavov. Moscow, 1968. [Multicomponent diffusion saturation of metals and alloys].
122. Li Yensin, Van Slomin // Jiechu rechuli. 1990. №4. pp. 34–40.
123. Kaidash N.G., Lutsenko L.I., Semenenko I.A., Nelyub M.G. Zakonomernosti obrazovaniya, struktura i svoistva boridokhromovykh diffuzionnykh pokrytii na stali // Diffuzionnoye nasyscheniye i pokrytiya na metallah. Kiev, 1977. P. 42–46 [The formation pattern, structure and peculiarities of boron-chrome diffusion coatings on steel].
124. Zavalishin A.N., Loskutov V.F., Permyakov V.G. et al. Vliyaniye kompleksnogo nasyscheniya borom i perekhodnymi metallami na svoistva stalei i chugunov // Zaschitnye pokrytiya na metallakh. Kiev, 1977. Iss. 11. pp. 76–78 [The influence of complex saturation by boron and transition metals on the properties of steel and cast iron].
125. Loskutov V.F., Khizhnyak V.G., Byakova A.V., Grinenko E.M. Mnogokomponentnye pokrytiya na stalyakh // Zaschitnye pokrytiya na metallakh. Kiev, 1980. Iss. 14. pp. 46–47 [Multi-component coatings on steels].
126. Borisenok G.V., Turov Yu. V., Kupriyanov I.L., Danilchenko E.P. Sostav dlya borirovaniya izdelii iz stalei i spechennykh materialov. Cert. № 1196415 C23C. Publ. 1985. №45 [The composition for boriding of details made from steels and baked materials].
127. Grinenko E.M., Loskutov V.F. Vliyaniye legiruyuschikh elementov na formirovaniye boridnykh pokrytii // Kiev Polytechnical Institute Bulletin. 1988. №25. pp. 94–97 [The influence of alloy elements on the formation of boride coatings].
128. Lisovskaya N.V., Kadenatskiy L.A., Kochatok L. Ya. Poroshkoobraznyi sostav dlya diffuzionnogo borirovaniya. Cert. № 1659530 C23C. Publ. 1991. №24.
129. Chocholowski M., Przybylowez K. // 5th Int. Congr. Heat. Treat. Mater. Budapest. 1986. Proc. V. 3. — Budapest: S.a. 1438–1444.
130. Galynskaya N.A., Kulikovskiy E.I. Osobennosti formirovaniya borokhromirovannykh sloev v alumotermicheskih smesyakh // Metallurgiya. Minsk, 1986. №20. pp. 41–42 [The formation peculiarities of boron-chromized layers in aluminum-thermal mixtures].
131. Kalner V.D., Karpman M.G., Kulazhenkov S.D. Posledovatelnoye borokhromirovaniye uglerodistykh stalei // FKHMM. 1986. 22, № 6. pp. 24–26 [The consecutive boron-chromizing of carbon steels].

132. Ma Zongfu, Tang Furong, Chen Daixun. Structural characterizationof vanadized boride layers on steel // Surface end Coat. Technol. 1991. V. 48, №3. P. 249–252.
133. Filonenko B.A. Kompleksnye diffuzionnye pokrytiya. Moscow, 1981. P. 136 [Complex diffusion coatings].
134. Hunger H.J, Trute G. Problem beam Borireren von Ni-Basis werstoffen // Hart.-Tech. Mitt. 1994. V. 49, №3. S. 215–218.
135. Zemskov G.V., Kogan R.L., Grankin P.I., Shevchenko I.M. Osobennosti formirovaniya boridnykh diffuzionnykh sloev na mertenstostareyuschikh stalyak // Zaschitnye pokrytiya na metallakh. Kiev, 1980. Iss. 14. pp. 44–45 [The peculiarities of boride diffusion layers' formation on martensite-ageing steels].
136. Golubin A.V., Banas I.P., Krukovich M.G., Shalkevich A.B., Goncharova G.I. Poverkhnostoye uprochneniye austenitnykh Fe-Ni-Cr splavov borirovaniyem // MIIT science proceedings. Moscow, 1991. Iss. 843. pp. 20–36 [The surface hardening of austenite Fe-Ni-Cr alloys with boriding].
137. Kroukovich M.G. Surface friction and wear of refractory steels after thermo-chemical treatment // Proceedings of EUROMAT 2000. Volume 1, 7–9 November, Tours, France. P. 609–614.
138. Carbucicchio M., Sambagna G. Influence of chromium on boride coatings produced on iron alloys // Thin Solid Films. 1985. V. 125, №3–4. P. 299–301.
139. Carbucicchio M., Palombarini G. Effecty of alloying elements on the growth of iron borid coatings // J. Mater. Sci. 1987. V. 6, №10. P. 1147–1149.
140. Gordiyenko L.K., Yakhnina V.D. et al. Poroobrazovaniye v borirovannom sloye na vysokolegirovannykh stalyakh // Fizika i khimiya obrabotki materialov. 1975. № 5. pp. 80–84 [The formation of pores in boronized layers in high-alloy steels].
141. Yakhnina V.D., Kozlov A.M. Borirovaniye khromistykh i khromonikelevykh nerzhaveyuschikh stalei // High education institutions bulletin. Mashinostroenie. 1975. №6 [Boriding of chrome and chrome-nickel stainless steels].
142. Kozlov A.M., Lobanov N.F. Vliyaniye kontsentratsii khroma v stali na osobennosti formirovaniya borirovannogo sloya // Tekhnologii remonta, vosstanovleniya i uprochneniya detalei maschin, mekhanismov, oborudovaniya, instrumenta i tekhnologicheskoi osnastki. Saint-Petersburg, 2007. pp. 121–124 [The influence of chrome concentration in steel on the peculiarities of the formation of boronized layers].
143. Gurvich O.S., Marmer E.N., Minkov O.B. Vysokotemperaturnoye iznosostoikoye pokrytiye dlya uzlov treniya vakuumnykh elektropechei // Sovremennoye elektrotekhnicheskoye oborudovaniye dlya termoobrabotki metallicheskikh materialov. Moscow, 1982. pp. 90–92 [High-temperature wear-resistance coating for friction points in vacuum electrical furnaces].
144. Minkov O.B., Murovannaya S.G. Sposob borirovaniya detalei. Cert. № 823456 C23C. Publ. 1981. №15 [The method of boriding of details].
145. Minkov O.B. Diffuzionnoye borirovaniye vysokolegirovannykh stalei s primeneniyem vakuumnoi termoobraborki. Moscow, 1984. 24 p [Diffusion boriding of high-alloy steels using vacuum treatment].
146. Sharivker S. Yu., Kaplina G.S. et al. Borirovaniye austenitnykh nerzhaveyuschikh stalei s predvaritelnym zhelezneniyem // MiTOM. 1974. №2. pp. 59–61 [Boriding of austenite stainless preliminarry-ironed steels].
147. Boyarsky V.N. Vosstanovleniye detalei selskokhozyaistvennoi tekhniki zhelezoboridnymi pokrytiyami. Moscow, 2000. 20 p. [The restoration of agriculture machinery details with iron-boride coatings].
148. Lyakhovich L.S. Diffuzionnoye uprochneniye metallov s predvaritelno nanesennymi metallicheskimi pokrytiyami // Zaschitnye pokrytiya na metallakh. Kiev, 1977. Iss. 11. pp. 18–20 [Diffusion hardening of metals with preliminary appied metallic coatings].
149. Lyakhovich L.S., Kukharev B.S., Devoino N.G., Levitan S.N., Pantelenko F.I. Diffuzionnoye uprochneniye metallov s predvaritelno nanessennymi metallicheskimi podlozhkami // Metallurgiya. Minsk, 1976. pp. 18–20 [Diffusion hardening of metals with preliminary applied metallic substrates].

150. Rashov Nikola. // Proc. 18-th Int. Conf. Heat. Mater. Detroit, Mich., 1980 (Metals, Park, Ohio), pp. 25–28.
151. Yanyuk N.F. O perenose veschestva elektricheskim polem pri diffuzionnykh protsessakh // Zaschitnye pokrytiya na metallakh. Kiev, 1980. Iss. 14. pp. 26–29 [On the transition of substances by the electric field during diffusion processes].
152. Shumagov A.I., Voronin V.G., Volobuev V.A., Ermolov V.I. Sposob borirovaniya stalnykh izdelii. Cert. № 1578227 C23C. Publ. 1990. №26 [The method of boriding of steel details].
153. Krukovich M.G. Printsipy postroyeniya mnogokomponentnykh diagram sostoyaniya v koordinatakh koncentratsiya-temperatura. Moscow, 1994. 24 p [The principles of constructing multi-component equilibrium diagrams in the coordinates concentration-temperature].
154. Lyakhovich L.S., Krukovich M.G., Turov Yu. I. Progressivnyi metod borirovaniya instrumenta i tekhnologicheskoi osnastki s ispolzovaniyem past. Minsk, 1977. 32 p [A progressive method of boronizing instruments and technological processing with the use of pastes].
155. Kidin I.N., Andryushechkin V.I., Volkov V.A. Elektro-khimiko-termicheskaya obrabotka metallov i splavov. Moscow, 1978. 320 p [Electro-chemical thermal processing of metals and alloys].
156. Andryushechkin V.I., Bashnin A.I. Borirovaniye stali 38XHM // MiTOM. 1981. №4. pp. 23–25 [Boriding of steel 38XHM].
157. Shalya M.A. Snizheniye khrupkosti borirovannogo sloya // MiTOM. 1969. №8. pp. 69–72 [The reduction of brittleness in boronized layers].
158. Andryushechkin V.I., Kidin I.N. Skorostnaya elektro-khimiko-termicheskaya obrabotka zheleza i stali // Zaschitnye pokrytiya. Kiev, 1967. Iss. 1. pp. 142–148 [Fast electrical-chemical-thermal processing of iron and steel].
159. Tkachev V.N., Polyakov B.N., Sidelnikova V.I. Mikrostruktura i svoistva boridnykh sloev, obrazuyuschikhsya pri induktsionnom nagreve stali // MiTOM. 1979. №1. pp. 2–5 [The microstructure and properties of boride layers formed at the induction heating of steels].
160. Shashina L.T., Makharov D.M. Sostav obmazki dlya borirovaniya stalnykh izdelii. Cert. № 1216249 C23C. Publ. 1986. №9 [The composition of a coating for boriding of steel details].
161. Борисенко А.П., Минина З.Д., Дубко В.Д. Борирование из паст деталей технологической оснастки // МиТОМ. 1976. №7. С. 39–40. Borisenko A.P., Minina Z. D., Dubko V.D. Borirovaniye iz past detalei tekhnologicheskoi osnastki // MiTOM. 1976. №7ю pp. 39–40 [Boriding from the pastes of details used for technological treatment].
162. Bayazitov M.I., Volkov V.A., Valiev A.A. Borirovaniye iz past v usloviyakh pechnogo nagreva // MiTOM. 1976. №5. pp. 53–54 [Boriding from pastes in the conditions of furnace heating].
163. Zozulya V.F., Legeida V.N., Zozulya L.V. Vliyaniye sostava borsoderzhaschei pasty na diffuzionnoye nasyscheniye stali 20X1M1ФTP borom // Kharkov Polytechnical Institute Bulletin. 1987. №250. pp. 51–63 [The influence of the composition of a boron-containing paste on the diffusion saturation of steel 20X1M1ФTP with boron].
164. Volkov V.A., Aliev A.A. Borirovaniye iz past v usloviyakh skorostnogo elektronagreva // Zaschitnye pokrytiya na metallakh. Kiev, 1975. Iss. 9. pp. 53–57 [Boriding in pastes in the conditions of speedy electrical heating].
165. Voronchikhina T.I., Rabek R.A., Poilov A.E., Tyutin N.I. Sostav dlya borirovaniya stalnykh izdelii. Cert. № 876774 C23C. Publ. 1981. №40 [The composition for boronizing steel details].
166. Voronchikhina T.I., Rabek R.A. Obmazka dlya borirovaniya. Cert. № 775174 C23C. Publ. №40 [The coating for boriding].
167. Pokhmurskiy V.I., Lizanets M.I., Kozub V.V., Uzyuma I.N., Tolstova S.V. Razrabotka tolstykh diffuzionnykh pokrytii dlya povysheniya abrazivnoi iznosostoikosti detalei mashin // Zaschitnye pokrytiya na metallakh. Kiev, 1975. Vol. 9. pp. 150–152. [The development of thick diffusion surfaces for the improvement of abrasive wear resistance in machinery details].

168. Gabinskaya G.D., Kolesnik N.M., Maryushkin L.G., Tovpyga V.N. Sposob diffuzionnogo borirovaniya detalei. Cert. № 937531 C23C. Publ. 1982. №23 [The method of diffusion boriding of details].
169. Woska Rainer. // Bander-Blech-Rohre. 1981. V. 22, №6. P. 154–155.
170. Tsipas D.N., Perez-Herz C. // J. Vater. Sci. Lett. 1982. Bd. 1, №7. S. 298–299.
171. Zemskov G.V., Vitchenko V.A., Guschin L.K. Povysheniye korrozionnoi stoikosti stali diffuzionnym nasyscheniyem iz past pri nagreve TVCH // Zaschita metallov. 1975. Vol. 11, №5. pp. 638–640 [The improvement of corrosion resistance in steels by diffusion saturation from pastes at HFC heating].
172. Castello Gonsales Iose, Garcia Dominignes Asdubal, Sitkevich M.V. // Constr. mat. 1987. V. 12, №2. P. 74–76.
173. Mitrokhovich N.I. Poverkhnostoye upravleniye press-form dlya litya pod davleniyem mednykh splavov // I Russian metallurgy science meeting proceedings. Part 2. Penza, 1993. pp. 38–40 [The surface control of press-forms for pressure molding of cuprum alloys].
174. Belskiy E.I., Sitkevich M.V. Sostav dlya borirovaniya stalnykh otlivok. Cert. № 755890 C23C. Publ. 1980. №30 [The composition for boriding of steel molds].
175. Sitkevich M.V., Liventsev V.E. Issledovaniye boridnykh pokrytii, poluchennykh v usloviyah povyshennykh temperatur // Metallovedeniye v mashinostroenii. Minsk, 1983. pp. 9–13 [The study of boride surfaces obtained in conditions of increased temperatures].
176. Lyakhovich L.S., Isakov S.A., Altshuler S.A. Intensifikatsiya borirovaniya iz past v tleyuschem razryade // Chernaya metallurgiya. 1986. №2. pp. 56–59 [The intensification of boriding from pastes in a glow current].
177. Khan Li Min, Liang Ling, Chi Yuang, Zhen Minxiong // Jiechu rechuli. 1991. №6. pp. 19–23.
178. Zhen Kaixian, Kyond Dezhuan, Pan Yong // Jiechu rechuli. 1990. №4. pp. 6–9.
179. Chi Yuong, Seoung Ikai, Yuang Lieu // Jiechu rechuli 1990. №11. pp. 35–37.
180. Poletika I.M., Borisov M.D., Krayev G.V., Meita V.P., Vaisman A.F., Tolkovskii M.G. Uprochneniye stali legirovaniyem v puchke relyativistskikh elektronov // Fizika. 1993. Vol. 36. №3. pp. 57–63 [The hardening of steel with alloying in relative electron beam].
181. Afanasyev A.A., Belikov A.M., Komarchev I.M. Sposob poverkhnostnogo uprocheninya detalei mashin. Patent 2133183 Russia, C, C23C. Publ. 1997. №34.
182. Zabelin S.F. Osnovy tekhnologii i kineticheskoi teorii protsessov diffuzionnogo nasyscheniya stalei v usloviyah termotsiklicheskogo vozdeistviya na material. Saint-Petersburg, 2004. 41 p [The foundations of the technology and kinetic theory of diffusion saturation of steels in the conditions of thermal-circular impact on the material].
183. Turov Yu. V. Issledovaniye osobennostei formirovaniya i svoistv boridnykh pokrytii na stalyakh. Minsk, 1974. 25 p [The study of peculiarities of the formation and properties of boride coatings in steels].
184. Voroshnin L.G., Asadula Sharif, Turov Yu.V., Krukovich M.G. Sostav dlya elektroliznogo borirovaniya stalnykh izdelii. Cert. № 876773 C23C. Publ. №38, 15.10.81.
185. Sokolova V.A., Vorobyev G.M., Klimenko F.A., Zelenskiy G.V. Sposob elektroliznogo borirovaniya. Cert. № 964023 C23C. Publ. 1980. №37. [The method of electrolysis boriding].
186. Afanasyev A.A., Venevtsev Yu. M., Korotkov V.D. Sposob elektroliznogo borirovaniya uglerodistykh stalei. Cert. № 768854 C23C. Publ. №37 [The method of electrolysis boriding of carbon steels].
187. Spiridonova I.M., Bodrikov V.P. Sposob elektroliznogo borirovaniya stalnykh izdelii. Patent 2004616 Russia C23C. Publ. 1993. №45–346 [The method of electrolysis boriding of steel details].
188. Cert. № 964023 C23C. Publ. 1980. №37 [The method of electrolysis boriding of alloy steels].
189. Afanasyev A.A. Issledovaniye borirovaniya i borotsirkonirovaniya postoyannym i reversivnym tokom. Voronezh, 1970. 25 p. [The study of boriding and boron-zirconium treatment with constant and reverse currents].

190. Electrolysis method of hardening areas of metal surfaces. Pat. UK №4531 S25D. Publ. 1976.
191. Korotkov V.D. Issledovaniye elektroliznogo borirovaniya metallov. Voronezh, 1968. 25 p. [The study of electrolysis boriding of metals].
192. Ляхович Л.С, Брагилевская С.С., Ворошнин Л.Г. Жидкостное борирование предварительно никелированных сталей // Докл. АН БССР, Серыя «Фізіка-тэхнічн. навук». 1967. Т. 11, №2. С. 162–165. Lyakovich L.S., Bragilevskaya S.S., Voroshnin L.G. Zhidkostnoye borirovaniye predvaritelno nikelirovannykh stalei // BSSR Academy of Science Report, 1967. Vol. 11. №2. pp. 162–165 [Liquid boriding of preliminary nickeled steels].
193. Lyakhovich L.S., Voroshnin L.G., Karpenko D.P., Kosachevskiy L.N., Dolmanov F.V., Krukovich M.G. Sostav dlya nizkotemperaturnogo elektroliznogo borirovaniya. Cert. № 261086 C23C. Publ. 1970. №4.
194. Lyakhovich L.S., Krukovich M.G., Turov Yu. V. Novoe v khimiko-termicheskoi obrabotke stalei // Ekspress informatsiya. Metalloobrabotka. Minsk, 1976. 32 p [New trends in chemical-thermal processing of steels].
195. Lyakhovich L.S., Voroshnin L.G., Vasilyev L.A., Krukovich M.G., Turov Yu.V. Nizkotemperaturnoye elektroliznoye borirovaniye // MiTOM. 1974. №2. pp. 54–55. [Low-temperature electrolysis boriding].
196. Artemchuk V.G., Klyuev V.N., Breev V.F., Solodukhin V.S. Sostav dlya elektroliznogo borirovaniya stalnykh izdelii. Cert. № 403613 C23C. Publ. 1987. №25. [The composition of electrolysis boriding of steels details].
197. Lyakhovich L.S., Kosachevskiy L.N., Turov Yu. V., Krukovich M.G., Voroshnin L.G. Sostav dlya elektroliznogo borirovaniya. Cert. № 261086 C23C. Publ. 1974. №16 [The composition for electrolysis boriding].
198. Krukovich M.G. Odnofaznoye borirovaniye stalei // Poverkhnostnye metody uprochneniya metallov i splavov v mashinostroenii. 1983. pp. 81–84 [One-phase boriding of steel].
199. Bomoni A., Habersaat R., and Bienvenu G. // Surf. Technol. 1978. V. 6. P. 313–319.
200. Matiasovsky K., Paucirova M.C., Felner P., and Makyta M. // Surf. Coat. Technol. 1988. V. 35. P. 133–149.
201. Ilyuschenko N.G., Chernov Ya. B., Anfinogenov A.I. et al. Sostav dlya borirovaniya izdelii iz chernykh i tsvetnykh metallov i splavov. Cert. № 953002 C23C. Publ. 1982. №31. [The composition for boriding of details made from black and precious metals and alloys].
202. Polikarpov E.M., Mikhalin V.M., Shipilov V.D. et al. Sostav dlya zhidkostnogo borirovaniya stalnykh detalei. Cert. № 931802 C23C. Publ. 1982. №20. [The composition for liquid boriding of steel details].
203. Smonikova E.A., Sarmanova L.M. Issledovaniye vozmozhnosti zhidkostnogo borirovaniya bystrorezhuschikh stalei // MiTOM. 1982. №1. pp. 30–32. [The study of the possibility for liquid boriding in cutting steels].
204. Polikarpov E.M., Mikhaliv V.M. et al. Sostav dlya zhidkostnogo borirovaniya stalnykh detalei. Cert. № 1021618 C23C. Publ. 1983. №21 [The composition for liquid boriding of steel details].
205. Hosokawa K., Yamashita T., Veda M., and Seki T. // Kinzoku Hyomen Gitjutsu. 1972. V. 23, №4. P. 211–216.
206. Smolnikov E.A., Sarmanova L.M., Knyshev E.A., Lipinskii I.E. Sposob podgotovki elektrodnoi vanny dlya bezelktroliznogo borirovaniya. Cert. №1206335 C23C. Publ. 1986. №3 [The method for preparing an electrode bath for non-electrolysis boriding].
207. Krukovich M.G, Stasevich G.V., Levchenko G.M. Sostav dlya zhidkostnogo borokhromirovaniya stalnykh izdelii. Cert. №901351 C23C. Publ. 1982. №4 [The composition for liquid boron-chromizing of steel details].
208. Krukovich M.G., Stasevich G.V. Vanna dlya zhidkostnogo borirovaniya stalnykh izdelii. Cert. № 773141 C23C. Publ. 1980. №39 [The bath for liquid boriding of steel details].
209. Pirizhnyak N.I. Vybor optimalnykh elektrokhimicheskikh vosstanovitelei dlya odnofaznogo borirovaniya // Metallurgiya. Minsk, 1986. Iss. 20. pp. 32–34 [Choosing optimal electro-chemical reducing agents for one-phase boriding].

210. Liu Lei, Lib, Ling. Jenshu rechuli. // Heat Treat. Metals. 1994. №1. P. 11–15.
211. Takeo Oki, Khideyuki Kanemazu, Motona Hamazaki // Pureningu To Kotingu. 1980. Vol. 10, №2–3. pp. 92–97.
212. Golubin A.V., Banas I.P., Goncharova G.I., Koloditskii M.E., Borisov V.A. Povysheniye iznosostoikosti detalei treniya iz splavov 58НХВКТБЮ borirovaniyem // Aviatsionnaya promyshlennost. 1992. №12. pp. 35–37. [The improvement of wear-resistance in friction details made from 58НХВКТБЮ alloys with boriding].
213. Atanassova I. // Int. Congr. Heat Treat. Metals, Warsaw, 1981, Part. 1, S. 236–238.
214. Arai Toru, Hendo Nozi. Boriding of steel. Patent. № 55–51510 C23C 9/04. Japan. Publ. 24.12.80.
215. Hill Vernon L., Stapleton Thomas F. Method of Boronizing /General Motors Corp.// Pat. USA, CL. 148-b, II (C23C), № 3201286, (15.03.62), Pub. 17.08.65.
216. Panich G.G. Zhidkostnyi metod diffuzionnogo nasyscheniya stalei nemetallicheskimi elementami. Minsk, 1968. 24 p [The liquid method of diffusion saturation of steels with non-metal elements].
217. The method of surface hardening of non-ferrous metals and alloys by boriding. Patent Japan 12A34 (C23C), № 721. Publ. 10.01.72.
218. Lyakhovich L.S., Kosacheskiy L.N., Krukovich M.G., Dolmanov F.V., Karpenko D. P. Sostav vanny dlya diffuzionnogo borirovaniya. Cert. № 345242 C23C. Publ. 1972. №22 [The compositions for baths for diffusion boriding].
219. Lyakhovich L.S., Kosachevskii L.N., Kucheryavyi A.G., Krukovich M.G., Turov Yu.V. Rasplav dlya zhidkostnogo bezelektroliznogo borirovaniya. Cert. № 398705 C23C. Publ. 1973 №38 [The smelt for liquid non-electrolysis boriding].
220. Lyakhovich L.S., Kosachevskii L.N., Kucheryavyi A.G., Krukovich M.G., Turov Yu.V. Sostav dlya zhidkostnogo borirovaniya. Cert. № 415337 C23C. Publ. 1974 №6 [The composition for liquid boriding].
221. Ilyuschenko N.G., Anfinogenov A.I., Plotnikova A.F., Belyaeva G.I. Sposob borirovaniya metallov i splavov. Cert. № 206270 C23C. Publ. 1968 №3 [The method of boronizing metals and alloys].
222. Chernov Ya.B., Anfinogenov A.I., Ilyuschenko N.G. Sposob khimiko-termicheskoi obrabotki stalnykh detalei. Cert. № 2004617 Russia C23C. Publ. 1993 №45–46. [The method of chemical-thermal processing of steel details].
223. Iosel Yu.A., Kaganov E.S., Strunskii M.G. Voprosy rascheta i modelirovaniya elektrokhimicheskoi antikorrozionnoi zaschity sudov. Leningrad, 1965. 271 p. [The issues of calculating and modeling electrochemical anticorrosion protection of vessels].
224. Pleskach A.M., Dubina I.D. Sostav dlya borirovaniya stalnykh izdelii. Cert. № 971911 C23C. Publ. 1982. №41 [The composition for boriding steel goods].
225. Duryagina Z.A., Pastukhova L.V., Kitsak M.I. Diffuzionnoe borirovaniye press instrumentov iz stali Di-22 // Zaschita metallov. 1987. Vol. 23, № 2. pp. 319–321 [Diffusion boriding of press tools made from steel Di-22].
226. Akulinicheva E.V. Mekhanizm obrazovaniya i strukturnye osobennosti borirovannykh sloev, poluchennykh pri kristallizatsii stalnykh otlivok // MiTOM. 1979. №1. C. 5–7 [The formation mechanism and structural peculiarities of boronized layers obtained while crystallizing steel moulds].
227. Pons M., Caillet M., Galerie A. La boruration superficiel du fer par faiseau laser // Mater. et techn. 1985. V. 73, №12. P. 699–708.
228. Fedosienko S.S., Brover G.I., Varavka V.N. Lazernoe legirovaniye metalloobrabatyvayuschikh instrumentov // Tekhnologiyya i organizatsiya proizvodstva. Kiev, 1988. №1. pp. 46–47 [Laser alloying of metal processing tools].
229. Bloyce A., Bell T. // Laser Treat. Mater. Eur. Conf. Bad Nauheiv. 1986. — Oberursel, 1987. S. 251–264.
230. Lakhtin Yu.M., Kogan Ya. D., Buryakin A.V., Fedin V.M., Zakharov B.V., Minkevich A.N. Poverkhnostnoye uprochneniye stali izlucheniem lazera // Poverkhnostnye metody

uprochneniya metallov i splavov v mashinostroenii. Moscow, 1983. pp. 65–67 [The surface hardening in steels by laser beams].

231. Lyakhovich L.S., Isakov A.A., Kartoshkin V.M. Opredeleniye uslovii borirovaniya stalei pri nagreve lazernym izlucheniem // MiTOM. 1985. №11. pp. 12–14 [The determination of boriding conditions for steels under laser heating].
232. Tanenko I.A., Levchenko A.A., Guiva R.T., Guiva V.A., Sittsevaya E.Yu. Lazernoye borirovaniye vysokoprochnogo chuguna // Fizika i khimiya obrabotki materialov. 1991. №5. pp. 89–95 [Laser boriding of high-resistant cast iron].
233. Bernstein A.M., Yandimirskiy E.M., Ermakova O.A., Somov B.B. Poverkhnostoye lazernoye legirovaniye stalei i chugunov metodom inzhektsii karbida bora // Elektricheskaya obrabotka materialov. 1991. №3. pp. 25–28 [Surface laser alloying of steels and cast irons with injections of boron carbide].
234. Devoino O.G., Kardapolova M.A., Panich G.G. Vliyaniye parametrov lazernogo legirovaniya na strukturu borotsementovannykh sloev // Lazernaya obrabotka stalei i titanovykh splavov. Perm, 1991. pp. 33–38 [The influence of laser alloying parameters on the structure of boron-cemented layers].
235. Rabitch K., Ebner R., Major B. Borid layers alloying of high-speed steel // Sci. Met. And Mater. 1994. 30, №11. S. 253–258.
236. Gavrilov V.B., Pelenev R.S. Lazernoye legirovaniye poverkhnostei stalei i titanovykh splavov. Perm, 1991. pp. 45–50 [Laser alloying of steel and titan alloy surfaces].
237. Malinov L.S., Kharlamova E.Ya., Kolechko A.A., Agarkov V.Ya., Lisakovich A.V. Lazernoye uprochneniye shtampovogo instrumenta // Tekhnologiya i organizatsiya proizvodstva. Kiev, 1986. №2. pp. 46–48 [Laser hardening of stamping tools].
238. Pogrebnyak A.D., Shumakova N.I. Modifikatsiya pokrytii pod deistviem nizkoenergeticheskogo silnotochnogo puchka // FiKHOM. 1999. №6. pp. 13–16 [The modification of surfaces under the influence of low-energy strong-flow beam].
239. Bakhtin S.V., Kozyr I.G., Sharshakov I.M., Shatov Yu.S. Formirovaniye karbidokhromovykh sloev na uglerodistykh stalyakh s ispolzovaniem elektronnogo puchka // FiKHOM. 1995. №3. pp. 35–38 [The formation of carbide-chrome layers in carbon steels while using electron beam].
240. Sizov I.G. Razrabotka nauchnykh osnov i tekhnologii elektronno-luchevogo borirovaniya zhelezouglerodistykh splavov s polucheniem na poverkhnosti boridov tugoplavkikh metallov. Dissertatsiya doktora tekhnicheskikh nauk. Moscow, 2003. 305 p [The development of scientific foundation and technology of electron-beam boriding of iron-carbon alloys aimed at obtaining high-melting point metal borides on the surface. Doctorate dissertation].
241. Sizov I.G., Smirnyagina N.N., Semenov A.P. Osobennosti elektronno-luchevogo borirovaniya stalei // MiTOM. 1999. №12. pp. 8–11 [The peculiarities of electron beam boriding of steels].
242. Sizov I.G., Prusakov B.A., Novakova A.A., Kornikova A.A. Sposob borirovaniya uglerodistoi stali. Patent RF № 2293789 [The method of boronizing carbon steel].
243. Krukovich M.G., Levchenko G.M. Sostav dlya borirovaniya stalnykh izdelii. Cert. № 876774 C23C. Publ. 1981. №40 [A composition for boriding steel details].
244. Krukovich M.G., Levchenko G.M. Sostav dlya khimiko-termicheskoi obrabotki stalnykh izdelii. Cert. № 885344 C23C. Publ. 1981. №44. [A composition for chemical-thermal processing of steel details].
245. Krukovich M.G. Sovershenstvovaniye tekhnologii polucheniya iznosostoikikh diffuzionnykh sloev // International MIIT Bulletin. Moscow, 1983. Iss. 724. pp. 91–95 [The development of obtaining wear-resistant diffusion layers].

246. Krukovich M.G. Issledovaniye zhidkostnykh bezelektroliznykh protsessov khimiko-termicheskoi obrabotki. Dissertaciya kandidata tekhnicheskih nauk. Minsk, 1974. 298 p. [The study of liquid non-electrolysis processes of chemical-thermal processing. Candidate of science dissertation].
247. Lyakhovich L.S., Kosachevskii L.N., Krukovich M.G., Turov Yu.V., Vasilyev L.N., Levitan S.N. Sostav dlya borirovaniya. Cert.№ 413115 C23C. Publ. 1976 №17 [The composition for boriding].
248. Lyakhovich L.S., Kosachevskii L.N., Krukovich M.G., Turov Yu.V., Levitan S.N. Sostav dlya zhidkostnogo borirovaniya. Cert. №445710 C23C. Publ. 1974 №37 [The composition for liquid boriding].
249. Lyakhovich L.S., Krukovich M.G., Turov Yu.V. Rasplav dlya borosilitsirovaniya. Cert. №548658 C23C. Publ. 1977 №8 [The smelt for boron-silicon treatment].
250. Lyakhovich L.S., Stasevich G.V., Krukovich M.G., Turov Yu.V. Sostav dlya boroalitirovaniya stalnykh izdelii. Cert. № 590369 C23C. Publ. 1978 №4 [The composition for boron-aluminizing of steel details].
251. Lyakhovich L.S., Kosachevskii L.N., Krukovich M.G., Turov Yu.V. Sostav rasplava dlya nizkotemperaturnogo borirovaniya. Cert. № 411167 C23C. Publ. 1974. №2 [The composition of the smelt for low-temperature boriding].
252. Lyakhovich L.S., Kosachevskii L.N., Krukovich M.G., Turov Yu.V., Voroshnin L.G. Sostav rasplava dlya nizkotemperaturnogo borirovaniya. Cert. № 405971 C23C. Publ. 1973. №45 [The composition of the smelt for low-temperature boriding].

Chapter 9
The Structure Compositeness as the Foundation for the Plasticity of Boronized Layers

Abstract The chapter describes the effective technologies used to improve the plasticity of layers based on metal borides which determine the composite structure containing a softer component of a-solid solution and inclusions of borides positioned freely or in a specific order. The layers with composite structure lack the main disadvantage of boride layers—high brittleness—due to the created plasticity reserve. A certain amount of borides (no less than 60%) preserves the high hardness and wear-resistance. The chapter also studies the perspective methods for obtaining composite structure which are based on the use of laser and electron-beam heating. The authors analyze the formation peculiarities of different morphology under the use of boron carbide and amorphous boron as well as the influence of different heating parameters on the structure and mechanical properties of boronized layers in different brands of steels and iron casts. The chapter also includes the description of formation mechanisms for boronized layers under laser and electron-beam heating from pastes and coatings and additional processing of preliminary obtained boronized layers (combined boriding). The unique structures of boronized layers, first obtained by the authors, are presented. The obtained data allows to significantly expand the usage of boronized layers and recommend the process to be used to harden the details of machines and tools working under the conditions of dynamic loadings, e.g. shafts, chain sheaves, gears, star wheels, cam gears, stamps etc.

One of the effective methods of improving the plasticity of iron boride layers is the formation of layers with a composite structure. The composite structure of boronized layers is the structure which includes, apart from borides, one or several additional phases located in the layer in a random or specific order. These additional phases normally have lower hardness in comparison with the borides. Layers, exclusively containing iron borides, usually have an increased deterioration level in conditions of shock load despite their hardness. The reason for this process is chipping of separate areas of the layer due to its high brittleness. Layers with a composite structure lack this disadvantage. The additional phases with lower hardness increase the general plasticity and the borides make the layer more wear-resistant, i.e. composite layers are formed according to the Charpy principle;

M.G. Krukovich et al., *Plasticity of Boronized Layers*,
Springer Series in Materials Science 237,
DOI 10.1007/978-3-319-40012-9_9

the surface acquires a micro-relief during a wearing process due to its structural heterogeneity. Moreover, composite boronized layers have a rather low stress state due to the relaxation of stresses on the border "borides-solid solution".

Thus, the composite structure in a boronized layer allows to increase its plasticity and preserve its high wear-resistance. The composite structure is acquired by different methods; the most perspective methods are based on laser and electron beam heating. The main purpose of using concentrated energy sources is the construction of a layer with divided borides. The compact needle or globular structures of borides should be replaced by the disperse structure with random orientation of borides. The use of these heating methods allows to carry out processing in the mode of thermal circulation which also leads to the formation of the composite structure according to the diffusion and diffusion-crystallization formation mechanisms.

One of the structural elements of composite layers is the eutectics component. The important stage of predicting the general structure of a layer and the occurrence of the eutectic component is obtaining true data on temperature-concentration conditions on the processed surface which are determined by polythermal sections of the multicomponent equilibrium diagram schemes (Chap. 5). The temperature of the eutectic reaction and eutectic concentrations of components in unknown multicomponent systems must be calculated with statistical methods (Appendix A, [1]) which involve the data on melting points of the phases comprising the eutectics.

According to the existing data [2, 3], the eutectics formation is a complicated physical-chemical process which results in the formation of the special colonial structure of crystals due to their simultaneous crystallization at certain temperature and the chemical composition of solid solution. The appearance and growth of colonies is controlled by the leading eutectics phase which determines the type of eutectics. The leading phase is characterized by multiple directed and other stronger bonds between atoms in crystals. The second and further phases are formed on the surface of the leading phase. Thus, the leading phase for the boride eutectics formation is the metal boride and the forming structure of eutectics in layers is determined by the crystal structure of borides, temperature-concentration conditions in certain layer's zones and cooling speed. The structure of the composite layer in general also depends on its state during treatment. In particular, a liquid crystal state of the layer (as a result of diffusion exchange) causes the simultaneous melting of new areas of the matrix and the boride crystallization which are spontaneously oriented in the liquid solution and solid solution (after complete crystallization). This creates conditions for disperse structures which provide the formed surface with a high reserve of plasticity, hardness and wear-resistance.

9.1 Laser Boriding

9.1.1 Peculiarities of Boriding with Laser Heating

The laser chemical-thermal treatment involves the surface alloying of a part of the metal in solid or melted state, i.e. under the diffusion or diffusion-crystallization formation of new alloys. Chemical-thermal processing, understood this way, is the topic of many studies on laser hardening [4–13]. The main conditions for the layer formation during chemical-thermal processing is the occurrence of diffusion and gradual transformation in chemical compositions (saturation element concentration) going from the surface to the core of the detail. This statement may be applied to both mechanisms of the surface layers' generation. Layers, obtained as a result of total melting of the surface areas, may be also attributed to the groups of surface alloying or hard facing [14]. According to the authors of [15], hard facing is a method of applying coatings; coatings are formed after the liquid phase solidifies. The liquid phase is obtained as a result of melting of the applied and basic material caused by the external source of energy. However in the case of laser hardening, it's important to remember about diffusion processes in the layer as a result of the concentration gradient of alloy element and in the zone of thermal impact as a result of alloy elements redistribution. These processes form the layer and sublayer and determine their structure, strain state, plasticity and exploitation conditions of the hardened detail.

9.1.2 The Influence of Laser Heating on the Structure of Boronized Layers

Studies on the formation of boride layers with reduced brittleness under laser heating have been carried out on different materials: carbon (45, У8, У10) and alloy (4Х4М2ФВС, 12Х18Н10Т, 40Х, 20Х13) steels, cast irons (СЧ20, ВЧ70), titan alloys (ВТ3-1) etc. [16–19].

The technology of laser boriding includes laser emission processing of the surface with boron-containing coatings. The brittleness reduction in boronized layers under laser heating is reached by creating a layer with heterogeneous (composite) structure [20–25] or its special alloying by introducing certain elements into the boron-containing coating [26–28].

Heterogeneous structures are obtained when laser heating causes the surface layer to fuse. It allows to get boronized layers of reduced brittleness in high-strength iron casts [4]. The X-ray diffraction analysis showed that laser boriding makes the surface layer to contain 4 phases: FeB, Fe_2B, α-phase and boron-cementite of $Fe_3(B, C)$ solid solution with alternating boron and carbon concentrations on the base of cementite. The primary crystals of boron-cementite in the layer are in the form of plate compounds—flat dendrites, which in perpendicular sections look as thin lines. The structural-free boride Fe_2B crystals are shown as stem crystals in different forms of squares, rhombs, triangles etc. (all possible sections of a tetragonal prism) in the cross-section.

Eutectic layer components are represented by two morphological types: independent zones in the layer (eutectic component) and combined with an excessive phase (pre-eutectic or over-eutectic component). The first type has a distinct cell substructure with typical coarsening of the cell border structure; the second type of eutectics is characterized by distinct basic plates of boron-cementite forming eutectic colonies. The eutectics can be characterized by different dispersability and different quantitative correlation between phases both as a part of one layer or in different layers. Apart from these eutectics, high cooling speeds may cause the formation of two more types of eutectics in local areas: the first one is double, the second—triple sectoral. The structure of the layers in all cases is heterogeneous—both in volume, width and depth. Despite a significant variety of heterogeneity in the layer's composition, there is a distinct dependence between the boron content reduction in deeper layers and the increase of carbon content going from the core to the surface-substrate which is manifested in the consequent appearance of structure elements. The mentioned heterogeneity of the structure combining solid and softer components decreases the layer's brittleness after laser boriding and creates good anti-friction conditions.

The heterogeneous structures after laser boriding are studied in work [5]. The laser boriding (like in studies [20–23]) was carried out in boron-containing coatings. The study samples were: ARMCO iron, steels 14XH3MA, 20X13, 40X, 45 and У8. Laser alloying was done by the stimulated emission CO_2-laser in the wide spectrum of processing modes: the voltage was changed from 1 to 3 kW, the speed of sample movement in the beam from 2.5 to 110 mm/s, the diameter of laser spot—from 1 to 6 mm, the coating width from 60 to 140 mkm. The X-ray diffraction analysis showed that laser boriding in the mentioned modes caused the formation of α-Fe, Fe_3B, Fe_2B and FeB in the hardened layer in ARMCO iron. Areas with low boron concentration in iron (determined by the mass-spectrometer analysis) obtain a boride eutectic structure where dendrites of α-solid solution of boron in iron grow. The eutectics has a plate composition; the plates' width is 0.08 mkm. The structure's microhardness is relatively low and is approximately 510–890 HV. It may decrease to 400–500 HV and increase up to 945–1000 HV in accordance with the boron and alloy elements concentrations in steel.

The increase of the boron content leads to the formation of the boronized layer of enhanced hardness. The microhardness of the hardened layer would be the boride eutectics where disperse light particles—iron borides—are distributed relatively evenly.

It was found out that all processing modes can be characterized by several basic structure combinations of eutectics and boron depending on the intensity of physical processes happening during the layer's formation.

- The borderline melting zone of the basic metal is characterized by eutectics; dendrites of α-solid solution of boron in iron are positioned in separate zones

between the border and the surface; Fe_2B boride of angled form is located closer to the surface. The microhardness of Fe_2B is 677–1010 HV (steel 20X13).
- The increase in boron concentration causes the transformation of Fe_2B: the form changes from angled to complicated branched.
- Maximum boron content leads to the formation of borides with a cellular form. Their brittleness and hardness (890–1790 HC) in steel 20X13 allows to conclude that they are FeB boride.

Some processing modes caused the formation of highly consistent, boron-containing areas with 1410–2570 HV microhardness.

The studies on the influence of chemical composition of the substrate in the layer's structure showed that main hardening phases in different steels are iron borides. The increase of carbon content in steel leads to the generation of boron cementite and the increase of chrome content causes the boride dispersability to decrease and change the structure of their boride eutectics. Simultaneously the decrease in brittleness takes place (in comparison with the boride layer obtained by the diffusion mechanism of traditional boriding).

Work [20] also mentions a special structure of the boride layer obtained by laser boriding. In comparison with the boronized layer generated with the diffusion mechanism, the structure of the boronized layer after laser boriding obtained by the diffusion crystallization mechanism consists of 3 characteristic zones. The boride zone including FeB, Fe_2B and high-boron phase (2500–2700 HV) is located near the surface. A specific gray-blue color of the phase inclusions, high microhardness and additional lines in the diffraction scheme of the boride zone indicated that this phase is a solid solution of iron in β-rhombohedric boron. Another zone containing FeB and Fe_2B borides (microhardness 1300–1600 HV) is located further. The borderline melting zone has a 10–25 mkm eutectic layer with 550–750 HV microhardness.

Thus, the use of laser heating is an effective instrument of obtaining boronized layers characterized by the heterogeneous structure and increased plasticity.

9.2 Electron-Beam Boriding

9.2.1 The Peculiarities of Boriding with Electron-Beam Heating

The peculiarities of laser boriding, described in Sect. 9.1.1, can be partly attributed to the electron-beam chemical-thermal processing. However it is necessary to take into account the specifics of electron-beam heating. Electron-beam heating, in comparison with laser heating, has a wider spectrum of impact: the penetration of surface layers by the electron beam increases its temperature and initiates diffusion processes in a given metal. This leads to the formation of layers based predominantly on diffusion processes. The regulation of electron-beam heating parameters can change the mechanism of layer formation from the diffusion to the diffusion-crystallization type.

Electron-beam chemical-thermal processing on metal surfaces causes different physical-chemical processes which are determined by the simultaneous influence of temperature, charged particles flux and alloy elements. The change in temperature is accompanied by the changes in the crystal lattice (the formation of point defects, changes in lattice parameters) which determine completely different (from the traditional chemical-thermal treatment) processes. Electron-beam chemical-thermal processing is carried out in the conditions of strong vacuum (10^{-3}–10^{-5} Pa) which also has a great significance for the procedure in relation to the formation of surface properties. The combination of temperature and point defects leads to the development of relaxing and diffusion processes, the formation of layers with specific structure which causes the special character of properties that are impossible to obtain in normal heating conditions.

The studies on electron-beam chemical-thermal processing show that the positive final result is not additive, i.e. is not a simple unity of effects from chemical-thermal and electron-beam processing. An integral effect of the influence on the structure and material's properties is achieved.

There are several advantages of electron-beam heating in comparison with laser heating [29–33]:

- the high ratio of electron beam absorption by the metal; it allows to effectively process the surface without using absorbing surfaces;
- the simplicity of speedy electron-beam scanning of the processed surface;
- the high working efficiency of the electron emitter—70–80%;
- the possibility of creating relatively compact electron-beam machines of dozens and hundreds kW capacity.

The electron beam which falls on the surface is partially reflected; the second part penetrates the detail causing its heating [34, 35]. For iron alloys, the energy volume transforming into heat in a thin layer of the detail (the beam falls vertically) is approximately 75%. Another part is lost with reflected electrons and is distributed to the x-ray and heat emission and secondary electrons [36, 37].

The electrons penetrating the surface layers cause, apart from heating, the formation of Frenkel couples—interstitials and vacancies [33, 36]. The formation of defects with crystal structure is determined by the parameters of the electron beam (accelerated stress, current density etc.). The range of experiments have showed [37–43] that the formation of multiple defects takes place at high accelerated stresses (hundreds of keV). The increase of current density (energy density) of the electron beam allows to accelerate the process of defect formation even at the accelerated stress of several dozens keV.

The study on the impact of low-energy strong-current electron beams on the structure and properties of diffusion chrome and chrome-nickel surfaces in steel CT3 was carried out on the machine for electron beam irradiation "Nadezhda-2" (a strong-current energy source; vacuum 10^{-3}–10^{04} Pa). The electron beam current density varied from 5×10^6 to 2.5×10^7 A/m^2; the impulse duration was 2 mks which correlated with the energy flow density between 4.5×10^4 and 1.2×10^5 J/m^2. The area of the beam section was significantly larger than the samples' size (about 25 cm^2) [44].

The electron-beam irradiation of burnt and unburnt samples with 1–8 mkm diffusion chrome coatings was accompanied by the insignificant reduction of the lattice's period in CCS solid solution (the occurrence of structural microstresses), splitting of line 1 (110) and redistribution of line intensities (accumulation of package defects). The result is the radiation-stimulated diffusion of elements going into the depth of the processed details accompanied by the formation of a diffusion layer.

The surface processing by the low-energy strong-current electron beam causes the formation of subgrains (1 mkm) inside the grains (the average size 6.1 mkm) [45]. The parameters necessary for this process are: the maximum of electron energy 15 keV, the energy density in the flow 5–8 J/cm^2.

The use of the electron flow (2.5×10^3 W/cm^2) during heating before soaking for 10 s increases the density of dislocation in comparison with furnace heating [46]. Thus, the improvement of electron beam current density (even at relatively small numbers of accelerated stress—20–30 kV) stimulates the defect formation.

Therefore electron-beam heating has better technological properties (in comparison with laser heating) for the formation of layers: apart from melting processes, the surface layers are characterized by the initiation of diffusion processes which determine the formation of composite layers according to the diffusion crystallization mechanism and diffusion mechanism (during thermal-circular processing).

9.2.2 The Influence of Electron-Beam Heating on the Structure of Boronized Layers

The use of electron-beam heating is possible at different stages of the process of layer formation:

- the direct use of electron-beam heating for boriding;
- the electron-beam processing of a preliminary generated boronized layer.

The direct use of electron-beam heating in boriding (electron-beam boriding). The study of the structure formation of boronized layers during the use of electron-beam heating was carried out in different compositions of saturation coatings applied on the processed surface [47].

The electron flow, directed at the surface, penetrates the surface zone of the sample through the coating layer. The penetration depth under the stress, accelerated up to 30 kW, reaches 4–5 mkm. 25% of the flow density is reflected, <1% is used for heat emission, <1% is used for x-ray emission, thermal-electron emission and secondary electrons, <10% is absorbed by the saturation coating. As a result of

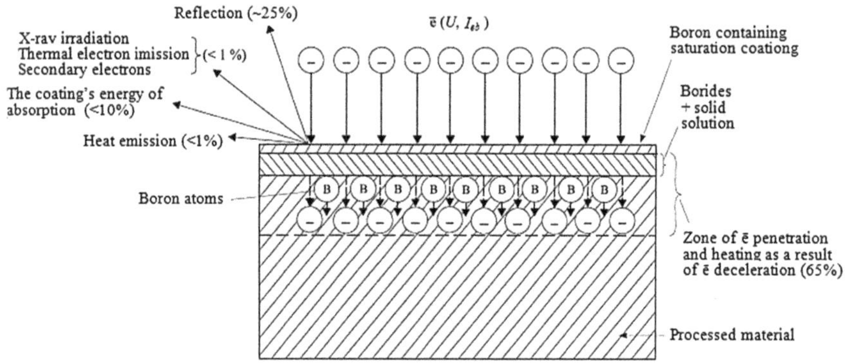

Fig. 9.1 The scheme of layers' formation under electron-beam boriding in vacuum (U—accelerated stress, meV; I_{eb}—electron beam current intensity)

the electron deceleration in the metal, their kinetic energy is transformed into a heat energy which leads to the temperature rise in the thin surface layer. The presence of crystal structure defects and high temperature encourage the boron atoms diffusion in the surface layers (Fig. 9.1).

The prolongation of electron-beam heating of the surface causes fusing of the surface. The electrons' kinetic energy activates the process of mixing between the thin fused metal layer and the coating layer. The total depth of the boronized layer increases and reaches 0.5–1.0 mm.

The study [47] used two-component powder mixtures B_4C + Cr and WC + B_4C in proportion B_4C/Cr = 7; 5; 3.5; 3; 2; 1.5 and WC/B_4C = 10; 8; 7; 6; 5; 4; 2; 1; three-component mixture WC + B_4C + CR was also used. The correlation in the three-component mixture was WC/B_4C = 7; 5; 4; 3; 2; 1; 2/3; 1/3; 1/6; 1/15, B_4C/Cr = 3 in all cases. The parameters of electron-beam processing: accelerated stress U = 1.4 meV, flow current I = 28 mA, reaming current I = 2.4 A. These irradiation parameters determined the flow energy absorbed by the steel surface as Q = 4.6 kJ/cm^2.

The study established that electron-beam alloying by clean boron carbide with the flux ($Na_2B_4O_7$) caused the surface layer's structure to be pre-eutectic and consist of primary austenite-martensite grains and ledeburyt. The X-ray diffraction analysis revealed the presence of α-phase and cementite Fe_3C; the lattice periods of Fe_3C are larger than the tabled data which can be explained by the partial replacement of carbon atoms by boron atoms and the formation of carbon-borides $Fe_3(C, B)$.

The introduction of chrome into the alloying mixture in proportion B_4C/Cr = 7 causes the increase of ledeburite amount. The x-ray diffraction analysis fixates the occurrence of $Cr_{23}(C, B)_6$ phase. A part of atoms is located in α-solid solution and, possibly, cementite.

The ratio B_4C/Cr = 5 signifies about the predominantly ledeburite structure with inclusions of over-eutectic structure with excessive carbon borides.

If the proportion is smaller (3.5), a typical over-eutectic structure with large primary carbon boride $Cr_{23}(C, B)_6$ elements is formed. Certain zones contain large volume parts of the second phase.

The number of carbon borides in the structure decreases if the correlation $B_4C/Cr = 3$; they become more disperse and evenly distributed in ledeburite.

The correlation $B_4C/Cr = 2.5$ leads to the appearance of small primary austenite grains in the structure (because of eutectics); $B_4C/Cr = 1.5$ causes the decrease of eutectics and makes the structure pre-eutectic. Simultaneously x-rays show the presence of the inter-metal connection line.

This mechanism of the structure forming may be explained by the border transition from one structural area of Fe-B-C-Cr diagram to another. The concentration border in the system was overcome at $B_4C/Cr = 3$.

The mentioned peculiarities of the structure have impact on mechanical properties (depending on the mixture composition): chrome causes the fall in hardness, wear resistance but improves the plasticity of the layer.

According to [48–52] there are two mechanisms of layer formation under boronizing: the diffusion and diffusion-crystallization mechanisms. The use of laser or electron-beam heating determines the possibility of predominant mechanism with layer formation while crystalizing the liquid phase [53–56] which is accompanied by diffusion processes.

The diffusion-crystallization mechanism on the surface's area or under-surface layer requires special conditions for the formation of liquid crystal state under the phase correlation which would sustain the detail's form and roughness of processed surfaces. Local fusing of the processed surface doesn't cause the risks of deformation that's why it's possible to use high processing temperatures. The surface roughness increases in these conditions. However the layer formation would be accompanied by the diffusion and crystallization. The processing temperature and the amount of the liquid phase generate either eutectic disperse structures of boronized layers or layers with eutectic structure. Eutectic structure layers are mainly generated under crystallization from the liquid phase [57, 58].

The structure of the layer obtained by the diffusion mechanism under electron-beam heating has 3 distinct zones.

Zone 1 is located on the surface; it is represented by light, pickling-resistant 5–7 mkm boride stripes.

Zone 2 is a 10–15 mkm dark zone which consists of the boride irregular-formed particles directed inwards. The study of these zones at large closing-ups (>X600) in the optical microscope showed that zone 2 continues zone 1. There are now distinct borderlines between them; the border reveals itself in a smooth change of the pickling ability.

Zone 3 is a transition zone. The structure and microhardness show that the foundation of this zone is α-solid solution which was obtained as a result of boron diffusion and iron and carbon atoms redistribution in the sublayer [58].

The diffusion-crystallization mechanism formation is accompanied by partial fusing of the surface without any significant changes in the surface form. The layer's depth is 200–250 mkm.

The complex impact of the electron flux in the conditions of high vacuum leads to the acceleration of diffusion and subsequently the formation of layers; this is especially important for the diffusion mechanism.

The structure of the layer, obtained through the crystallization of the liquid phase with diffusion processes under electron-beam heating, contains a main layer (up to 500 mkm wide) and a sublayer (35–40 mkm). The main layer has a distinct eutectic structure. The layer's formation leads to the total loss of the surface roughness which requires an obligatory mechanical treatment. The diffusion-crystallization mechanism is again used; the boronized layer consists of the separate disperse borides distributed in soft solid solution, it has a typical eutectic structure.

The analysis of the study results led to the development of electron-beam boriding which allows to obtain a composite structure of the boronized layer [59, 60]. It involves the use of saturation coating applied on the processed surface; then it is heated by the electron beam (2–2.5) × 10^3 W/cm^2 in vacuum 2 × 10^{-3} Pa for 4−5 min. The change of saturation coating composition and parameters of electron-beam heating allows to control the process of structure formation for the boride layer, change the layer's width and obtain layers with different structural components.

Boronized layers acquire eutectic structure when they are formed from coatings with average boron potentials at the temperatures exceeding eutectic (T_{eut} ~ 1174 °C for Fe-B system, 1100 °C for Fe-B-C system, 1022–1035 °C for Cr-Ni-Fe-B system) (Fig. 4.7). The X-ray diffraction phase analysis revealed that the layers comprise of FeB, Fe_2B iron borides and α-solid solution. The formation of layers from boron carbide (providing a high boron potential) leads to obtaining the layers containing round particles located on the surface and forming the eutectics structure.

Round particles are primary Fe_2B boride crystals. This fact correlates with the entropy criterion of sustainability for crystals of limited shape during the process of crystallization in the equilibrium conditions. According to the criterion, the entropy of fusion $\Delta S \leq 8.35$ J/(mol K) means that crystals have a round form. Fe_2B iron boride's entropy of fusion is $\Delta S = 6.29$ J/(mol K). However the conditions of the boriding crystallization are not equilibrium, due to the presence of boron concentration gradient, temperature gradient and directed heat withdrawal. This finally causes borides to obtain a slightly prolonged shape.

The boride layer formed from coating on the base of amorphous boron has a different structure (Fig. 9.2c, d). It comprises of differently shaped particles: rhombic, prism, dendrite. In this case the shape of the crystal is determined by the type of Fe_2B iron boride lattice. This boride has a body-centered tetragonal crystal lattice. Presumably, iron boride crystallites obtain a form of elementary cells during the formation of a layer from amorphous boron containing coating. As a result, primary boride crystals have rhombic, parallelogram etc. shapes, established by different incline angles of the crystal lattice (prism) to the surface of the lap [61].

Layers obtained by this method are not thick (up to 400–500 mkm). However they don't preserve roughness and sometimes the form of the processed surface. It requires additional mechanical treatment [62, 63].

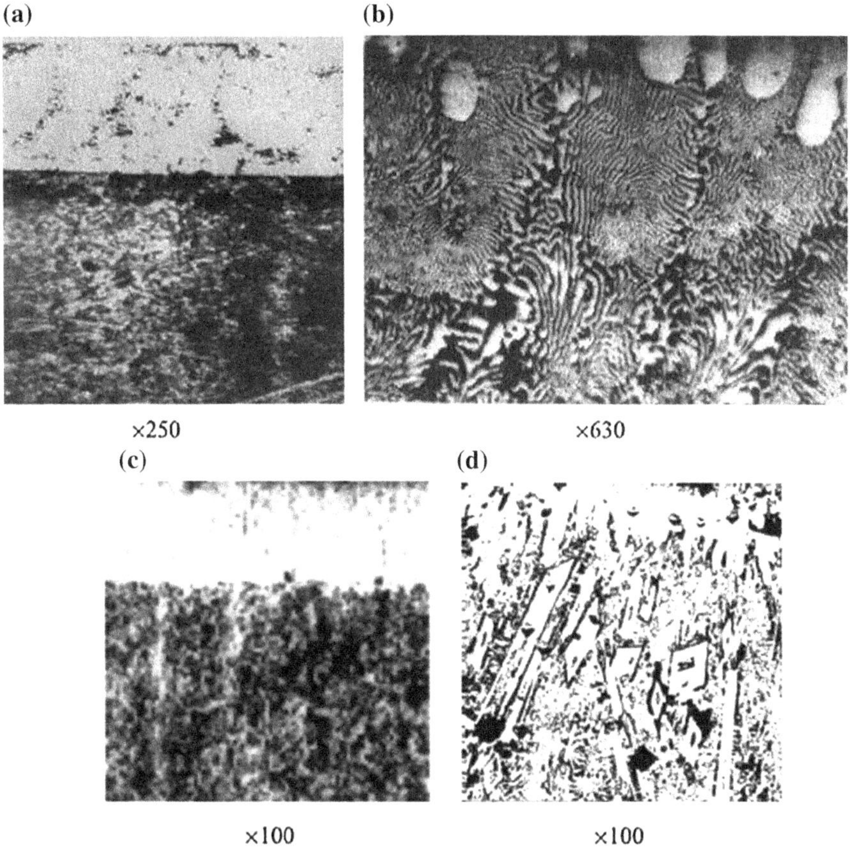

Fig. 9.2 The microstructure of boronized layers formed in steel 45 under electron-beam boriding from coating on the base of boron carbide B_4C (**a**, **b**) and amorphous boron (**c**, **d**)

The composite structure is obtained by boronized layers in tool and high-speed steels. However the layer's structure would be more compact with the inclusion of borides.

Traditional methods of boriding are not very useful for these steels as boride layers are fused under further heating for quenching (T_q = 1280 °C for P18, T_q = 1220 °C for P6M5) [64, 65]. Lower temperature quenching is not appropriate as it doesn't realize full potentials of high-speed steels.

The authors of [66–68] proposed a method of low-temperature boriding of high-speed steels at 550–727 °C. In their opinion, the use of this method is the most appropriate decision if it is combined with high-temperature tempering. The width of the boride layer in this case is 3–7 mkm. The wear resistance of the processed tools and details, produced from high speed steels, increases 2–2.5 times after low-temperature boriding.

Thus, the wear resistance of end mills, made from steel P6M5, increases twice between sharpenings as a result of electron-beam boriding at T = 560 °C for 4 h. The wear resistance of similar disc mills for treating flat spots increases by 2–3 times between sharpenings [58].

The intensity of the electron beam determines the temperature in the processed zone and the level of diffusion stimulation, therefore different intensity levels form different layer structures. In particular, steel P6M5 can generate the following structures.

- The surface power density of the electron beam—$P_d = 2.8 \times 10^3$ W/cm^2—for 4 min causes the formation of a layer as a result of deep fusing which determined its construction (Fig. 9.3d). The directed crystallization is obvious; it manifests itself in the orientation of primary and secondary axis of linear dendrites towards the heat withdrawal. The construction of dendrites (a chain of separate globules) demonstrates the interrupted manner of their formation. The layer's width reaches 100–110 mkm.
- At the surface power density of electron beam—$P_d = 2.5 \times 10^3$ W/cm^2—the layer predominantly consists of star dendrites (Fig. 9.3c) at the width 70–80 mkm.
- $P_d = 2.2 \times 10^3$ W/cm^2 causes the surface layer to also contain star dendrites but their amount isn't predominant (Fig. 9.3b). The layer's width is 40–50 mkm.

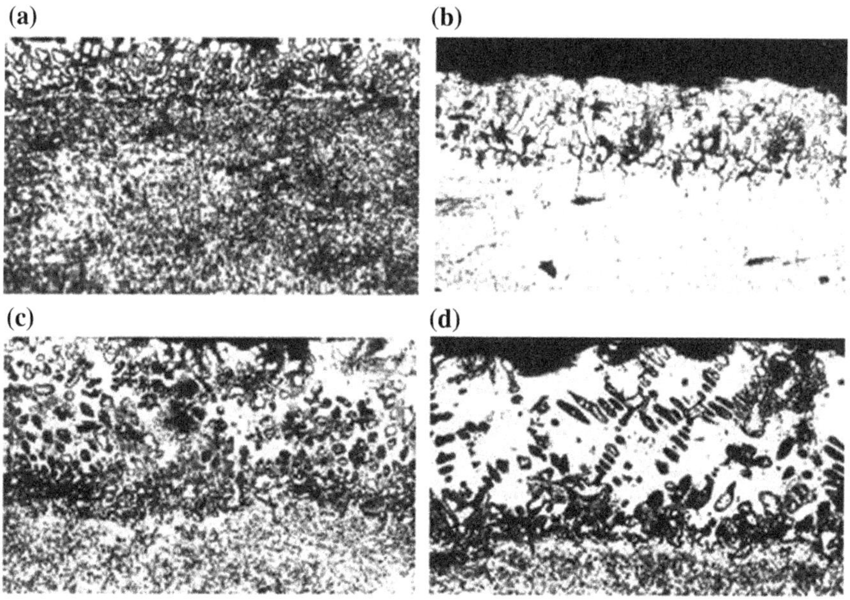

Fig. 9.3 The layer's structure in steel P6M5 after electron-beam boriding (×320) [58]: **a** $P_d = 2.0 \times 10^3$ W/cm^2; **b** $P_d = 2.2 \times 10^3$ W/cm^2; **c** $P_d = 2.5 \times 10^3$ W/cm^2; **d** $P_d = 2.8 \times 10^3$ W/cm^2

Fig. 9.4 The microstructure of the cutter (steel P6M5) after electron-beam boriding (×100)

- $P_d = 2.0 \times 10^3$ W/cm^2 makes the layer's composition to combine separate boride and carbon boride inclusions (Fig. 9.3a). The distribution zone of inclusions is 20–23 mkm.

The microstructure of the layer in steel P17, obtained at electron-beam boriding of the cutter without surface fusing, is demonstrated in Fig. 9.4. A 8–20 mkm layer is formed on the surface after processing P18 plates with applied boron-containing coatings. It has the even width spread across almost all the processed surface. The processing time is 3 min.

These results of electron-beam boriding were obtained in high-speed steels after the final thermal processing (quenching and triple tempering). The measuring of matrix hardness near the layer showed that it doesn't change after boriding and is 61–63 HRC. It is explained by short-term heating of the surface during boriding, a certain depth of electron penetration and with low thermal conduction of high-speed steels. Finally the sublayer's zone which high-temperature tempering regime exceeded limits turned out to be minimal and didn't affect the hardness change.

Composite layers can also be obtained in cast irons after electron-beam boriding. Iron borides are the foundation of the layer (Fig. 9.5). The saturation coating consisted of amorphous boron and bonding material. The layers formation went according to the diffusion and diffusion-crystallization mechanisms.

The boride layer, obtained by the diffusion mechanism (Fig. 9.5a), doesn't have a distinct border. It has separate small graphite inclusions; their number increases towards the surface.

The boride layer, obtained by the diffusion-crystallization mechanism (Fig. 9.5b), has a distinct borderline «layer-matrix» and consists of borides of different shapes: needle, prism etc. The layer's width reaches 250 mkm.

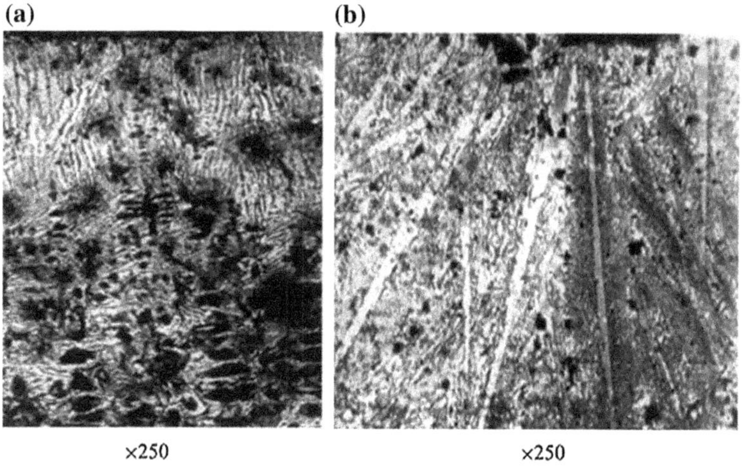

Fig. 9.5 The layer's microstructure after electron-beam boriding of the gray cast iron according to the diffusion mechanism (**a**) and diffusion crystallization mechanism (**b**) [58]

According to the X-ray diffraction phase analysis, the layers include iron borides, boron solid solution in α-iron and cementite Fe_3C.

The processing of preliminary formed boride layers with the electron beam (combined boriding). There are few published papers on the influence of electron beams on the structure and properties of boronized layers preliminary formed with traditional boriding methods. Presumably it may be explained by the fact that these studies require significantly precise analysis methods. The phase transformation in boride zones must be studied with Mössbauer spectroscopy [69, 70] due to the small depth of electron penetration at studied regimes.

Preliminary boriding of control samples, produced from steels Ct3 and 45, was carried out in the powder saturation mixture, containing 97% B_4C and 3% KBF_4 in a container with fusible lock at 940 °C for 3 h. The obtained layers were 80–90 mkm wide. The next step was the treatment of the surface with the electron beam in vacuum ($P = 2 \times 10^{-3}$ Pa) at the surface power density 2.9×10^4 W/cm^2 for 15–50 s [71]. The irradiation of samples at normal power density $(2.0–2.8) \times 10^3$ for 3–5 min didn't cause any significant changes in the boride layers' structure. Therefore the surface processing was done at higher density which allowed to reduce the process duration.

The X-ray diffraction phase analysis showed that the boride layer consisted of two phases: FeB and Fe_2B—which correlates with the results of other authors [72–74].

The use of Mössbauer spectroscopy allowed to specify the phase composition of the boride zone and estimate the influence of the electron beam on the phase composition. The study involved the use of the non-destructive method of Mössbauer converse spectroscopy which doesn't require a preliminary chemical treatment and polishing of the sample. This method presupposes that Mössbauer spectrum is filmed in the geometry of reverse reflection from the sample surface using a special running proportional detector [75, 76].

This detector can film the Mössbauer spectrum during the change of energy modes of registration and variation of gaseous atmosphere by registering either the converse resonant electrons or the resonant characteristic x-ray emission. If electron spectrums are registered, the information on the phase composition of the surface layer 0.1–0.2 mkm deep is obtained (depending on the level of roughness); a resonant x-ray spectrum gives information on the layer 20 mkm deep.

The Mössbauer spectrum of studied samples were obtained on the experimental machine with an electrodynamic vibrator which allows to move a resonant absorber by the source in the range between −10 mm/s to +10 mm/s depending on the set amplitude. The source of resonant γ-emission with E = 14.4 keV was an active isotope Co^{57}, obtained in cyclotrope, implanted with the use of tempering into the metallic matrix. This method of producing sources allows to get a single Mössbauer emission line close to the natural level width.

The detection of γ-quants was done by a scintillating detector with a 0.15 mm crystal NaJ(Tl). The registered impulses go into different channels of the analyzer so that the number of the channel is proportional to its movement speed. The obtained Mössbauer spectrum was fixated on the display of the multichannel analyzer; the numerical information was digitally printed and put into the computer memory where it was mathematically processed in UNIVEM software [77].

The analysis showed that the spectrum of the boronized sample, filmed near the surface 0,3 mkm layer (Fig. 9.6a), has a complex form which signifies about the ordered magnetic iron boride phases formation in the surface layer. The Mössbauer parameters of these phases are thoroughly studied [78] and differ significantly which allows to carry out a reliable phase analysis. The decomposition of the spectrum into compound components demonstrated that it presents itself as the superposition of sub-specters of ordered magnetic phases (FeB, Fe_2B, Fe_3B and a solid solution $FeB_{(1+x)}$ with x < 0,4) and the widened paramagnetic duplicate which correlates with the quasi-amorphous disordered solid solution of $FeB_{(1+x)}$ where X > 0.4. The spectrum of the same sample for the near-surface 20 mkm layer (Fig. 9.6b) significantly differs in form: the central paramagnetic part almost disappears and the redistribution of magnetic boride phases' intensities takes place.

The Mössbauer specters of the preliminary boronized sample under electron-beam heating during 10 s are demonstrated in Fig. 9.7. As seen in the picture, the spectrum of the near-surface 3000 Å area abruptly changes after irradiation in comparison with the spectrum of the normally boronized sample (Fig. 9.6a). The spectrum begins to resemble the spectrum of a deep (20 mkm) layer of a normally boronized layer. The spectrium of the layer ~20 mkm of the irradiated sample insignificantly differs from the spectrum of the surface layer.

The mathematical processing of the Mössbauer specters showed significant differences in the phase composition of layers obtained after solid-phase and combined boriding (Table 9.1).

The table demonstrates that the most significant changes take place in the thin surface layer after electron beam processing: a relative amount of Fe_2B phase

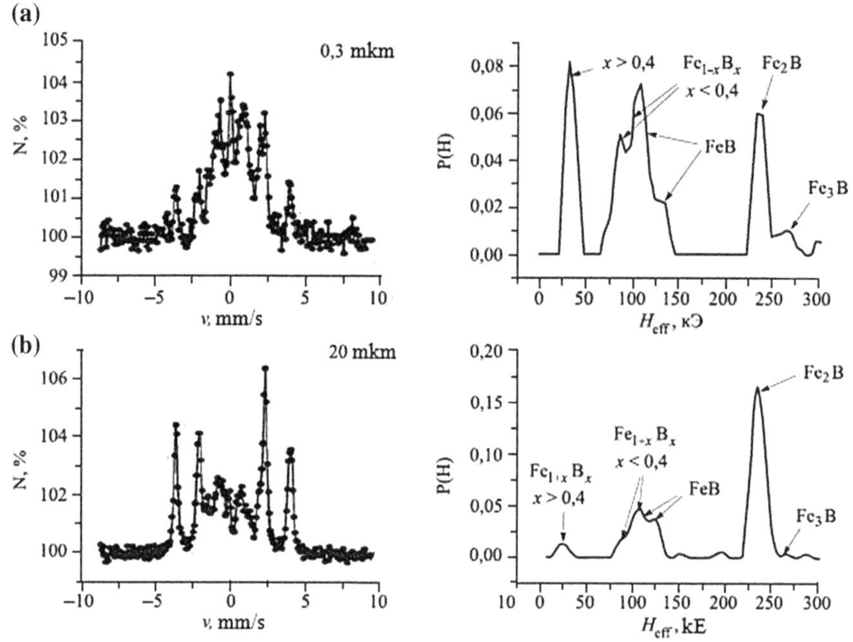

Fig. 9.6 The Mössbauer converse spectrum of boronized samples obtained from the surface layers 0.3 mkm (**a**) and 20 mkm (**b**) wide. N—intensity of spectrum lines, P(H)—intensity of magnetic-split components

sharply increases, the amounts of other boride phases decrease differently. The largest fall in the amount is seen in the most boron-rich disordered phase $FeB_{(1+x)}$ (x > 0,4). The changes are preserved at 20 mkm depth but they are weaker.

The result of these processes is the change in the boronized layer's structure (Fig. 9.8).

Electron-beam heating at the set power density of electron flux (2.9×10^4 W/cm^2) causes complex processes in boride layers; they are explained by the complex impact of the electron flux on the surface and are accompanied by the rise of temperature, formation of multiple point defects and the occurrence of directed diffusion flows. The result of this impact is the change in the phase correlation: the decrease in boron-rich phases on the surface, the increase of boron-deficient phases (Table 9.1.).

The significant changes in the structure are seen in the zone of the sublayer where conditions for a liquid crystal state occurrence may be observed (as a result of alloy elements and carbon redistribution).

The process of the structure transformation in accordance with the microstructures (Fig. 9.8) goes through several stages:

- The radioactive and thermal dispersion of borides in the layer's surface and needle points accompanied by the formation of the poly-channel structure,

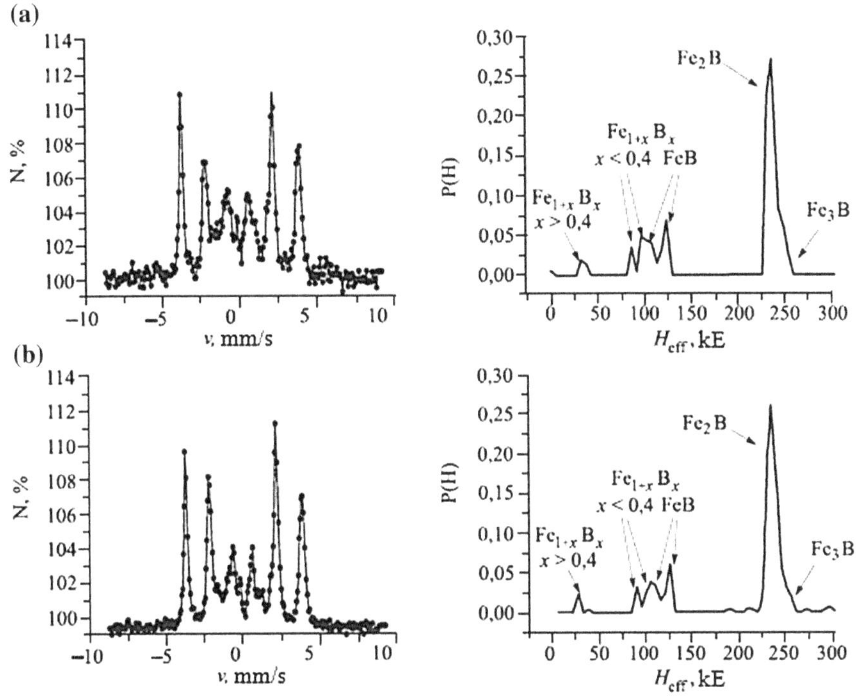

Fig. 9.7 The Mössbauer converse spectrum of boronized samples (treated with the electron beam) obtained from the surface layers 0.3 mkm (**a**) and 20 mkm (**b**) wide

predominantly parallel to the temperature, radiation and concentration gradients. The formation of this structure should be connected to the formation of multiple point defects in borides and the occurrence of two diffusion flows of boron and iron atoms from the middle part of the boride layer to the borderlines of the boride part with the environment and main metal (substrate). It is accompanied by a partial withdrawal of boron atoms into the environment and boron atoms diffusion from the boride needle points inwards the substrate. Also the borderline between the layer and substrate smoothens out as there is no evident growth of the poly-channel structure in a certain crystallographic direction (the texture character of the boride layer is eliminated at the acquired temperature, concentration and radiation gradients).

- Carbon is replaced and alloy elements are accumulated in the substrate; this process is accompanied by the formation of alloy borides and carbon borides. The surface layers in the zone of the poly-channel structure hold the natural selection of crystallites which are favorably oriented towards the surface radiation impact and ascending diffusion flows of iron and boron atoms. A favorable orientation manifests itself in the cubic form of crystallites, typical for the tetragonal crystal structure of Fe_2B structure which in cross-section has a shape of square with the sides parallel and perpendicular to the diffusion flows. Those

Table 9.1 The results of processing Mössbauer specters

Analyzed layer's width, mkm	The content of a boronized layer's phases after processing (vol. %)									
	Solid phase boriding					Combined boriding				
	FeB	Fe2B	Fe3B	Fe1 + Bx (x < 0.4)	Fe1 + Bx (x > 0.4)	FeB	Fe2B	Fe3B	Fe1 + Bx (x < 0.4)	Fe1+Bx (x > 0.4)
0.3	21	14	17	26	22	10	62	2	23	3
20	18	59	4	19	4	15	65	5	13	2

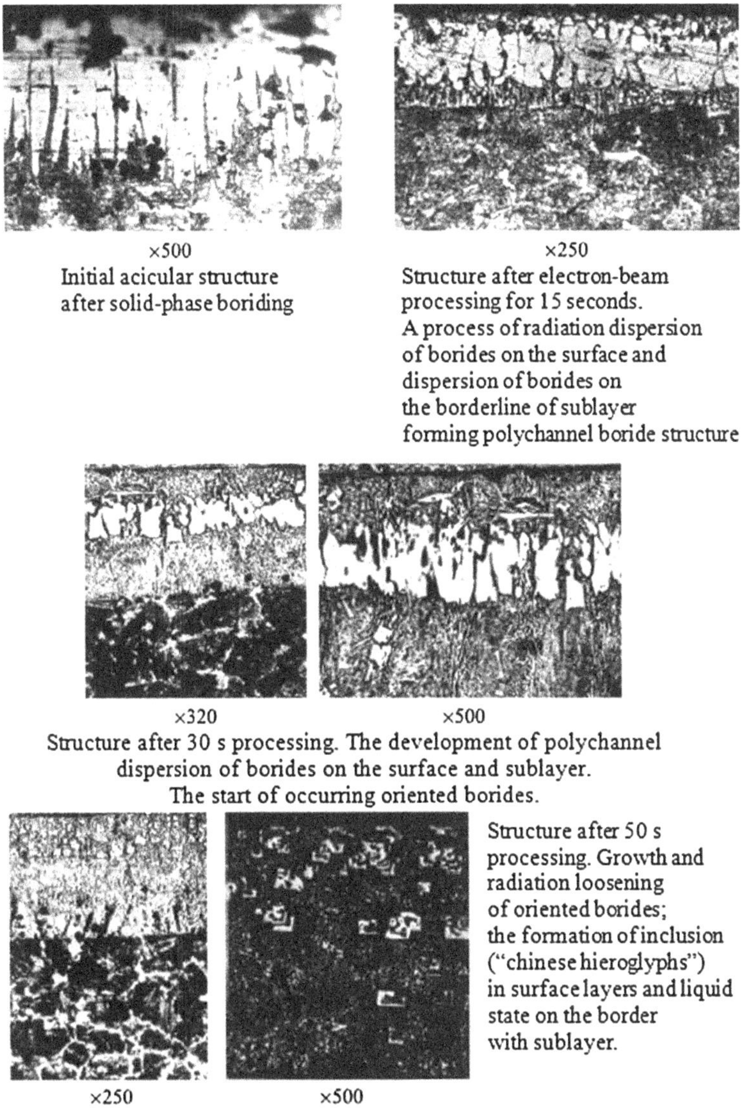

Fig. 9.8 The microstructure of the layer after combined boriding (P = 2.0×10^4 Vt/cm^2, T = 1050–1100 °C)

crystallites, which are oriented in another direction, go under further dispersion and create conditions for the formation of favorably oriented crystals. High-boron FeB phase is replaced by Fe_2B phase which is also oriented rightly in the structure of the boride layer.

- The growth of favorably oriented boride crystallites in the near-surface zone provided by the diffusion flows from the deeper layers and their radiation loosening from the surface. It manifests itself in the formation of crystallite structure with an open area where the radiation flow affects it; these crystallites are so called "Chinese hieroglyph" borides. The formation of these borides at the constant radiation impact is accompanied by the generation of porousness in the surface layers which is well seen in the pictures as dark areas. The porousness functions as oil-retaining pockets which provides greasing of the surfaces in joints while preserving the high hardness and plasticity of the boronized layers. There are no sufficient conditions for forming "Chinese hieroglyph" borides in the poly-channel zone of the boronized layer next to the substrate. It is explained by the same direction of radiation and diffusion flows and insufficient boron concentration. The conditions for a liquid state occurrence are possible to acquire in the sublayer (by forming areas with increased carbon and alloy elements concentration).

- Further radiation boride layer's loosening on the surface and formation of α-solid solution of boron in iron (during the boron atoms withdrawal into the environment). The development of the layer's structure from the inside and the sublayer is determined by the alloyness of the substrate, carbon content and process temperature. However it is also connected with the formation of carbon borides and an α-solid solution of boron and alloy elements in iron.

The character of spreading hardness in the layer at every stage is different which allows to form boronized layers of different plasticity by changing the duration of the process or setting a thermal circulation mode. The manner of changes in the structure and properties allows to classify boronized layers after a combined targeted processing:

- the improvement of the wear-resistance of the details while preserving high hardness of borides requires the use of layers after 30 s radiation impact;
- shock loads and intensive deterioration require plastic layers of ~1400–1450 kgf/mm^2 after 30–50 s radiation impact;
- exploitation conditions at alternating loading, spinning or bending require plastic layers after 50–75 s radiation impact or layers obtained directly after electron-beam boriding.

The dominant role of the electron beam can be proved by the fact that the furnace heating of boride layers doesn't cause the formation of layers with this structure. The heating of boronized layers in the furnace at 900–1200 °C for 5 min leads to a slight growth of the layer (1000 °C) due to the resorption of a high-boron phase; then it is replaced by the decrease of the boride zone width due to the prevailing corrosion of the surface (Fig. 9.9).

Simultaneously the growth of grains in the surface zone under boron influence is observed.

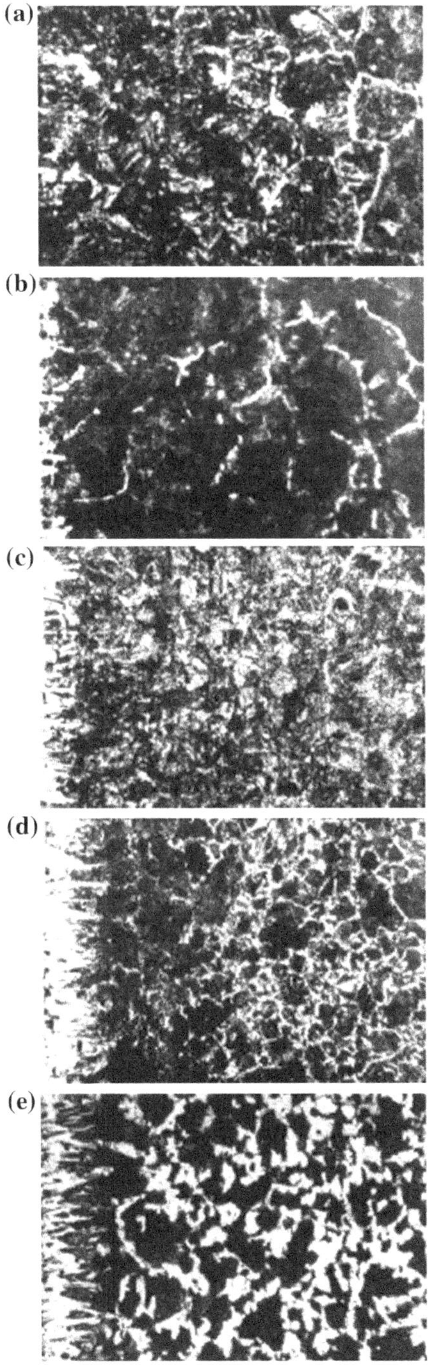

Fig. 9.9 The influence of the 5 min furnace heating on the structure of the boride layer (×250): **a** initial state; **b** 900 °C; **c** 1000 °C; **d** 1100 °C; **e** 1200 °C

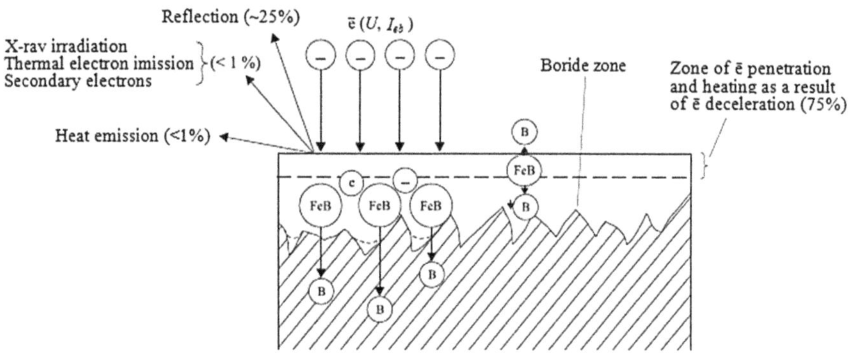

Fig. 9.10 The process scheme of the layers' formation under combined boriding

Generally the process of electron-beam influence causes the kinetic energy to transform into a heat energy in the processed zone (Fig. 9.10). During this procedure, the contact zone of the electron beam and the surface becomes a place for the elastic collision between electrons and sample's iron atoms. It results in point defects (vacancies and interstitial defects) while only light boron atoms are ionized and replaced from the structural positions in the crystal lattice. A strong heating of the surface layers under a falling electron flow (T $\sim$ 1000 °C) causes boron atoms to leave the crystal lattice and withdraw from the surface into the environment. The point defects diffuse into the deeper layers forming new defect zones and the poly-channel layer's structure. The high temperature leads to the formation of another flow of boron atoms directed towards the core of the metal going by inter-grain borders with accelerated speed under the influence of the electron beam; this causes the formation of another poly-channel structure in this zone.

The withdrawal of boron atoms from the surface layer under the influence of the electron beam means the abrupt reduction of the surface-based boron-saturated disordered FeB_{1+x} phase. The result of all these processes is the boron deficit in the near-surface area. Prevailing boron and iron atoms flow provides the formation of crystallites oriented towards the surface. The radiation impact from the surface makes crystallites open towards this impact. Formed borides acquire a "chinese hieroglyphs" shape, open towards the surface.

9.2.3 The Mechanical Properties of Boronized Layers with Composite Structure

Microhardness. The electron beam processing of boronized layers (combined method) causes the change in the microhardness distribution mechanism along the depth. The microhardness of the boride zone decreases; the level of decrease directly depends on the duration increase. Moreover the occurred structural

Table 9.2 The mechanical properties of boronized layers

No.	The type of processing	Microhardness HV, kgf/mm^2	σ_{sp}, mPa	ε_{ult}	Bend loading for crack formation, kgf	Brittleness point
1	Solid-phase boriding	1950–2050	245	1.13	155	4
2	Electron-beam boriding	650–1450	484	1.65	245	3
3	Combined boriding	560–1650	414	1.38	210	2

heterogeneity makes the microhardness change in a certain interval; the microhardness of the transition zone increases. The optical microscope study didn't reveal any distinct structural changes in the transition zone. Presumably, the increase of microhardness is connected to the diffusion of boron atoms from the boride zone and the increase of its content in α-solid solution (Table 9.2).

Brittleness. The studies show that the use of the electron beam at different stages of the layer formation allows to reduce brittleness and increase plasticity [79]. The influence of the electron-beam on the boride layers' properties was experimentally studied by measuring split tension and calculating the ultimate strain. According to [80], split tension σ_{sp} was used as a criterion of brittleness; it is measured under the impact of the diamond pyramid on the surface and is calculated by the formula

$$\sigma_{sp} = 0.174\,P/(2 \cdot l^2 + c \cdot l)'$$

where P—the loading on the indenter while measuring microhardness; l—the minimal distance from the centre of the imprint to the end of the sample which doesn't allow splitting; c—the diagonal of the imprint. The split tension is an integral characteristics of the layer's brittleness which spontaneously includes the physical-mechanical properties of borides (microhardness, elasticity module), phase composition, phase correlation, their dispersability and mutual position in the layer and on the surface, stress state of the layer and its plasticity. High split tension signifies about low brittleness and therefore high plasticity. Table 9.2 contains the indexes of σ and brittleness points of layers in steel 45.

The plasticity was evaluated by the index of the ultimate strain [81]. The experiments were carried out on the machine PMT-3 at 50 g loading. The pyramid indentation forms angles—points of stress concentrations which cause breaks—cracks; it accelerates the moment of destruction. Two processes take place—the plastic deformation (the imprint's diagonal) and the destruction of the material (the crack's length). Their correlation serves as an indicator of ultimate plasticity of the material. The plasticity of the material (structure) depends on the index of ultimate plasticity:

$$\varepsilon_{ult} = D_{imp}/L_{cr}$$

where D—the imprint's diagonal, L—the crack's length between imprints.

Table 9.2 contains the indexes of ultimate plastic deformation of the layers obtained with different methods of boriding.

The analysis of microhardness, split tension and ultimate plastic deformation shows that the composite layers obtained under boriding combined with electron-beam heating are characterized by the reduced microhardness in comparison with solid-phase boriding layers, but they are superior in the plasticity.

The increased plasticity of composite boronized layers is also proved by the occurrence of cracks on bends in these layers requires larger loadings (Table 9.2).

Adhesive strength. Adhesive strength was calculated by a sclerometer [74, 82]. The experiments were done with the sclerometer (МЭИ-С2) which allowed to automatically register the diagram of deformation of the surface layer by the horizontal movement of the pointed diamond indenter (standard diamond Vickers pyramid) on the depth of 5 mkm. The diagram helped to determine the quantitative characteristics of mechanical properties of the surface material which indicate its destruction resistance.

The important criteria of adhesive strength (wear resistance) of the surface are the hardness HGV and the energy density of the local destruction ω [83].

The hardness HGV during scratching characterizes the resistance to the local destruction of the surface layer in given conditions and is calculated by the formula:

$$HGV = 0.076\,P_{scr}^{max}/t^2$$

where P_{scr}^{max}—maximum scratch load; t—depth of the indenter's intrusion.

The energy density of local destruction ω is calculated according to the formula:

$$\omega = W/V,$$

where W—the absolute energy of destruction (J) which is estimated by the area limited by the scratch diagram; V—the volume of the material replaced during scratch;

$$V = FL_{scr}^{max},$$

where F—square of the cross section of a crack, L_{scr}^{max} —the maximum length of the scratch.

Figure 9.11 shows the diagrams of scratch of the boride layers.

The analysis of the diagrams shows that the boride layer's electron beam processing with $P_y = 2.9 \times 10^4$ W/cm^2 for 15 s leads to the increase of local destruction energy ω from 112,038 to 123,118 MJ/m^3 and a slight reduction in

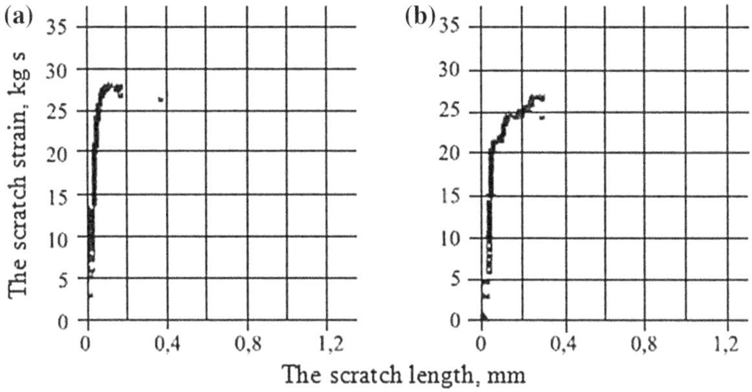

Fig. 9.11 The scratch diagrams of the initial boride layer (**a**) and the boride layer processed by the electron beam (**b**)

hardness at scratching HGV5 from 83,914 to 80,561 MPa. The prolongation of the processing from 30 to 50 s reduces both ω and HGV5: from 82,232 to 51,604 MJ/m^3 and from 70,919 to 51,438 MPa.

Apart from the mentioned mechanical properties, other characteristics of boronized layers change. Thus, for instance, the heating of the layers obtained with traditional contact boriding from gaseous mediums beginning with 700 °C causes the formation of cracks (Fig. 9.12a). The crack formation begins at the surface. The heating temperature rise causes the inward growth of cracks and their opening. This manner of cracks distribution characterizes primarily only these layers and conditions of their formation. Layers with a composite structure which combine solid (brittle) and more plastic structural components don't generate cracks under these conditions (Fig. 9.12b).

Thus, the boride layers with composite structure are characterized by the slightly reduced microhardness in comparison with the traditional layers but are superior by plasticity characteristics. This allows to use boriding for hardening machinery details and tools working under dynamic loadings (stars, ratchets, eccentrics, stamps etc.).

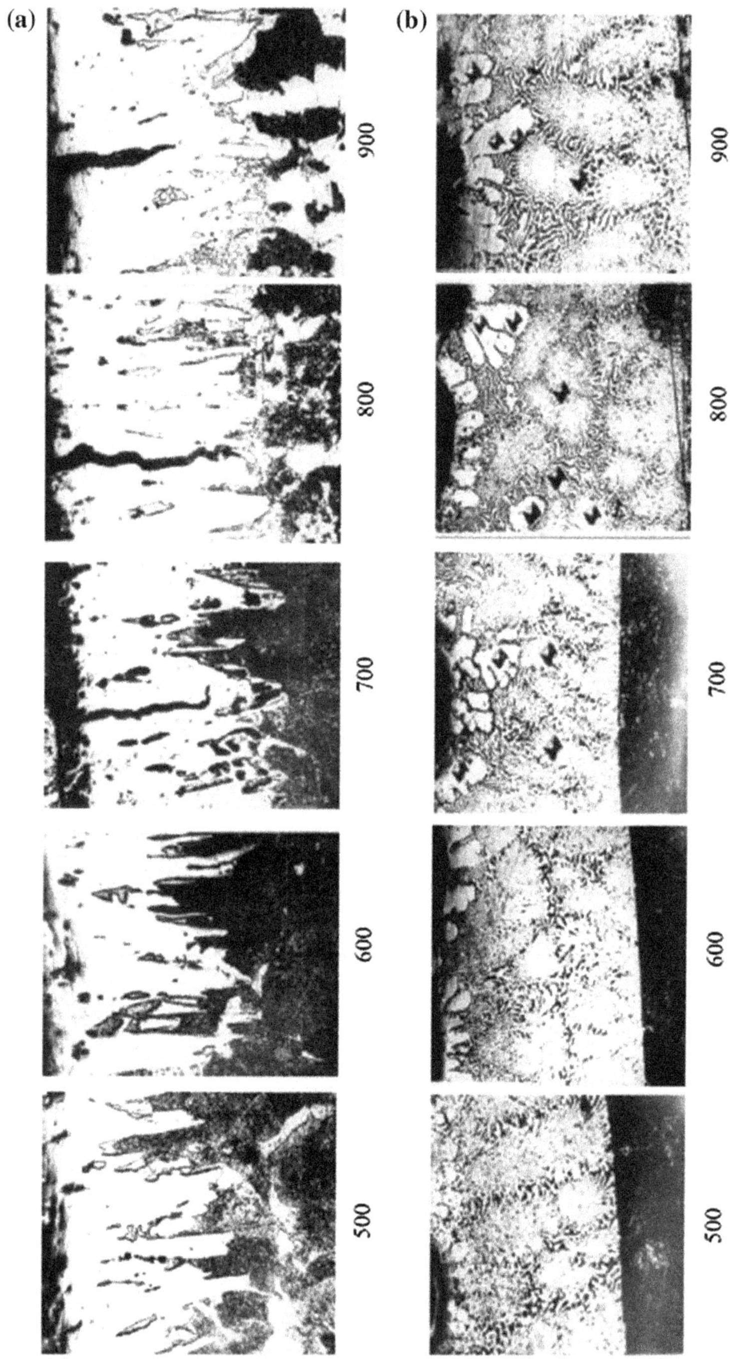

Fig. 9.12 The influence of the secondary heating temperature (temperature, °C) on the microstructure of boronized layers in steel 45: **a** after solid-phase boriding (×500): **b** after electron-beam boriding (×500)

References

1. Krukovich M.G. Raschet evtekticheskikh kontsentratsii i temperatury v dvukh i mnogokomponentnykh sistemakh // MiTOM. №10. Pp. 9–17 [The calculation of eutectic concentrations and temperature in two- and multi-component systems].
2. Taran Yu.T., Mazur V.I. Struktura evtekticheskikh splavov. Moscow, 1978. 455 p. [The structure of eutectic alloys].
3. Novikov I.I. Metallografiya. Laboratornyi praktikum. Moscow, 1984. 125 p. [Metallurgy. Laboratory lessons].
4. Chudina O.V. Poverkhnostnoye legirovaniye zhelezouglerodistykh splavov s ispolzovaniem lazernogo nagreva // MiTOM. 1994. №12. C. 2–7 [Surface alloying of iron-carbon smelts using laser heating].
5. Draper C.W., Poate J.M. Surface Modificate and Alloy by Laser, Lin and Electron Beams // Proc. NATO Adv. Study Inst., Trevel, 27–28 Aug; 1981. New York, London. 1983. P. 385–404.
6. Postnikov V.S., Belova S.A., Erofeeva E.M. Obrazovaniye struktury pri lazernom karboborokhromirovanii stali X12M // Metallovedeniye i termicheskaya obrabotka metallov. 1995. №12. Pp. 13–15 [The structure formation during laser carbon-boron-chrome treatment of steel X12M].
7. Lysenko A.B., Kozina N.N., Miroshnichenko I.S., Borisova G.V. Osobennosti obrazovaniya struktury v stalyakh, podvergnutykh lazernomu poverkhnostnomu legirovaniyu // Metallovedeniye i termicheskaya obrabotka metallov. 1995. №12. Pp. 10–12 [The peculiaritis of structure formation in steels under laser surface alloying].
8. Паршин А.М., Кириллов Н.В. Физические и структурные аспекты обработки сплавов концентрированными источниками энергии // Металлы. Parshin A.M., Kirillov N.V. Fizicheskiye i strukturnye aspekty obrabotki splavov kontsentrirovannymi istochnikami energii // Metally. 1995. №3. Pp. 21–26 [Physical and structural aspects of processing alloys with concentrated energy sources].
9. Kolesnikov Yu. V. Sostav dlya lazernogo boroazotirovaniya. Cert. 1482975. 30.05.89. Bul. №20 [The composition for laser boron-nitrating].
10. Ariely S., Shen J., Bamberger M., Dausiger F. Lazer surface alloying of steel with TiC // Surface and Coat Technol. 1991. V. 45, №1–3. P. 403–408.
11. Dorozhkin N.N., Kardapolova M.A., Krupskaya M.M. Metallograficheskiye i duyrometricheskiye issledovaniya poverkhnostnogo iznosa, uprochnennykh lazernoi obrabotkoi // Inzhenerno-fizicheskii zhurnal. 1998. Vol. 71, №1. Pp. 112–115 [Metallographic and durametric studies of the wear of surfaces hardened with laser treatments].
12. Lysenko A.B. Vliyaniye tekhnologicheskikh faktorov na sostav, stroenie i svoistva zony lazernoi khimiko-termicheskoi obrabotki // FiKHOM. 2001. №2. Pp. 25–30 [The influence of technological factors on composition, structure and properties of a laser chemical thermal processing zone].
13. Lakhtin Yu. M., Kogan Ya. D., Buryakin A.V. Poverkhnostnoye nasyscheniye stali borom pri vozdeistvii izlucheniya lazera // MiTOM. 1985. №11. Pp. 9–11. [Surface saturation of steel with boron during the impact of laser irradiation].
14. Lakhtin Yu. M., Kogan Ya.B., Zelenov A.E., Chudina O.V. Sposob lazernogo legirovaniya poverkhnosti metalla. Cert. 1557193.15.04.90. Bul. №14 [The method of laser alloying of the metal's surface].
15. Kershenbaum V.Ya., Prusakov B.A., ed. Mnogoyaychnyi tolkovyi slovar' 'Metally. Stroeniye. Svoistva. Obrabotka'. Moscow, 1999. 710 p. [Multi-language explanatory dictionary 'Metals. Structure. Properties. Processing'].
16. Abilsitov G.A., Safonov A.N., Baskov A.F. et al. Razrabotka protsessov borirovaniya stalei s pomoschyu nepreryvnogo CO-2 lazera // DAN USSR. 1989. Vol. 305, №2. Pp. 351–354 [The development of boriding processes for steels using discontinuous CO-2 laser].

17. Postnikov V.S., Tagirov M.N. Lazernoye borirovaniye titanovykh splavov // MiTOM. 1994. №1. Pp. 14–15 [Laser boriding of titanium alloys].
18. Tananko I.A., Levchenko A.A., Guiva R.T., Guiva V.A., Sittsevaya E.Yu. Lazernoye borirovaniye vysokoprochnogo chuguna // FiKHOM. 1991. №5. Pp. 89–95 [Laser boriding of resistant cast iron].
19. Safonov A.I., Tarasova T.V. Protsessy lazernogo legirovaniya stalei nemetallicheskimi komponentami // MSTU Bulletin. Machinery construction. 1997. №2. Pp. 69–77 [Processes of laser alloying of steels with non-metallic components].
20. Lyakhovich L.S., Ivanov S.A., Kartoshkin V.M. et al. Opredeleniye uslovii borirovaniya stali pri nagreve lazernym izlucheniem // MiTOM. 1985. №11. Pp. 12–14 [The determination of conditions for steel boriding under laser irradiation heating].
21. Bely A.V., Makushok E.M., Pobol I.L. Poverkhnostnaya uprochnyayuschaya obrabotka s primeneniem kontsentrirovannykh potokov energii. Minsk, 1990. 79 p. [Surface hardening treatment involving concentrated energy flows].
22. Bernstein A.M., Yandimirkin E.M. et al. Poverkhnostonye lazernoye legirovaniye stalei i chugunov // Elektronnaya obrabotka materialov. 1991. №3. C. 25–28 [Surface laser alloying of steels and cast irons].
23. Tananko I.A., Levchenko A.A., Guiva R.T., Guiva V.A., Sitsevaya E.Yu. Lazernoye borirovaniye vysokoprochnogo chuguna // FiKHOM. 1991. №5. Pp. 89–95 [Laser boriding of highly resistant cast iron].
24. Gordienko A.I., Ivashko V.V. Polucheniye boridnykh pokrytii na Ti-splavakh s pomoschyu izlucheniya nepreryvnogo CO2-lazera // Zaschitnye pokrytiya na metallakh. 1990. №24. Pp. 66–68 [Obtaining boride surface on Ti-smelts using discontinuous CO2-laser irradiation].
25. Safonov A.I. Osobennosti borirovaniya zheleza i stalei s pomoschyu nepreryvnogo CO2-lazera // MiTOM. 1998. №1. Pp. 5–9.
26. Grigoryants A.G., Safonov A.N. Metody poverkhnostoi lazernoi obraborki. Moscow, 1987. 191 p. [Methods of surface laser processing].
27. Mitrokhovich N.M. Obosnovaniye vybora sostava dlya borirovaniya v pastakh // Termicheskaya obrabotka i fizika metallov. Sverdlovsk, 1982. Iss. 7. Pp. 144–147 [The foundation for choosing a composition for boriding in pastes].
28. Postnikov V.S., Kalashnikova M.S. Issledovaniye struktury poverkhnostnogo sloya posle lazernogo legirovaniya nizkouglerodistykh konstruktsionnykh stalei // FiKHOM. 1999. №4. Pp. 47–51 [The study of the structure in surface layers after laser alloying in low-carbon construction steels].
29. Tosto S., Nenci F. Surface cladding and alloying of AISI 316 stainless steel on C45 plain carbon steel by electron beam // Met. et etud. sci. Rev met. 1987. V. 84, №6. P. 311–320.
30. Semenov A.P., Kovin I.B., Petrova I.P. et al. Metody i sredstva uprochneniya poverkhnostei detalei mashin kontsentrirovannymi potokami energii. Moscow, 1992. 404 p [Methods and means of hardening surfaces of machinery details with concentrated energy flows].
31. Benshou R.F. Istoriya sozdaniya i razvitiya elektronno-luchevykh tekhnologii // Vvedeniye v tekhnologiyu elektronno-luchevykh protsessov. Moscow, 1965. Pp. 7–22 [The history of creating and developing electron-beam technologies].
32. Rykalin N.N., Uglov A.A., Kokora A.N. Lazernaya obrabotka materialov. Moscow, 1975. 296 p [Laser processing of metals].
33. Pobol I.L. Ispolzovaniye elektronno-luchevogo vozdeistviya v tekhnologiyakh vtorogo pokoleniya poverkhnostei obrabotki metallicheskikh materialov // Treniye i iznos. 1993. T. 14, №3. Pp. 524–531 [The use of electron-beam impact in technologies of the second generation on the surfaces of metal materials].
34. Rykalin N.N., Zuev I.V., Uglov A.A. Osnovy elektronno-luchevoi obrabotki materialov. Moscow, 1978. 239 p. [Basics of electron-beam processing of materials].
35. Vladimirskiy A.R., Livshits V.B., Payuk V.A. O vozmozhnosti ispolzovaniya oblucheniya elektronami vysokikh energii dlya legirovaniya materialov // USSR Academy of Science Bulletin. Metals. 1988. №5. P. 128–133 [On the possibility of using high energy electron irradiation for alloying of materials].

36. Pobol I.L. Elektronno-luchevaya termoobrabotka metallicheskikh materialov // Itogi nauki i tekniki. 1990. Vol. 24. Pp. 90–166 [Electron-beam thermal processing of metallic materials].
37. Popov V.F., Gorin Yu. I. Protsessy i ustanovki elektronno-ionnoi tekhnologii. Moscow, 1988. 254 p. [Processes and settings of electron-ion technologies].
38. Ryazanov M.I., ed. Vzaimodeistviye zaryazhennykh chastits s tverdymi telami. Moscow, 1979. 73 p. [The interaction of charged particles with solid bodies].
39. Platonov P.A. Deistviye oblucheniya na strukturu i svoistva materialov. Moscow, 1971. 40 p [The impact of irradiation on the structure and properties of materials].
40. Konobeevsky S.T. Deistviye oblucheniya na materialy. Vvedeniye v radiatsionnoye materialovedeniye. Moscow, 1967. 401 p. [The influence of irradiation on materials. Introduction to the radiation materials science].
41. Didenko A.N. et al. Vozdeistviye puchkov zaryazhennykh chastits na poverkhnost metallov i splavov. Moscow, 1987. 187 p. [The influence of charged particles beams on the surfaces of metals and alloys].
42. Kirsanov V.V. et al. Protsessy radiatsionnogo defektoobrazovaniya v metalle. Moscow, 1985. 272 p. [Processes of radiation defect-formation in metals].
43. Лейман К., Кристиан Д. Взаимодействие излучения с твердымтелом и образование элементарных дефектов. — М.: Атомиздат, 1979. — 296 с. Leiman K., Kristian D. Vzaimodeistviye izlucheniya s tverdym telom i obrazovaniye elementarnykh defektov. Moscow, 1979. 296 p. [The interaction of irradiation and solid bodies and the formation of elementary defects].
44. Pogrebnyak A.D., Shumakova N.I. Modifikatsiya metallichekikh pokrytii pod deistviem nizkoenergeticheskogo silnotochnogo puchka // FiKHOM. 1999. №6. Pp. 13–16 [The modification of metallic surfaces under the influence of low energy strong-flow beam].
45. Klimenov V.A., Ivanov Yu. F., Perevalova O.B., Proskurovskiy D.I., Torgunakov Yu. B., Rotshtein V.P., Andrien S.P. Formirovaniye struktury i mekhanizmy uprochneniya poverkhnostnykh sloev nerzhaveyuschei stali, obrabotannoi nizkoenergeticheskim silno-tochnym elektronnym puchkom // FiKHOM. 2000. №2. Pp. 41–47 [The formation of the structure and hardening mechanisms for surface layers in stainless steels processed with a low-energy strong-flow electron beam].
46. Pobol I.L. Issledovaniye i razrabotka metodov elektronno-luchevogo nagreva i poverkhnostnogo uprochneniya metallicheskikh izdelii. Minsk, 1987. 23 p [The study and development of methods of electron-beam heating and surface hardening of metallic goods].
47. Poletika I.M., Borisov M.D. Tverdost' i iznosostoikost' stali posle oblucheniya puchkom relyativistskikh elektronov // MiTOM. 1996. №12. Pp. 16–19 [Hardness and wear-resistance of steels after irradiation with relative electrons flow].
48. Sizov I.G., Smirnyagina N.N., Semenov A.P. Elektronno-luchevoye borirovaniye zhelezouglerodistykh splavov // ESSTU papers. Technical science. Ulan-Ude. 1999. Iss. 7. Pp. 95–106 [Electron-beam boriding of iron-carbon smelts].
49. Krukovich M.G. Razrabotka teoreticheskikh i prikladnykh aspektov upravleniya strukturoi i svoistvami borirovannykh sloev i ikh ispolzovanie pri proizvodstve transportnoi tekhniki. Moscow, 1995. 45 p. [The development of theoretical and applied aspects of controlling the structure and properties of boronized layers and their use in producing transport machinery].
50. Galynskaya N.N., Panich G.G. Mekhanizm formirovaniya diffuzionnykh pokrytii pri posledovatelnom nasyschenii stalei borom i khromom. Minsk, 1998. 7 p [The mechanism of the formation of diffusion surfaces during continuous boron and chrome saturation of steels].
51. Acimovic Sorica. Mehanizam obrazovania i rasta boriranoy sloja // Zast. mater. 1996. V. 37, №3. P. 119–120.
52. Trusova G.V. Mekhanizm formirovaniya, struktura i svoistva boridnykh sloev na stali. Tomsk, 1996. 17 p. [The formation mechanism, structure and properties of boride layers in steels].
53. Burakov V.A., Burakova N.M., Semenov A.S. Fenomenologicheskaya model lazernogo legirovaniya // Rostov-na-Donu, 1991. 22 p. [The phenomenological model for laser alloying].

54. Gemelli H., Gallerie A., Caillet M. Improvement of resistance to oxidation by laser alloying on tool steel // Sc. Mater. 1998. V. 39, №10. P. 1345–1352.
55. Verfanren rum Laserlegieren von Metallteilen mit Zubuhrung der Legierungs komponenten: №196, №196502586. 4.12.96. —. 10.6.98.
56. Sizov I.G., Smirnyagina N.N., Semenov A.P. Issledovaniye struktury pokrytii na osnove boridov, poluchennykh s primeneniem elektronno-luchevogo nagreva v vakuume // IV International Conference on Vacuum Technologies Proceedings. Kharkiv, 2001. Pp. 88–92 [The study of the structures of surface on the basis of borides obtained as a result of electron-beam heating in vacuum].
57. Semenov A.P., Smirnyagina N.N., Sizov I.G. Obrabotka poverkhnosti stali elektronnym puchkom i formirovaniye boridnykh sloev // IIIrd International Symposium "Advanced Technologies and Equipment" Proceedings. Kharkiv, 1999. Pp. 101–105 [Processing of steel surfaces with electron flow and the formation of boride layers].
58. Sizov I.G. Razrabotka nauchnykh osnov i tekhnologii elektronno-luchevogo borirovaniya zhelezouglerodistykh splavov s polucheniem na poverkhnosti boridov tugoplavkikh metallov. Dissertatsiya doktora tekhnicheskikh nauk. Moscow, 2003. 305 p [The development of scientific foundation and technology of electron-beam boriding of iron-carbon alloys aimed at obtaining high-melting point metal borides on the surface. Doctorate dissertation].
59. Semenov A.P., Sizov I.G., Smirnyagina N.N., Korobkov N.V., Tselovalnikov B.I., Vandanov A.G. Sposob elektronno-luchevogo borirovaniya stalei i chugunov. Patent №2186872 [The method of electron-beam boriding of steels and cast irons].
60. Smirnyagina N.N., Semenov A.P., Sizov I.G. Elektronno-luchevoye modifitsirovaniye i borirovaniye zhelezo-uglerodistykh splavov // Fizika i khimiya obrabotki materialov. 2000. №3. Pp. 45–49 [The electron-beam modification and boriding of iron-carbon alloys].
61. Sizov I.G., Smirnyagina N.N., Semenov A.P. Osobennosti elektronno-luchevogo borirovaniya stalei // MiTOM. 1999. №12. Pp. 8–11 [The peculiarities of electron beam boriding of steels].
62. Smirnyagina N.N., Sizov I.G., Semenov A.P. Issledovaniye struktury poverkhnostnykh sloev posle elektronno-luchevogo borirovaniya zhelezouglerodistykh splavov // 6th International Conference "Films and coatings 2001" Proceedings. Saint-Petersburg, 2001. Pp. 556–561 [The study of the structure of surface layers after electron-beam boriding in iron-carbon alloys].
63. Sizov I.G., Smirnyagina N.N., Semenov A.P. Comparativ estimation of boron layers produced by the electron beam and traditional borating // The proceedings of VI China-Russian symposium «New materials and technology». Beijin. 2001. P. 506.
64. Smolnikov E.A., Sarmanova L.M. Issledovaniye vozmozhnosti zhidkostnogo borirovaniya bystrorezhuschikh stalei // MiTOM. 1982. №11. Pp. 30–32 [The study of the possibility of liquid boriding of high-speed steels].
65. Safroshenkov A.F., Safroshenkov I.A. Tekhnologiya borirovaniya bystrorezhuschikh stalei // Univ. studies. Chernaya metallurgiya. 1997. Pp. 67–70 [The technology of boronizing high-speed steels].
66. Voroshnin L.G. Borirovaniye promyshlennykh stalei i chugunov. Minsk, 1981. 205 p. [Boriding of industrial steels and cast irons].
67. Lyakhovich L.S., Krukovich M.G., Turov Yu. V. Novoye v khimko-termicheskoi obrabotke stalei // Ekspress informatsiya. Minsk, 1976. 32 p. [New trends in chemical-thermal processing of steels].
68. Lyakhovich L.S., Voroshnin L.G., Vasilyev L.A., Krukovich M.G., Turov Yu. V. Nizkotemperaturnoye elektroliznoye borirovaniye // MiTOM. 1974. №2. Pp. 54–55 [Low-temperature electrolysis boriding].
69. Novakova A.A., Sizov I.G., Gvozdover R.S., Golubok D.S., Kiseleva T.Yu., Semenov A.P., Smirnyagina N.N., Prusakov B.A. Strukturnyi analiz borirovannykh sloev na poverkhnosti malouglerodistoi stali do i posle elektronno-luchevoi obrabotki // Poverknost. Rentgenovsiye, sinkhronotropnye i neitronnye issledovaniya. 2003. №3. Pp. 99–103 [The structural analysis of boronized layers on the surfaces of low-carbon steel before and after electron-beam processing].

70. A.A. Novakova, I.G. Sizov, D.S. Golubok, T.Yu. Kiseleva, P.O. Revokatov. Electron-beam boriding of low-carbon steel // Journal of Alloys and Compounds. 2004. V. 383. P. 108–112.
71. Sizov I.G., Smirnyagina N.N., Semenov A.P., Prusakov B.A., Novakova A.A. Sposob kombinirovannogo borirovaniya stalei. Patent Russian Federation № 2210617 [A method of combined boriding of steels].
72. Acimovic-Stojadinov Zorica Fashi sastav borizang sloja // Zast. Mater. 1996. V. 37, №4. P. 176–178.
73. Kukharev B.S., Phisenko S.P., Shabunya S.I. Matematicheskoye modelirovaniye borirovaniya zheleza // Enginerring-physical of iron. 1996. Vol. 69, №2. Pp. 187–193 [Mathematical modeling of iron boriding].
74. Matyunin V.M., Volkov P.V., Saidakhmedov R.Kh. et al. Opredeleniye mekhanicheskikh svoist i adgezionnoi prochnosti ionno-plazmennykh pokrytii sklerometricheskim metodom // MiTOM. 2002. №3. Pp. 36–39 [Determining mechanical properties and adhesive resistance of ion-plasma surfaces with a sclerometer method].
75. Kiselev A.A., Kuzmin R.N., Novakova A.A. // Pisma v ZHTF. 1986. Vol. 12. C. 32–36.
76. Novakova A.A., Kuzmin R.N. Messbauerovskaya konversionnaya spektroskopiya i ee primenenie. Moscow, 1989. 72 p. [Mössbauer conversion spectroscopy and its use].
77. Novakova A.A., Kiseleva T. Yu. Metody messbauerovskoi spektroskopii v fizike tverdogo tela. Messbauerovskii praktikum. Moscow, 2000. 30 p. [Method of Mössbauer spectroscopy in the physics of solid bodies. Mössbauer practice].
78. Sanchez F.H., Budnick J.I., Zhang J.D., Hines W.A., Choi M., Hasegawa R. The study of boride phase by Messbaur method // Phys. Rev. B. 1986. V. 34, №7. P. 4738–4731.
79. Sizov I.G., Semenov A.P., Smirnyagina N.N. On the increase of plasticity of boron layers using electron-beam boron plating // The proceedings of 5-th International Conference on Modification of Materials with Particle Beams and Plasma Flows. — Tomsk, 2000. V. 3. P. 262–266.
80. Grigorov P.K., Katkhanov B.B. Metodika issledovaniya khrupkosti borirovannogo sloya // Povysheniye nadezhnosti i dolgovechnosti detalei mashin. Rostov-na-Donu. 1972. Iss. XVI. Pp. 97–99 [The method of studying the brittleness of a boronized layer].
81. Skudnov V.A., Grigoryev I.N., Evdokimov S.V., Gavrilov L.A. Sposob otsenki plastichnosti uprochnennogo metalla. Patent №2085902. Publ. 1997. №18 [The method for evaluating the ductility of hardened metals].
82. Matyunin V.M. Metody i sredstva bezobraztsovoi ekspress-otsenki mekhanicheskikh svoistv konstruktsionnykh materialov. Moscow, 2001. 93 p. [Methods and means of non-sample express-evaluation of mechanical properties of construction steels].
83. Matyunin V.M. Metody i sredstva bezobraztsovoi operativnoi otsenki mekhanicheskikh svoistv materialov i elementov konstruktsii i mashin. Moscow, 1993. 36 p. [Methods and means of non-sample express evaluation of mechanical properties of materials and elements of constructions and machines].

Chapter 10
The Connection Between the Plasticity of Boronized Layers and the Mechanical and Exploitation Properties of Boronized Steels

Abstract Boronized layers are a multi-phase system which can have different combinations of structural components (acicular, layered etc.) of different dispersability; their plasticity can broadly vary depending on the external conditions of processing and the type of material. The chapter analyzes the technological methods used to control mechanical and exploitation properties as well as the nature and formation patterns for most important characteristics of boronized layers: the hardness of homogeneous and heterogeneous layers; stress state after boriding and further thermal treatment; brittleness and its functional relation with microhardness; tribological properties with structural classification and usage recommendations in different conditions of wear; thermal fatigue and hear resistance which are necessary for improving the exploitation resistance of details and tools working in high-temperature conditions; corrosion and corrosion-mechanical properties providing the period of effective protection; mechanical and physical-mechanical properties. The chapter demonstrates the influence of different factors on the hardness values of boronized layers: the amount, size, form and mutual position of boride phases; texturedness and stress state of layer phases; the level of alloyness of the layer; the microstructure of sublayer and metal base. The methods used for calculating the microhardness of boronized layers with heterogeneous, eutectic and pseudo-eutectic structures are given. The authors suggest that the most favourable distribution of residual stresses in the boride layer is observed in one-phase structures. The level of residual stresses in boronized layers with composite structure is at minimum.

The plasticity of boronized layers is the complex characteristics which depends on the mechanical properties and interaction of structural components of layers and properties of the main metal. Since boronized layers are multi-phase systems which have all kinds of combinations of components obtained with different technological methods, the plasticity of layers changes in a wide range. The structural components of boronized layers may be compact and non-compact, acicular and layered, with different level of dispersibility (pre-eutectic, eutectic, over-eutectic, pseudo-eutectic and of different morphology of the interior of inclusions). Boronized layers

M.G. Krukovich et al., *Plasticity of Boronized Layers*,
Springer Series in Materials Science 237,
DOI 10.1007/978-3-319-40012-9_10

and the sublayer's structure determine the final properties of boronized steels and their exploitation characteristics. The increased plasticity of boronized layers slows down the formation and development of brittle microcracks and provide the boronized details with the higher bearing capacity.

10.1 The Generation Pattern for the Hardness of One-Phase and Multiple-Phase Structures in Boronized Layers

Transition metals are characterized primarily by the high hardness of their borides. At the same time, the microhardness of diffusion boronized layers is influenced by other factors: the amount, size, shape and mutual position of boride phases; the texture characteristics and stress state of layer's phases; the level of the layer's alloyedness; the structure of the sublayer and the main metal.

The microhardness of borides in ARMCO-iron, established by the frequency curve of distribution, is 2125 HV for FeB and 1625 HV for Fe_2B. The results are influenced by the loading. If the loading on the hardness tester's indenter is reduced, the interval of microhardness indexes increases towards higher numbers. The interval of indexes tells about the hardness anisotropy of borides in different crystal-graphic directions. The most appropriate loading for measuring hardness of boride phases is 100 gf [1, 2].

Boriding of ingot iron and low-carbon steels slightly reduces the microhardness in phases. The increase of carbon content in steel reduces the microhardness of FeB boride but Fe_2B doesn't significantly change (Table 10.1). This manner of microhardness change is connected to the change in the amount and sign of the forming residual stresses.

The residual stresses also influence the hardness of phases and depend on the layer's phase composition. Mostly it concerns Fe_2B borides. Their hardness is higher in two-phase layers than in one-phase layers (in the same conditions).

Table 10.1 The influence of carbon content in steels on the microhardness of boride phases

The medium content	Layer's phases	The microhardness of phases (HV) at different carbon content in steel, mass%			
		0.07	0.15	0.45	0.75
70% $Na_2B_4O_7$ + 30% B_4C	FeB	1950	1940	1900	1760
	Fe_2B	1380	1380	1400	1450
100% $Na_2B_4O_7$ (electrolysis)	FeB	2100	2030	1860	1750
	Fe_2B	1450	1440	1450	1430

Note Non-electrolysis boriding was carried out at 950 °C for 6 h. Electrolysis boriding—at 950 °C, for 3 h, at current density 0.25 A/cm^2

Alloying of borides from the saturation mixture or from the foundation material by tungsten, vanadium, zirconium and chrome improves the values for FeB phase; cuprum, nitrogen, aluminum and nickel reduces them. The increase of Fe_2B phase's microhardness is mostly observed when alloying is done with molybdenum, aluminum, chrome and zirconium; the decrease—under the alloying with manganese, nitrogen and nickel. Alloying with other elements doesn't show any distinct correlation with the microhardness due to the presence of multiple influence factors. For instance, the impact of nickel and molybdenum depends on the saturation temperature and has a reverse effect. The boronized layers obtained at 800 °C are positively influenced by nickel and molybdenum (Fe_2B microhardness increases); the microhardness decreases in the layers obtained at saturation temperatures >900 °C (Fig. 10.1). It is presumably connected to the increase of diffusion mobility of molybdenum and nickel at high temperatures and their dissolution in borides. The measurement of the microhardness of boride phases in standard tool steels showed that it lies in the range 1750–2400 HV for phase FeB and 1450–1750 HV for Fe_2B [1, 3].

The complex alloying treatment of the boronized layers obtained from the saturation compact material (coating) and the substrate's material during boron-nitrating, boron-siliconizing, boron cuprum treatment and boron chromizing of alloy steels leads to a certain deceleration of the layer's growth and significant improvement of their plasticity. This improvement is connected to the micro-alloying of the layer's borides and the change in their hardness (Table 10.2)

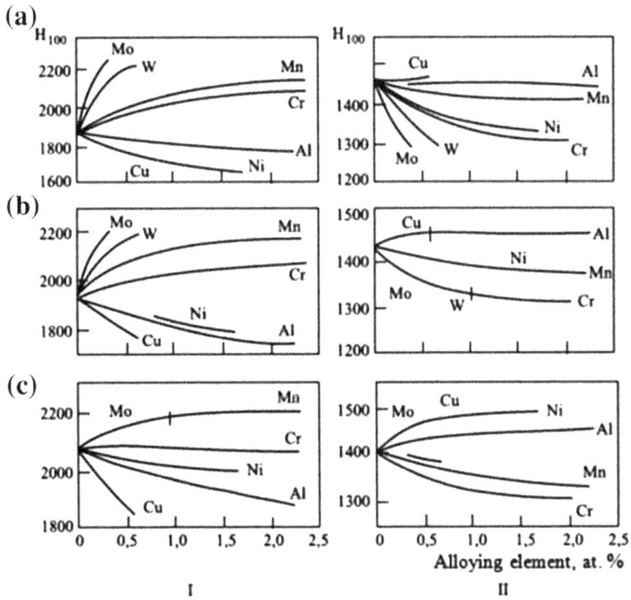

Fig. 10.1 The influence of alloy elements on the hardness of boride phases (steel 0.4% C, τ = 3 h, j = 0.25 A/cm^2); *I* FeB; *II* Fe_2B; **a** 950 °C; **b** 900 °C; **c** 850 °C [1]

Table 10.2 The influence of the brand of steel on the microhardness of the layer's boride phases [3]

The type of processing	Hardness of boride phases (HV) in different steels					
	5XHM		7X3		5X3B3MΦC	
	FeB	Fe_2B	FeB	Fe_2B	FeB	Fe_2B
Boron-nitrating	1730	1440	1800	1400	1700	1280
Boron-cuprum treatment	1800	1420	1780	1160	I860	1300
Boron-siH conizing	–	1280	1800	1500	1980	1500
Boron-chromizing	1950	1450	2060	1600	2080	1600
Bonding	1870	1430	2000	160	1820	1400

and also with the redistribution of phases in the layer, the formation of the sublayer and the generation of the favorable stress state. Hardness indexes are obtained at 100–200 mkm width of the boronized layer in steels 5XHM and 7X3 and at 50–60 mm width—in 5X3B3MΦC.

Carbide-forming elements in the layers of studied steels are distributed almost equally. These elements ensure micro-alloying of both borides, i.e. their amount in borides correlates with their content in steels. However the borders of boride phase needles contain more boride than the steel itself.

In particular, the boride needles' borders of steel 7X3 contain 5–8% chrome; the zones between needles—2–2.5%. The boride needle's borders of steel 5X3B3MΦC contain 4% chrome, 2.5% vanadium, 1.5% molybdenum; these amounts all exceed their content in the steel itself [3].

Due to the replacement of carbon by a growing layer of borides, carbide-forming elements on the border of the Fe_2B boride generates an alloyed carbon boride zone. If a steel contains sparingly soluble carbides (e.g. WC), they can be located as separate inclusions even in the boride layer and sublayer.

The distribution of microhardness along the width of the boride layer has a gradual character in most steels (Fig. 10.2). The pseudo-eutectic structure of the

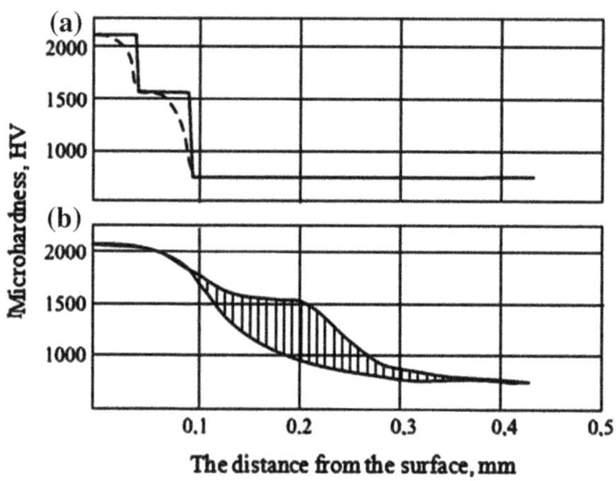

Fig. 10.2 The pattern of a microhardness change along the width of boronized layers in steel У8: **a** in boride phases and transition zones (*dotted line*—average microhardness); **b** the combination of boride phases on the surface and a pseudo-eutectic structure in the sublayer [2]

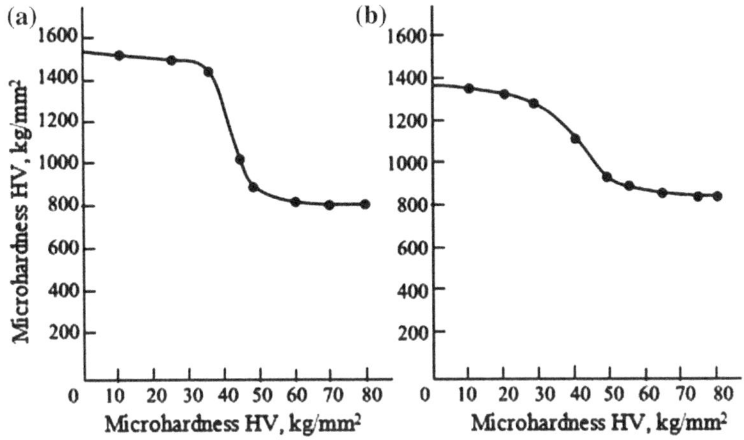

Fig. 10.3 The pattern of a microhardness change along the width of boronized layers in steel X12M: **a** after contact gas medium boriding; **b** after electron-heat processing of the same layers (combined boriding)

sublayer in steel У8 was formed at 1100 °C as a result of the replacement of carbon with a boride layer and setting the sublayer temperature higher that solidus temperature. At the same time microhardness >2200 HV and the layer's width >80 mkm abruptly increase the unfused boride layer's tendency to spontaneous chipping.

The general tendency of changes in microhardness after combined boriding, connected to further electron-beam heating of the boride layer, is the hardness reduction on the surface, thickening of the boronized layer and the increase of plasticity (Fig. 10.3) [4].

The microhardness of boronized layers with heterogeneous structure follows the rule of addictiveness (in first approximation):

$$HV = q_1 HV_1 + q_2 HV_2 + \cdots + q_n HV_n \tag{10.1}$$

where HV—the total calculated microhardness; HV_1, HV_2… HV_n—the microhardness of phases comprising the layer's structure; q_1, q_2… q_n—a specific amount of each phase.

This calculation of hardness should be done for heterogeneous areas of boronized layers with the interrupted FeB layer on the surface, in transition zones from FeB to Fe_2B, from Fe_2B to the sublayer and from the sublayer to the structure of the main metal (due to the different levels of acicularity of borides and sublayer's structure) (Fig. 10.4).

The calculation is even more appropriate for eutectic and pseudo-eutectic structures (Fig. 10.4c) as the distribution of width microhardness has a wide oscillation interval (Figs. 8.15 and 8.18).

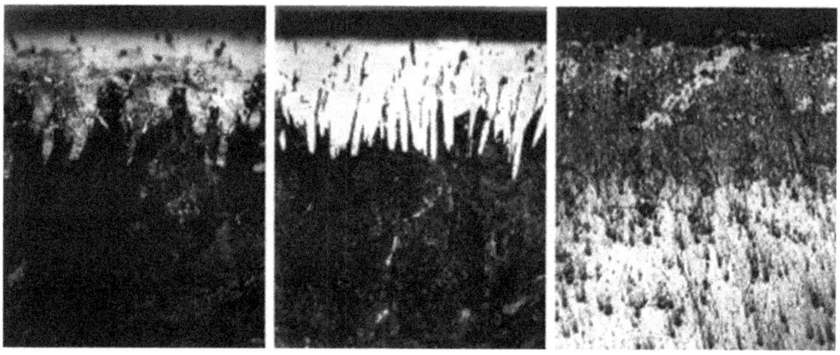

Fig. 10.4 The microstructures of boronized layers of different acicularity and heterogeneity: **a** a two-phase layer obtained in the bath containing 70% $Na_2B_4O_7$ and 30% B_4C (950 °C, 6 h, steel 45, ×320); **b** a two-phase layer obtained in the bath containing 70% $Na_2B_4O_7$ and 30% silicon zirconium (950 °C, 6 h, ×320): **c** a pseudo-eutectic layer obtained in the bath containing 96% $Na_2B_4O_7$ snd 4% MgB_{12} (1050 °C, 2.5 h, smelt ЭП877, ×100, not polished section)

Pseudo-eutectic composite structures are formed during the boronizing of high-alloy steels and smelts by the diffusion-crystallization method. Thermal and concentration conditions of processing are determined by the 4-component equilibrium diagrams and their polythermal cross sections.

For instance, the microhardness of structural components of plastic pseudo-eutectic boronized layers in the smelt ЭП718 varies from 440 to 1620 HV. The layer consists of light non-pickled crystals Cr_2B and CrB with 1350–1620 HV microhardness, yellow crystals Ni_3B (1100–1300 HV) and dark structural composites (450–650 HV) which represent Fe-Ni solid solution.

The calculated variable gives a more complete characterization of real properties of layer's areas than the curve of microhardness width distribution built on measurements. It is used to forecast the calculations of wear-resistance and work durability for boronized details.

10.2 The Pattern for the Stress State Formation in Boronized Layers

Iron borides in boride layers are oriented relatively to the saturation surface (textured) and naturally have the anisotropy of thermal expansion coefficients. This anisotropy significantly impacts the formation of temporary and residual stresses in boronized steels. The evaluation of main thermal expansion coefficients for FeB boride (α_a, α_b, α_c) and Fe_2B (α_a, α_c) was carried out by differentiating the curves of thermal dependency of transformation in their crystal lattices' parameters.

For FeB it is:

$$\alpha_a = \frac{1}{a}\frac{da}{dt};\quad \alpha_b = \frac{1}{b}\frac{db}{dT};\quad \alpha_c = \frac{1}{c}\frac{dc}{dT}. \tag{10.2}$$

For Fe_2B, which has a tetragonal crystal lattice, $\alpha_a = \alpha_b$.

The thermal expansion coefficients (parallel to the diffusion front) determined the type of stress:

$$\alpha_{||} = \frac{\alpha_a + \alpha_b}{2}. \tag{10.3}$$

Changes in FeB, Fe_2B and Fe borides, occurring after the change of temperature after boriding (<800 °C) and further cooling, show that compression stresses must be formed at Fe_2B-Fe phase borders. The border between FeB-Fe_2B should be characterized by a significant decrease of these stresses or even a transition to tensile stresses on the surface.

The thermal expansion coefficient (perpendicular to the diffusion front; the coefficient is similar to α_c) is responsible for the strength of the bond between the boride layer and the substrate and determines the layer's tendency to spontaneous chipping. For Fe_2B boride, this coefficient is permanent ($\sim 7 \times 10^{-6}$ 1/K); for FeB it changes from 7×10^{-6} to 20×10^{-6} 1/K at 20–1000 °C. This determines an abrupt reduction of compressive stresses on the FeB-Fe_2B border with their transformation into tensile stresses on the processed surface [5, 6]. This pattern of stress distribution leads to the non-equiaxial total compression if the wrong processing mode was set or the appropriate choice of material and layer's width combination and phase correlation was made. This, in combination with the high anisotropy of mechanical properties of crystal lattices in borides, the natural hardness and brittleness of FeB, explains the crack formation in two-phase layers and spontaneous flaking and the formation of separate shears on the edges and tops and even flat planes. This effect is specifically observed at high content of FeB.

The average volume thermal expansion coefficient (α_v) of boride phases and iron (Fig. 10.5), calculated according to the formula (10.3), openly reveals the pattern of stress state formation:

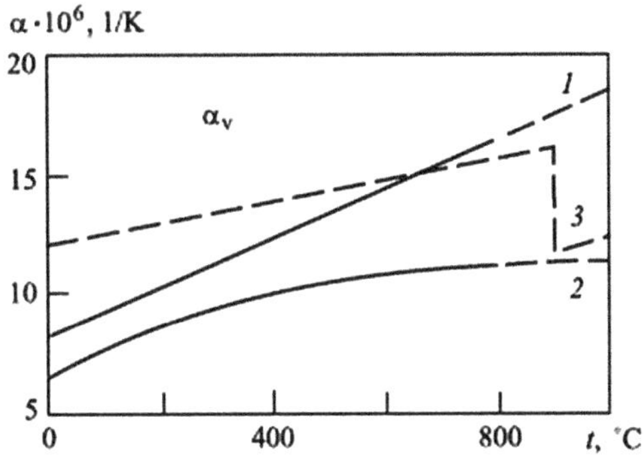

Fig. 10.5 The pattern of change for the volume thermal expansion coefficient of boride phases and iron: *1* FeB; *2* Fe_2B; *3* Fe

$$\alpha_v = \frac{\alpha_a + \alpha_b + \alpha_c}{3}. \tag{10.4}$$

Thus, Fe_2B-based one phase layers at 550–1100 °C saturation and further slow cooling are expected to generate residual compression stresses. Two-phase layers ($FeB + Fe_2B$) generate residual stresses after saturation at temperatures <800 °C and in the presence of a slight amount of high-boron phase FeB.

The residual stresses of boronized layers with different phase compositions were estimated according to the method described in [7]. It involved the use of boronized samples in the form of plate (50 × 10 × 2 mm) which were exposed to one-sided electrolysis bleeding. The boriding process was carried out at temperatures 800 and 950 °C. The process duration was chosen to obtain equal 100 mkm layers. The used methods and compositions for boriding (Table 10.3) ensured the generation of layers with different phase compositions. The calculation of residual stresses was done according to the recommendations [8].

The obtained results reveal that the most appropriate distribution of residual stresses in the boride layer is observed at its one-phase structure (Fig. 10.6, curves 3–4).

The level reduction of compression stresses on the layer's surface, obtained in composition 3, was caused by the influence of manganese. The presence of FeB phase (with a high thermal expansion coefficient) on the surface of the boronized layer creates a new borderline and leads to the reduction of compression stresses. The increase in its content leads to its more significant decrease. The maximum of compression stresses shifts deeper into the boride layer. This leads to a significant

Table 10.3 The methods and compositions used for obtaining different boride layers in order to calculate residual stresses [9]

No.	Boriding method	Composition of saturation mixture	Phase composition of the layer
1	Electrolysis method (salt bath)	100% $Na_2B_4O_7$ (j = 0.15 A/cm^2)	$FeB + Fe_2B$
2	Non-electrolysis (salt bath)	70% $Na_2B_4O_7$ + 30% B_4C	$FeB + Fe_2B$
3	Non-electrolysis method (salt bath)	70% $Na_2B_4O_7$ + 30% silicomanganese	$(Fe, Mn)_2B$
4	Gaseous contact method in powder medium	50% B_4C + 45% Al_2O_3 + 5% KBF_4	$FeB + Fe_2B$
5	Gaseous contact method in powder medium	45%CaB_6 + 45% CaO + 7% NaF + 2% KBF_4 + 1% S (rod)	$Fe_2(B, S)$

Note Boriding in compounds 4 and 5 was carried out in containers for powder saturation; the fusible lock used (B_2O_3 material)

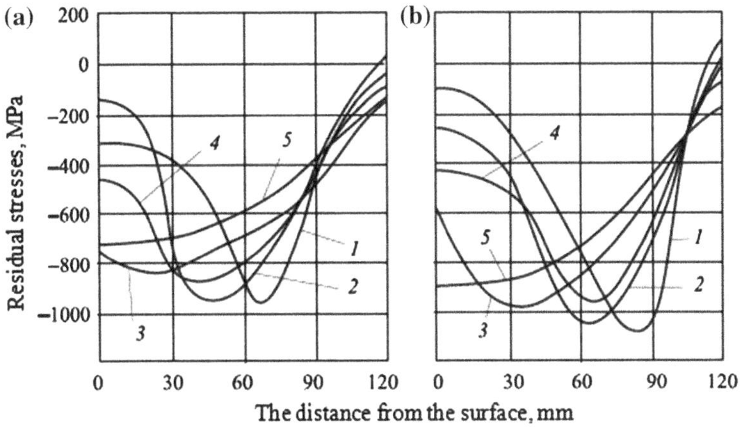

Fig. 10.6 The distribution of residual stresses in boronized layers (numbers near curves correlate with compositions from Table 10.3): **a** steel 45; **b** steel У8

drop of residual stresses which negatively effects the exploitation properties of boronized details.

These patterns are typical for carbon and low-alloy steels. The boride layers in medium- and high-alloy steels have their own peculiarities of the distribution of residual stresses in the layer. In particular, the presence of 2% Si, 3% Cr and 3% Ni in a borided medium-carbon steel leads to an abrupt fall of compression stresses; they are completely eliminated or transformed into tensile stresses with the due consequences. The influence of silicon and nickel in this case is explained by the formation of a transition zone of α-solid solution in the sublayer and an interrupted or constant new border sublayer-substrate. Bigger amounts of alloy elements determine their influence by change in the layer and the sublayer's structures and the level of solubility of these elements in borides which leads to the change in correlation of linear expansion coefficients among the borides, sublayer and steel. For instance, α-solid solution of silicon in iron has a bigger thermal expansion coefficient than iron. This leads to the formation of tensile residual stresses in the sublayer and increased tendency of the layer to peeling. A certain influence is done by the change of phases' thermal conductivity.

Accelerated cooling after boriding allows to prevent the precipitation of borides, iron carbon borides and alloy elements in the sublayer (they cause a negative redistribution of stresses in boronized layers); also it allows to avoid a significant deformation of boronized details.

The amount of residual stresses is effected by the boride layer's width and scale factor. Thickening of the boride layer reduces compression stresses on the surface. A negative value of the scale factor (ε) abruptly increases the layer's tendency to spontaneous shearing:

$$\varepsilon = h_s / h_{sample} \quad (10.5)$$

where h_c—the boride layer's width: h_{sample}—the width or diameter of the sample or detail.

Faultless two-phase layers (FeB + Fe_2B) are generated in steels which contain:

$$0.2\%\,C \text{ at } \varepsilon \leq 0.02$$
$$0.45\%\,C \text{ at } \varepsilon \leq 0.016$$
$$0.8\%\,C \text{ at } \varepsilon \leq 0.012.$$

At the same time, a boronized surface of details with a two-phase layer may generate a grid of small cracks due to the realization of residual stresses which doesn't reduce exploitation characteristics of boronized details and tools.

One-phase boride layers have a certain reserve of plasticity and allow certain improvements of the scale factor. For instance, for steel 45 it is ≤0.023. The scale factor manifests itself mainly during boriding of small details and tools. Boriding of construction medium- and high-alloy steels requires the reduction of the scale factor and acknowledging the sublayer's structure in order to obtain faultless layers.

Immediate quenching after boriding leads to the reduction of residual compression stresses (Fig. 10.7).

The level of compression stresses increases again after tempering. These changes totally correlate with changes in the volume of the substrate during quenching and further tempering.

The chance of cracks grid formation on the surfaces with two-phase boride layers may be reduced by the decrease of the phase stresses level. It can be achieved

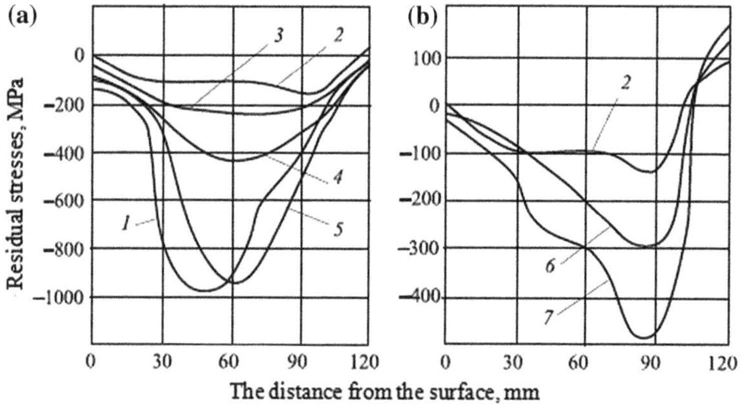

Fig. 10.7 The influence of a cooling speed (**a**) and a tempering temperature (**b**) on the distribution pattern of residual stresses in steel 45 after boriding in composition 2 (Table 10.2): *1* cooling in furnace; *2* quenching in water; *3* isothermal quenching in a bath (300 °C); *4* same (500 °C); *5* air cooling; *6* tempering at 200 °C; *7* tempering at 400 °C

by using isothermal quenching. A high level of residual compression stresses in a boronized detail can also be guaranteed by quenching accompanied by high frequency current heating. The pattern of residual stress distribution would be the same as in one-phase boronized layers.

The residual stresses in boride layers have an efficient temporal stability. Three months after soaking, the residual stresses decrease by 15%. During further soaking they change insignificantly [1]. The change of residual stresses is observed at the beginning of boronized details' work cycle and depends on external conditions.

The stress state of composite structures is determined by the presence of a continuous zone of α-solid solution, the amount and level of inclusions' dispersability. However even an insignificant difference in thermal expansion coefficients provide the opportunity for relaxing temporary compression and tensile stresses due to the plasticity of α-solid solution. The formation of boride or carbon boride inclusions with a coherent bond with α-solid solution (ensuring disperse solidifying) in boronized layers hasn't been observed. Therefore the level of residual stresses in boronized layers with composite structures has minimal values.

10.3 The Pattern for the Brittleness Formation in Boronized Layers

The brittleness of boronized layers is determined by their phase composition, stress state, type and structural state of the substrate's material. Since boronized layers are multiphase structures with alternating width hardness and alternating stress state, the brittleness of these layers is also alternating. Therefore widely used characteristics of brittleness in boronized layers (microbrittleness, brittleness point, brittleness factor) are only comparative values and are quite subjective. Nevertheless these comparative characteristics have practical significance and allow to choose the most efficient coating.

The most advanced method is microbrittleness tests [1, 10]. The total value of the brittle fracture for boride phases in ARMCO-iron demonstrates the advantage of a Fe_2B phase (Fig. 10.8).

The studies on the microbrittleness of one-phase layers revealed that the increase in width and hardness leads to the increase in microbrittleness. The alloy elements, changing the boride layer's width, the patterns of stress state and the properties of the Fe_2B boride itself, slightly transform this dependency. However the general regularity of microbrittleness growth and the boride layer's hardness is preserved in alloy steels.

A large number of studies show that the total value of brittle fracture of the FeB boride in steel 40 is twice and the brittleness factor is 6 times higher than for the Fe_2B boride. Thus, the brittleness of a boride layer in general is determined by the brittleness of a surface FeB phase and its amount [1].

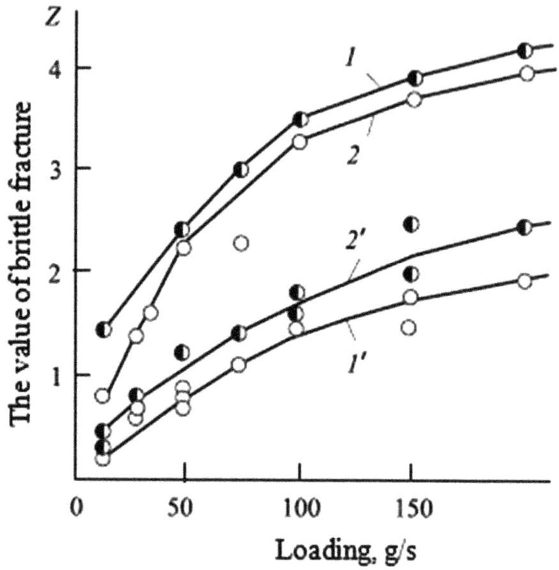

Fig. 10.8 The influence of the boriding method on the total value of brittle fracture in boride phases of ARMCO-iron: *1*, *2* FeB; *1′*, *2′*—Fe_2B; *1*, *1′*—boriding in the bath of 35% B_4C and 65% $Na_2B_4O_7$ (1000 °C, 8 h); *2*, *2′*—electrolysis boriding in the bath 100% $Na_2B_4O_7$ (1000 °C, 4 h, 0.3 a/cm^2)

The evaluation of the brittleness value of boride layers according to the method for measuring brittleness in nitrated layers (Chap. 6) demonstrates a comparatively low level of the boride layer's brittleness (Table 10.4) obtained in the positive combination of the layer's width, material and processing conditions.

The calculated average brittleness value doesn't exceed 1.6. One-phase boride layers got 1 point (out of 4) in all studied modes of saturation (i.e. the layers weren't brittle). The shown results allow to make a conclusion that the appropriately chosen

Table 10.4 The influence of steel composition on the microhardness and the brittleness value of boride layers [1]

Evaluation characteristics of the brittleness value and microbrittleness	The distribution of imprints by the brittleness values and microhardness (HV) for different steel brands							
	40	40Г2	40Х2	40Н2	40Д (0.65)	40В2	40М (0.62)	40Ю
Non-brittle	14	11	10	12	10	10	9	10
Slightly brittle	5	7	10	3	6	4	3	5
Brittle	2	2	–	–	–	3	3	–
Medium brittle	1.44	1.55	1.50	1.25	1.38	1.59	1.6	1.33
Microhardness	1860	2150	2110	1680	1750	2260	2270	1760

Note The conditions for the generation of boronized layers: electrolysis boriding 950 °C, 3 h, j = 0.25 a/cm^2. Experiments were carried out at 10 kgf loading. The duration of experiment—30 s saturation mode leads to establishing a comparative low level of brittleness and ensures a good connection with the substrate's material.

The simultaneous silicon and boron saturation reduces the brittleness factor of FeB boride by 30%, Fe_2B doesn't change; cuprum and boron leads to the reduction of the hardness and brittleness factors of the high-boron phase and the oscillation in the margin of error in Fe_2B boride; chrome and boron—for FeB the brittleness factor increases, for Fe_2B—doesn't change [3].

The energy of interatomic interaction in melted borides is comparatively big that's why they have high values of elasticity module (for FeB—5.5×104 kgf/mm^2; for Fe_2B—3×104 kgf/mm^2). If the maximum critical shearing stress, which determines the resistance of crystal borides against plastic deformation, is 0, 1 from shear modulus, the high deformation resistance of borides is observed. However the properties of borides in diffusion layers are influenced by the technological factors, growth peculiarities, the level of dispersability, quantitative phase correlations, the layer's compactness etc.

The widely-used evaluation characteristics for brittleness in boronized layers are subjective and don't reflect their real properties. The most appropriate brittleness characteristic is the shear stress (σ_{sh})—the minimal stress causing a shear while measuring microhardness at different distances from the sample's edge. This stress integrally includes the natural properties of the layer's phases (microhardness, elasticity modulus) and its state (phase composition and their correlation, dispersibility and mutual position in the layer and on the surface, stress state, texture properties, ductility reserve etc.) [2].

The transition from FeB phase on the surface to the mixed zone of FeB + Fe_2B and further to Fe_2B phase and to the sublayer is accompanied by the reduction of brittleness in boronized layers. Moreover the disconnected distribution of borides reduces the total microhardness in eutectic and pseudo-eutectic composite structures which leads to the brittleness reduction (Fig. 10.9).

The given data shows a distinct correlation between the microhardness and the shear stress. Thus, the processing of a particular detail may be appointed to

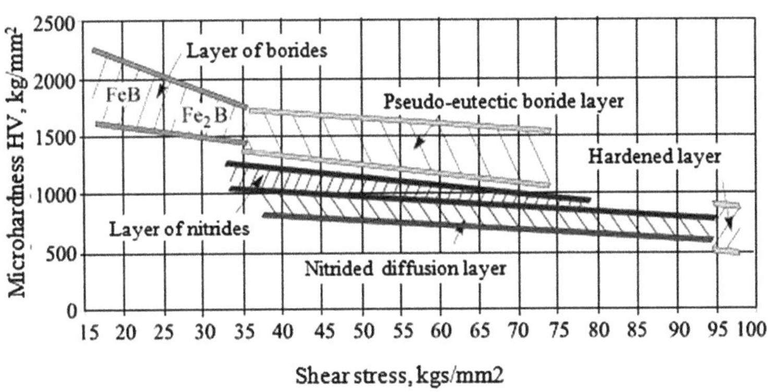

Fig. 10.9 The correlation between microhardness and shear stresses of boride compact, pseudo-eutectic boronized and nitrated layers [2]

Table 10.5 The influence of processing conditions on the brittleness of boronized layers

Boriding type		HV, kgf/mm^2	σ_{sh}, kgf/mm^2	ε	P, kgf	Brittleness value
Gaseous contact		1950–2050	24.5	1.13	155	4
Electron-beam		650–1450	48.4	1.65	245	3
Combined	a	1680–900	40.5	1.32	195	2
	b	1550–750	41.4	1.38	200	1.6
	c	1350–550	43.1	1.41	210	1.4

Note HV—the microhardness; σ_{sh}—the shear stress; ε_{def}—the ultimate deformation (the correlation between the imprint's diagonal to the length of the crack between the imprints); P—the bending stress of the first cracks' formation. The conditions for combined (consequent) boriding at 1050–1100 °C: **a** $P_{str} = 2.1 \cdot 10^3$ W/cm^2, $\tau = 6$ min; **b** $P_{str} = 2.4 \cdot 10^3$ W/cm^2, $\tau = 5$ min; **c** $P_{str} = 2.8 \cdot 10^3$ W/cm^2, $\tau = 3$ min. The results of gaseous contact boriding didn't meet the optimal combination of phase composition, layer's width, material and processing conditions (100% B_4C, 940 °C, 3 h, 80–90 mkm)

particular exploitation conditions which would ensure the maximum effect, i.e. guarantee a favorable combination of hardness and plasticity reserve. In particular, abrasive wear conditions require boronized layers on the basis of borides while composite layers (eutectic, pseudo-eutectic, non-compact etc.) are more appropriate for alternating loading [11].

The change in shear stresses in the boronized layer allows to make a relatively reliable forecast on the durability of the hardened detail if the wear process forecast is needed.

The comparison of the boronized layers' properties in steel 45 obtained by contact gas boriding, electron-beam boriding and consequent combined processing (boriding + electron-beam heating) demonstrates the high plasticity of composite structures (Table 10.5).

10.4 The Wear-Resistance of Boronized Layers

A high level of wear-resistance in boronized layers manifests itself in the conditions of liquid borderline friction and grease-free friction during rolling, sliding and abrasive wear. The high productivity of boronized details and tools is demonstrated under fretting corrosion, at low and high temperatures, in vacuum and aggressive mediums [1]. The wear-resistance of boronized layers is practically the same as the wear-resistance of carbide layers and solid smelts. However boronized layers are wider than carbide layers and may be obtained in any metal materials therefore their use is more effective. In comparison with solid smelts, boronized steel tools are much more economically efficient; due to their foundation, these tools are very productive at stress loadings. Moreover the wear-resistance of many solid smelt tools may be increased by boronizing.

A relatively high wear-resistance of boronized layers can be explained from the point of the microgeometry of interacting surfaces during grease-free sliding friction [2]:

- Due to the absence of the phase transformation in borides and the complexity of diffusion processes, the precipitated friction heat accumulates in the surface layers of borides as the coefficient of thermal conductivity for borides is higher along the surface than in the perpendicular direction. It is estimated as ~20 W/(m K) for FeB boride and ~26 W/(m K) for Fe_2B boride. The precipitated surface heat is used for heating and oxidation.
- The revealed good conformability of boride layers is explained presumably by filling the cavities of boronized surface by the products of cutting the counterbody's more plastic metal due to the significant difference in hardness of the frictional couple and also by filling the cavities with oxidation products.
- The increased hardness of the counterbody material or the work of the frictional couple borides-borides fills the cavities with boron oxides and alloy elements. Boron oxides in solid state have a scaly structure and reduce the friction coefficient; they also play the part of a half-liquid lubricant material when softened (~450 °C).

The multiple studies on the wear-resistance of boronized layers don't offer any universal approach to the effective argumentative choice of the layer type for particular exploitation conditions. The tribotechnical properties determine the following classification of diffusion boronized layers:

1. Layers with the foundation in the borides of the processed material. Their peculiarity is the acicular construction of borides with a continuous zone on the surface. This structure has either homogenous (FeB or Fe_2B based) and heterogeneous (on the base of different combinations of boride phases and combination of Fe_2B and α-solid solution). Similar layers are normally obtained on low-alloyed, low- and medium-carbon steels and iron. They have stable tribological properties of the continuous layer and alternating properties for transitioning the acicular structure from one phase to another and for boride transitioning to the structure of the sublayer.
2. Alloy boronized layers based on the borides of the processed metal (material). The process of boride alloying may be determined by the saturation mixture or by the alloy elements of the processed smelt, or by the combination of these two factors. The structure of these layers is characterized by decreasing acicularity of borides and the presence of the sublayer in the form of solid solution important for the formation of stress state and final tribological properties of boronized layers in general.
3. Alloy boronoized layers based on alloy solid solutions with separate inclusions of boride needles. These layers have a heterogeneous structure with changing tribological properties in the layer. They can be obtained at simultaneous boron and silicon, boron and aluminum saturation or at consequent saturation (silicon and aluminum and then boron). This type of structure sometimes has layers

obtained according to the diffusion-crystallization mechanism with pseudo-eutectic and eutectic structures and layers after combined boriding (boriding with further electron-beam processing). Separated boride needles are obtained when the boron concentration is uneven along the processed surface.

4. Boronized layers based on the solid solution of boron and alloy elements in the processed material. These layers may be obtained on ferrite high-alloy steels during their processing in mediums with a low saturation capacity (a low boron potential) and processing of other alloy steels accompanied by the formation of layers based on solid solutions with boride and carbon boride inclusions along the grain borders. Tribological properties of these layers are close to the properties of a hardened steel and are characterized by a relative homogeneity.
5. Boronized layers based on solid solution of boron and alloy elements with inclusions of boron of different level of dispersity. These composite boronized layers may be obtained during welding, fusing of galvanic, electrochemical and other coatings, high frequency current saturation and electro-beam processing and diffusion-crystallization saturation under furnace heating [12]. The properties of these layers depend on the amount, form and mutual position of boride inclusions and the properties of α-solid solution. The tribological properties of these layers change insignificantly along the width.

Thus, the diversity of structures in boronized diffusion layers presupposes a more attentive approach to the development of the evaluation criteria for their exploitation characteristics. The tribological properties and durability of such layers depend on the phase composition, width, stress state and mutual position of structural components. The pattern of interaction between hardened rubbing surfaces is determined—on the one hand—by the kinetics and mechanism of friction and wear and—on the other hand—by the state of the coating's surface layers. This state is explained by different processes going on the surface of the friction: diffusion, adsorption, adhesion, thermal changes, substructure transformations etc. For friction and wear, the most important one is the pattern of destroying an adhesive bond on the contact spot of the frictional couple.

The destruction of the microwelding junction determines the damage of the microcontact according to the fatigue mechanism at the gradual structural loosening of the active layer. The friction process should be estimated as stable in the conditions of external friction. The deformation is concentrated in the thin surface layer. If the destruction of the adhesive bond takes place deeper accompanied by the tear-out of the metal, then the wear goes as a range of single catastrophic acts accompanied by the formation of wearing particles. The friction process wouldn't be stable.

The rule of the positive gradient of mechanical properties concerns the difference between the strength of adhesion bonds with the counterbody and the strength of lower metal layers. The more wear-resistant material would not be the roughest material but the material with a sufficient plasticity reserve which is capable of self-healing defects and tolerant to plastic deformation. The loss of this capability leads to the formation of cracks and fracture. Eutectic, pseudo-eutectic and complex boronized layers with a transformed boride morphology (Chinese hieroglyphs) meet most of these requirements.

Diffusion boronized layers with a wide range of structures are able to establish a reliable work of details in different exploitation conditions. Therefore processes taking place on friction surfaces of boronized details must be analyzed from the point of view of different wear mechanisms.

The fatigue wear at the initial stage is accompanied by the gradual accumulation of defects in the boride layer and its further fracture due to the loss of plasticity and the formation of cracks. The presence of grease material reduces contact stresses and decelerates the crack formation's process. The peculiarities of fraction of strong boride layers is that friction areas of contact surfaces have a typical fractography with a net of fatigue cracks. These cracks begin to develop at structural concentrators (pores, inclusions, boride needle points etc.) and go deeper into the metal forming voluminous "cells". The formation of cells is preceded by the processes of mechanical surface and near-surface layers loosening with the gradual accumulation of substructural defects of different kinds.

After the continuous layer of the brittle FeB phase is worn out, the process is carried out with a larger plasticity reserve and in the conditions more favorable from the point of view of the brittle wear. Normally this zone of the boronized layer has compression stresses which complicate the generation and development of fatigue cracks.

The friction coefficient for such materials is a structure-sensitive characteristic which depends on the structure peculiarities of the active layer on the surface which in turn is determined by the structural type of diffusion boronized layers. The presence of porous functioning as lubrication pockets ensures the generation of stable third body in contact zones and decrease in the friction. Boride layers are positively influenced by a less significant (than in other materials) reduction of wear resistance accompanied by increase of unit pressures. The increase of sliding speed normally causes the initial decrease of friction coefficient and then stabilization.

The mechanical-chemical wear of materials (with tendency to tribo-chemical and adsorbent loosening) causes the appearance of covers of secondary structures on the friction surfaces; they are formed as a result of mechanical deformation of thin surface layers accompanied by the simultaneous active interaction with aggressive components of the environment. The process of wear is controlled by the formation and fracture speeds of secondary structures. The friction speed in this case forms a temperature mode on the friction surface and the type of oxide covering. Depending on the structural type of boride layers, oxide coverings may be based on B, Fe, Cr, Ni etc. Boron oxides, having a layer structure and free bonds, are solid grease materials that guarantee the reduction of the friction coefficient.

The homogenous layer's structure makes a wear process dependable on the boride properties, the heterogeneous structure—on the crystal structure of borides and solid solutions. A certain function is given to the texture properties of homogenous boride layers and compression stress level. In particular, the accelerated experiments in Skoda-Savin machine on homogeneous boride layers obtained at different temperatures in sliding friction grease-free conditions revealed

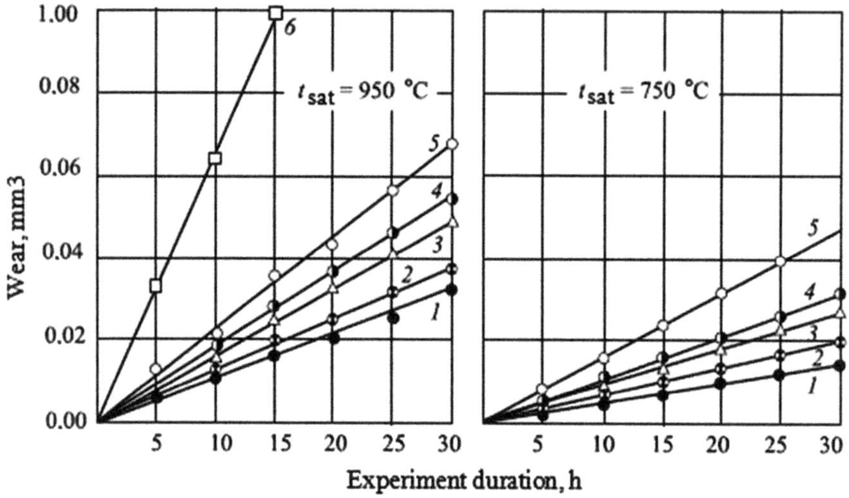

Fig. 10.10 The influence of an experiment duration on the wear of one- and two-phase boride layers in steel 45 obtained in baths with different electrochemical reducing agents: *1* B_4C; *2* ligature P70; *3* CaC_2; *4* Silicomanganese; *5* SiC; *6* steel 45 (after 800 °C quenching and low tempering)

a higher wear-resistance of layers obtained at lower temperatures and having better texture properties and compactness (Fig. 10.10). Two-phase boride layers had more advantages for wear-resistance.

When areas of α-solid solution of the sublayer withdraw on the friction surface in the conditions of the boride layer presence or when the heterogeneous complex structure layers wears off, it is necessary to choose basic materials of the frictional couple with different crystal structures, chemical composition and solubility to prevent the setting.

The abrasive wear determines the detail's durability by hardness of surface layers and correlation between the hardness and the abrasive. In particular, for the homogeneous structure of surface layers, a relative wear-resistance may be calculated according to the formula:

$$U = b \cdot \mathrm{HV}, \tag{10.6}$$

where b—proportion coefficient [13]; HV—hardness.

However even the homogeneous structure hardness is not universal for calculations as it doesn't account for the plasticity reserve and the stress state of the layer, its crystal structure and texture properties. The mechanical calculation of inclusions' size and their distribution density in the heterogeneous structure at set values of elasticity modulus [14] doesn't take into account the important real peculiarities of the structure of boronized diffusion layers either.

In all cases wear-resistance is more or less determined by the fatigue processes going in the diffusion layer, i.e. the generation and development of fatigue micro-

and macro-cracks. They depend on micro- and macro-barriers decelerating dislocation processes and the effects of their expansion.

The expansion deceleration may be reached by structural changes in diffusion layers, i.e. creating structural barriers of micro- and macro-levels or creating prolonged zones of compression stresses and cracks with blunt tops. One of the perspective directions is the structural type of cracks expansion blockage by forming a heterogeneous structure with optimal heterogeneity along the surface section and creating coherent microinclusions at saturation and friction (developing the effect of spontaneous ageing). Another effective way is changing the growth direction of cracks up to its turning into counter-direction or its swirl. It may be done by shaping borides into an acicular or plate form and their free deorientation in the layer. These structures of boronized layers are obtained during the diffusion-crystallization mechanism formation in high-alloy steels containing elements of boride generators. The formed plate pseudo-eutectic boronized layers are heterogeneous with boride inclusions which are evenly distributed in α-solid solution. The change in the boronized layer's morphology by adding electron-beam processing is also appropriate for a significant increase in the plasticity reserve and wear-resistance.

10.4.1 The Wear Process Pattern for Boronized Layers in Sliding Friction Conditions

The wear-resistance of boride layers is determined in many ways by the width and correlation of FeB and Fe_2B phases in the layers. Each steel brand has a critical boride layer width; the excess would lead to the increase of brittleness and spontaneous shearing on edges, tops and certain areas of flat plates.

The wear of two phase boride layers with loadings <20 kgf/mm^2 doesn't cause a significant stress in a contact zone and the wear process has a normal oxidation character. The friction surface is smooth and shining. In these conditions two phase layers have an advantage in comparison with one phase boronized layers. The increase of loading from 20 to 80 kgf/mm^2 at slow sliding speeds (<0.5 m/s) leads to the intensive growth of stresses in the contact zone, brittle fracture and coloring of separate boride particles which function as an abrasive and accelerate the wear process of the whole layer. The friction surface has marks of brittle fracture and the wear reaches its maximum value, i.e. they are worn out faster than one-phase layers.

The changing phase composition and the structure of boronized layers (beginning from the surface) defines the changing nature of wear processes. The smooth change of the wear manner happens in FeB-Fe_2B, Fe_2B-sublayer, sublayer-substrate borderlines.

The increase of sliding speeds leads to a temperature rise in the contact zone, the acceleration of corrosion and the formation of boron oxides which function as solid lubricants or in certain conditions soften and function as liquid lubricants. The friction coefficient and wear value decreases, the state of the surface improves [1].

Table 10.6 Wear processes of different boride layers in steel 45 [15]

Type of boronized layer	The value of wear, mkm/s, at different sliding speeds, m/s					
	0.05	0.1	0.5	1	3	5
Two-phase (FeB + Fe_2B)	49	40	8	3.5	2	1
One-phase (Fe_2B)	22	12	6	3	2	1
Eutectic (FeB + Fe_2B) + a-solid solution	8	6	1.5	1.2	1	1

Notes Pressure—10 kgf/mm^2; counterbody—VK-2 smelt

The typical peculiarity of the wear process for boronized layers is the absence of setting centers in the wide range of sliding speeds and loadings.

One-phase boride layers with a larger plasticity reserve (in comparison with two-phase layers) are less sensitive to stresses emerging at slow sliding speeds (up to 0.4 m/c). A certain positive role is given to the residual compression stresses which can be found on the surface and in the volume of Fe_2B borides. Faster sliding speeds even out the wear intensities of one-phase, two-phase and eutectic boronized layers (Table 10.6). The obtained patterns of wear process correlate with experiments done in other wear conditions at sliding grease-free friction [15].

Brittle fracture conditions increase the wear-resistance of composite layers to the maximum. The increase of sliding speeds changes the wear mechanism from the mechanical to the corrosion form; it is accompanied by the generation of secondary structures which unifies the wear for all types of boronized layers. At the same time this wear is several times lower than the wear of hardened steels and other different wear-resistant layers (nitrated, cemented, nitrocemented etc.).

Experiments in sliding grease-free friction conditions of layers obtained after a simultaneous saturation with boron and other elements showed that the wear-resistance of steel 45 increases twice or 2.5 times after boron-siliconizing; 1.9–2.0X after boron-chromizing; 1.2X after boron-cuprum treatment; after boron-tungsten and boron-molybdenum treatment—1.8–2.5X; boron-titanizing and boron-zirconium treatment—1.4X in comparison with boronized layers. The data on wear-resistance were obtained at the pressure of 1.5 kgf/mm^2 and 0.42 m/s sliding speed [3]. In these conditions the minimal wear-resistance was observed in steel 5X3B3MΦC as a result of coloring the boronized layers. This was due to the presence of large amount of FeB phase (up to 50%), the presence of an unhardened zone in the sublayer and the depression of the boride layer and the formation of tensile residual stresses with a big amplitude of values in FeB-Fe_2B transition zone.

The comparative studies of boronized layers of different morphology in their boride structure under conditions of sliding friction showed the advantage of the layers obtained in combined (consequent) boriding (Fig. 10.11). The experiments were done on the boronized layers with acicular structure, eutectic structure and separate round inclusions of borides on the surface and layers with boride inclusions (Chinese hieroglyphs), localized in a certain pattern relative to the surface. The study was carried out on the machine СМЦ-2. The studied samples were given as

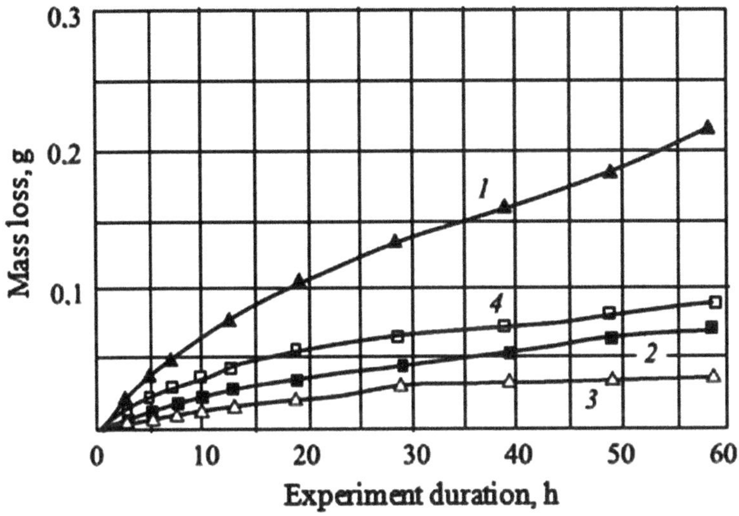

Fig. 10.11 The influence of experiment duration on the mass loss in boronized samples produced from steel Ст3: *1* initial state; *2* gaseous contact boriding in powder mixtures; *3* combined boriding; *4* electron-beam boriding

segments. The counter body was a disk made from hardened steel 45 (hardness 58HRC). The speed of the counter body rotation—85 cps. The loading—1.5 kgf [4].

The experiments on the pseudo-eutectic boronized structures in superalloys revealed that the load bearing capacity and wear-resistance of smelts increase. The study was carried out on the Ш-17 machine (ИМАШ-ВИАМ construction) in

Table 10.7 The results of the comparative experiments on superalloys after different treatments [2, 12, 16]

No.	Metal brand	Treatment	Speed of wear, V × 10^{-2} mm/min	Corrosion-abrasive wear, %
1	ВНС-45	Quenching + ageing boriding	1.4–1.6 0.13–0.20	83 29
2	ЭП 718	~	1.66–2.50 0.147–0.208	90 38
3, 4	ЭП 877	Quenching + ageing aluminizing boriding	1.6–2.3 0.25–0.26 0.13–0.20	85 95 30
5	8Х4В9Ф2	Quenching + triple tempering	0.147–0.208	150
6	90Х16НМ2Д	Quenching + ionic nitriding	1.8–2.2	142
7	13Х3НВМ2Ф	Cementation	0.8–1.03	155
8	14Х16Н2АМ	~	–	100

Note The wear is expressed in % in correlation with the frictional couple (cemented steel-resin 3862) in the corrosion-abrasive environment. After boriding, aluminizing and cementation, quenching and ageing were carried in the normal mode for the substrate's material rolling friction conditions with grease-free slipping at contact stresses $\sigma_c = 540$ MPa and temperature <400 °C (Table 10.7) [16].

The wear-resistance of boronized layers is in many ways determined by the sublayer's hardness. The creation of the solid quenched substrate under the boride layer prevents the depression and coloring of the boride layer during exploitation. Hard frequency current heating quenching (1.5 mm deep) increases the wear-resistance of boronized steel 45 twice or 2.5 times. Low tempering of the boronized steel after voluminous quenching also leads to the increase of wear-resistance by increasing compression stresses in the layer [1].

The influence of alloy elements on the wear-resistance of boronized layers manifests itself in the layer's structure, correlation of boride phases in the layer and microhardness, and also through the change of value and character of residual stresses' distribution in the layer.

10.4.2 The Behavior Pattern of Boronized Steels in Abrasive Wear Conditions

The wear-resistance of boronized steels in abrasive wear conditions is determined by the phase composition and hardness of the layer and also its stress state. The hardness of abrasive particles also has a significant importance on the wear speed.

The experiments with fastened abrasive particles are done at the side friction of the samples along the carborundum grinding sand-paper of certain granulosity. The two-phase boronized layers have the maximum wear-resistance (Fig. 10.12).

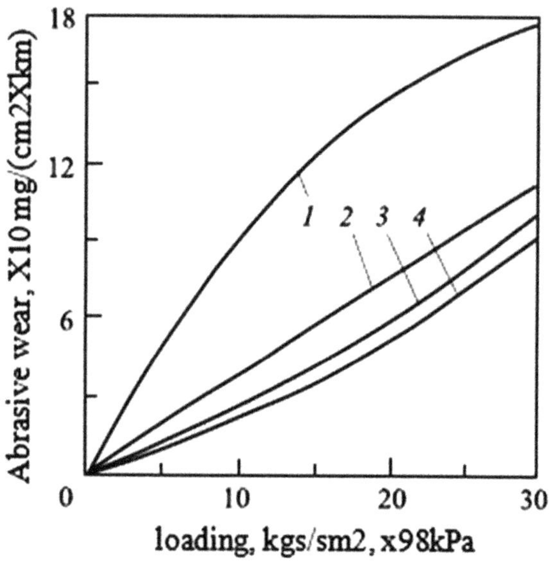

Fig. 10.12 The influence of loading on the abrasive wear of steel 45 after different treatments: *1* after quenching and low tempering; *2* after one-phase boriding; *3* after two-phase boriding; *4* after chromizing

Table 10.8 The abrasive wear-resistance of phases in the layer of carbon steel [1, 2]

No.	Treatment	Steel brand	Phase composition of a layer	Wear-resistance	
				tg α	ε
1	Two-phase boriding in the bath	45	FeB	0.020	6.75
			Fe_2B	0.060	2.24
		10	FeB	0.027	5.0
			Fe_2B	0.065	2.07
		У8	FeB	0.022	6.10
			Fe_2B	0.076	1.76
2	One-phase boriding	20	Fe_2B	0.051	2.65
		У8	Fe_2B	0.054	2.49
3	One-phase boriding	20	$(Fe, Mn)_2B$	0.030	4.3
		У8	$(Fe, Mn)_2B$	0.027	5.0

In comparison with hardened and low-tempering state of steel 45, one-phase boriding increases its abrasive wear-resistance 2–3 times, two-phase boriding—3, 5–6 times (Table 10.8).

The tests with unfastened particles demonstrated the 6–8X increase of wear-resistance. For instance, in the conditions of clay solution flowing with inclusion of abrasive particles.

High resistance to abrasive wear processes is characteristic for boronized layers in steel 45 in pressuring conditions for a large-grain anthracite blend [17]. Moreover, this boronized steel is more abrasive wear-resistant in the given conditions than boronized steels 20X, 30ХГСН2А and Х12Ф1.

Thus, the abrasive wear-resistance of boride layers is completely determined by the structure and phase composition. Following the correlation of the boride layer and the material width and the right choice of technological parameters preserve the abrasive wear resistance at a high level in the interval of loadings from 10 to 30 kgf/cm^2 and sliding speeds from 5 to 30 m/min. Recurrent boriding is possible on the surface of details and tools after the boride layer is worn out.

10.5 The Thermal Fatigue Resistance and Scale Resistance in Boronized Layers

Boride layers at heating temperatures up to 800 °C increase the air corrosion resistance in carbon and low-alloy steels for some time. Two processes take place in the layers: the corrosion of the surface and the layer's resolution. Corrosion is accompanied by the formation of boron and iron oxides which at temperatures <750 °C fully cover the surface and protect it from the air influence for some time. The increase of temperature causes the formation of boron anhydride drops, the exposure of the surface areas and the acceleration of corrosion. Finally, after 3–4 h the surface generates a layer of clean iron. The oxidation of the layer is accompanied by the decrease of FeB phase content and thickening of the boride layer. The temperature 900 °C and soaking >6 h lead to the total destruction of boride layers and don't prevent the corrosion.

The presence of alloy elements (Cr, Mn, Si, Mo, W) in medium-carbon steels (1–2%) doesn't significantly effect the kinetics of the boride layers' corrosion. A significant corrosion deceleration is observed in high-alloy steels [1] or during additional alloying of the layer in the saturation mixture. In particular, alloying of two-phase boronized layers with manganese increases their scale resistance in double [2].

One-phase layers based on Fe_2B have a high boron diffusion coefficient therefore they resolute faster, i.e. they are less scale resistant than two-phase layers.

A limited corrosion protection, presented by boronized layers, determines their rare use for increasing the scale resistance while the scale resistance results are significant for further thermal treatments.

Table 10.9 The influence of the number of cycles on the length of fatigue cracks [2]

Марка стали	Treatment	The number of cycles, n	Fatigue cracks length l, mm
3X2B8	Standard thermal treatment	3200	0.11
	Cementation	3300	0.11
	Boriding (FeB, Fe_2B)	3900	0.09
	Boriding [(Fe, $Mn)_2B$]	4800	0.06
	Chromizing	4200	0.08
3X3M3ФC	Standard thermal treatment	2600	0.13
	Cementation	2600	0.12
	Boriding (FeB, Fe_2B)	3200	0.11
	Boriding [(Fe, $Mn)_2B$]	3600	0.09
	Chromizing	3500	0.09
5XHB	Standard thermal treatment	2000	0.21
	Cementation	2400	0.2
	Boriding (FeB, Fe_2B)	3000	0.16
	Boriding [(Fe, $Mn)_2B$]	4100	0.1
	Chromizing	3500	0.13

Notes Two-phase boriding was carried out in the bath containing 70% $Na_2B_4O_7$ and 30% B_4C. One-phase boriding was carried out in the bath containing 70% $Na_2B_4O_7$ and 30% silicomanganese. After boriding all samples were exposed to quenching at 1050 °C (to prevent fusing) and high tempering at 560 °C. For steel 5XHB the heating temperature for quenching was 860 °C. Chromizing was carried out as a separate process with the fusible lock at 1000 °C in metal-thermal mixture containing 49% Cr_2O_3, 20% Al, 5%AlF_3, 26% Al_2O_3

When boronized layers are used at increased temperatures with a simultaneous mechanical impact, their durability increases. These conditions are specific for hot-stamp tools.

The determination of boride layers' durability in conditions of increased temperatures and multiple changes of temperature was carried out for two-phase and one-phase layers in stamp steels 5XHB and 3X2B8. The heating took place in a melted lead bath at 600 °C for 20 s; further cooling was carried out in boiling water. The evaluation of the sample damaging as a result of thermal circulation was carried out by the total length of thermal fatigue cracks (l) and by the number of cycles (b) before obtaining first 25–30 mkm deep fatigue cracks. The obtained results (Table 10.9) demonstrate a 1.5–2 times increase of thermal fatigue resistance in comparison with non-boronized samples produced from stamp steels and exposed to standard thermal processing.

The given results demonstrate a favorable influence of manganese on thermal fatigue resistance in boronized layers. An additional fast evaluation of thermal fatigue resistance was done by high frequency current heating of samples (lamp generator W = 10 kW, f = 440 kHz) up to 850 °C for 3 h (heating speed 285 °C/s). Cooling was done on air. After 10 changes of temperature, the non-alloy boride layers contained a layer of oxides on the surface; the manganese alloyed layers darkened insignificantly. Further 15 changes of temperature led to the formation of deep fatigue cracks along the whole surface and sides of the samples in the non-alloy boride layers. The manganese alloyed boronized layers generated a small number of small surface cracks.

Thus manganese alloying of one-phase and two-phase boride layers, apart from increasing the plasticity, improves their thermal fatigue resistance and scale resistance which makes this method appropriate for hardening stamp tools with hot and cold pressure treatment.

10.6 The Corrosion and Corrosion-Mechanical Fracture of Boronized Layers

Currently boronized layers are not used as corrosion-resistant layers as their effective protection has a limited durability [2]. The multiple effects of corrosion-resistance improvement, studied in different works, are explained by a short-term character of tests and convenient choices of objects for comparison which aren't normally used as corrosion-resistant materials. At the same time certain details of low-alloy steels with boride layers may function as substitutes of high-alloy corrosion-resistant steels for short-term protection from corrosion. It can be justified economically but anyway it should recognize its limitedness. Details must be changed or reboronized after a certain period of time.

The maximum of corrosion-resistance is provided by homogeneous protective coatings. The choice of protective coating (especially heterogeneous) should be done from the position of its corrosion-resistance and also—electronegativity in relation to the substrate's material and connected details, density, level of stress state and fracture pattern (which in many cases turns out to be a decisive factor). In particular, the corrosion sublayer fracture may be accompanied by the peeling of FeB boride layer's areas as a result of residual stresses' impact [11].

Electrochemically heterogeneous boronized layers of eutectic or pseudo-eutectic structure, separate boride inclusion in α-solid solution after combined or electron-beam boriding are not corrosion-resistant enough. The most appropriate choice of corrosion-resistant coating requires the systematic analysis of corrosion and corrosion-mechanical fracture of boronized layers.

The intensity of corrosion and corrosion-mechanical fracture is determined by the electrochemical heterogeneity of the system which ensures spontaneous functioning of galvanic elements in conditions of applied diffusion multiphase coatings on the surface and their work with heterogeneous materials in the presence of liquid or solid electrolytes (mediums with ionic conductivity). The difference of components' electrode potentials in the galvanic couple and their mutual position in the range of stresses assembled for electrolysis of the environment set up the speed of

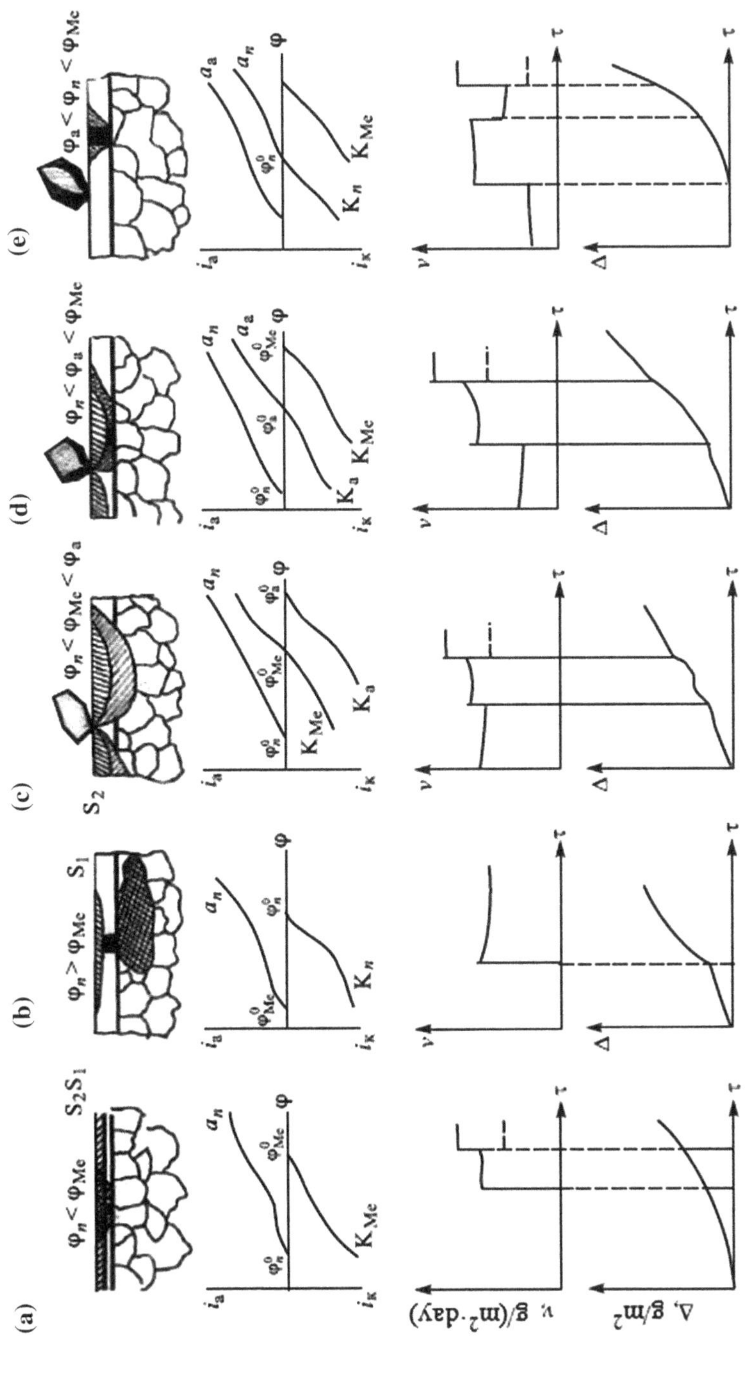

Fig. 10.13 The pattern of corrosion and corrosion-mechanical fracture of boronized layers (a–e)

redox-reaction process; its progress kinetics is characterized by polarization diagrams (Fig. 10.13). The possibility of accelerated corrosion process is more real at higher differences.

The corrosion fracture leads to the anode resolution process taking place on the anode surface; the main metal is fully protected and doesn't depend on the defectiveness of the diffusion layer (Fig. 10.13a). The minimal speed of an evenly spread corrosion fracture is observed on the continuous homogeneous faultless layer; the corrosion speed abruptly increases if areas of the layer's second phase or the sublayer expose themselves or there are defects (like penetrating pores) (Fig. 10.13i_j(1)). The further layer's resolution speed is determined by the correlation of the phase surfaces, however the main metal still stays under protection.

The cathode coating (Fig. 10.13b) and the faultless layer set the minimal corrosion speed at the first period, i.e. before exposing areas of the second phase and metal base. When areas of the second phase and metal base begin to expose, the intensive resolution of the second phase and the sublayer begins due to the electrochemical heterogeneity of the surface, i.e. the undersurface corrosion develops. It is accompanied by the formation of deep-laid cavities. The speed of the undersurface fracture would be determined by the speed of reaction products withdrawal from the interaction zone.

The pattern of corrosion fracture of boronized levels was demonstrated in 10% water solution of H_2SO_4; the pattern of corrosion-mechanical fracture—in the same solution with an added abrasive material powder (SiC).

The recreation of stable electrode potentials' stress range for the chosen solution (Fig. 10.14) showed that boride layers for iron are cathode coatings. Moreover, depending on the phase composition and the correlation between phases and the level of their alloyeness (Table 10.10), they have different values of stable electrode potentials. The stable electrode potentials were determined with the express method involving the use of carbon-oxygen electrons for comparison.

In accordance with the location of layers in the range of stresses, minimal corrosion speed in the continuous faultless layer (before exposure of the second phase areas) is expected to be generated in the non-alloy two-phase layers obtained at the electrolysis saturation and the silicon alloyed two phase layers obtained in baths (calcium-manganese ligature used as a reducing agent). These layers are

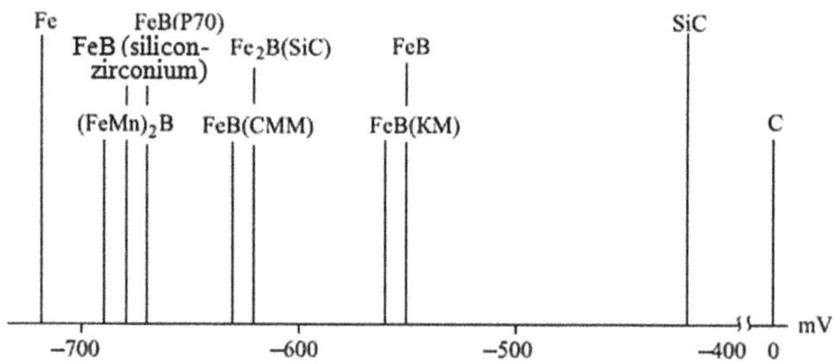

Fig. 10.14 The range of stresses in boride layers in 10% solution of H_2SO_4

Table 10.10 The shorthand signs and phase compositions of studied boronized layers [2]

No.	Shorthand of a layer in the range of stresses	The composition of saturation medium	Phase composition of a layer	Alloy elements for a layer
1	(Fe, Mn)$_2$B	60% $Na_2B_4O_7$, 10% NaCl, 30% silicon-manganese	(Fe, Mn)$_2$B	Mn, Si
2	FeB (P70)	60% $Na_2B_4O_7$, 10% NaCl, 30% P70 (ligature containing 70% P3M)	FeB + Fe_2B	Si, P3M
3	FeB silicon-zirconium	65% $Na_2B_4O_7$, 10% NaCl, 25% silicon-zirconium	eB + Fe_2B	Si, Zr
4	FeB (CMM)	70% $Na_2B_4O_7$ и 30% CMM (silicon-mishmetal)	FeB + Fe_2B	Si, P3M
5	Fe_2B (SiC)	70% $Na_2B_4O_7$ и 30% SiC	Fe_2B	Si
6	FeB (KM)	60% $Na_2B_4O_7$, 10% B_2O_3, 30% KM (calcium-magnum ligature)	FeB + Fe_2B	Si
7	FeB	100% $Na_2B_4O_7$ (electrolysis)	FeB + Fe_2B	–

characterized by the lowest electronegativity in the studied electrolyte in comparison with other boronized layers.

In cases when the continuity of boronized layers is interrupted (by exposing the sublayer's areas), the minimal speed of the undersurface corrosion must be provided by the layers which are characterized by the minimal difference of stable electrode potential with iron.

According to the range of stresses, these layers are one-phase boronized layers based on (Fe, Mn)$_2$B, obtained in the bath of non-electrolysis saturation with silicon-manganese. The tests proved this statement (Fig. 10.15).

A significant change of mass loss (Fig. 10.15) and spasmodic change of the corrosion speed (Fig. 10.13f–j) is explained by the exposure of the metal's sublayer areas. The further undersurface corrosion fracture's speed correlates with the difference of stable electrode potentials of the boride layer and iron, the level of polarization of interactive materials and the speed of reaction products withdrawal from the contact zone. In all the cases boronized layers, obtained at lower temperatures, are more appropriate by the corrosion resistance due to their high density and faultlessness.

The microstructure of surface layers after corrosion fracture illustrates the undersurface character of this fracture (Fig. 10.16). It is impossible to predict the detail's durability in given conditions. It is quite dangerous for important details and joints.

To obtain comparable results, the equal width (100 mkm) of boronized layers was acquired by varying saturation duration.

The structure of the sublayer also influences the development of corrosion fracture. In particular the increase of carbon content in the sublayer of steel 45

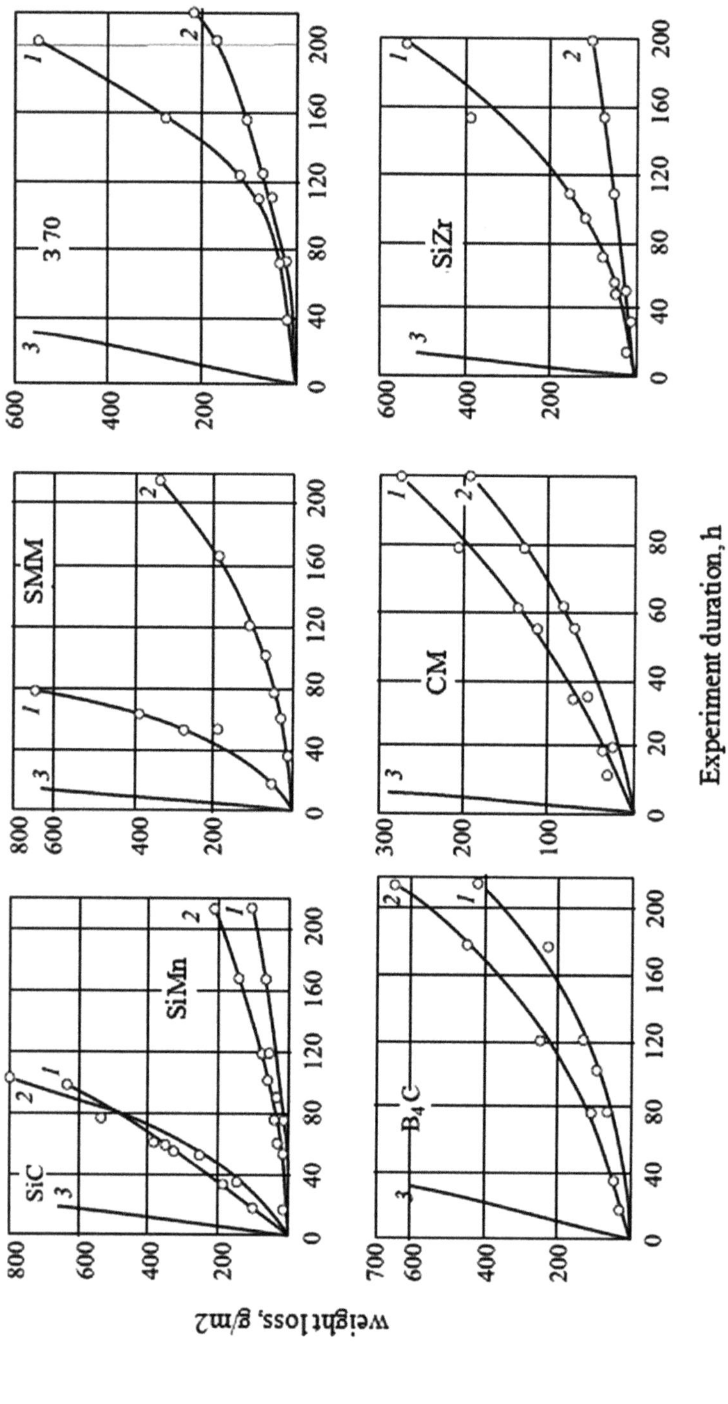

Fig. 10.15 The kinetics of corrosion fracture for boronized layers in steel 35, obtained in borax smelts with the addition of electrochemical reducing agent powders (named on the diagrams): *1* saturation temperature 950 °C; 2 saturation temperature 800 °C; 3 steel 35

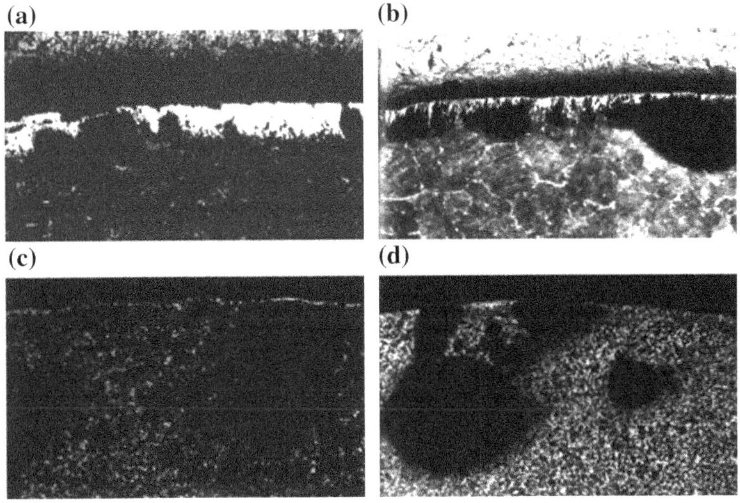

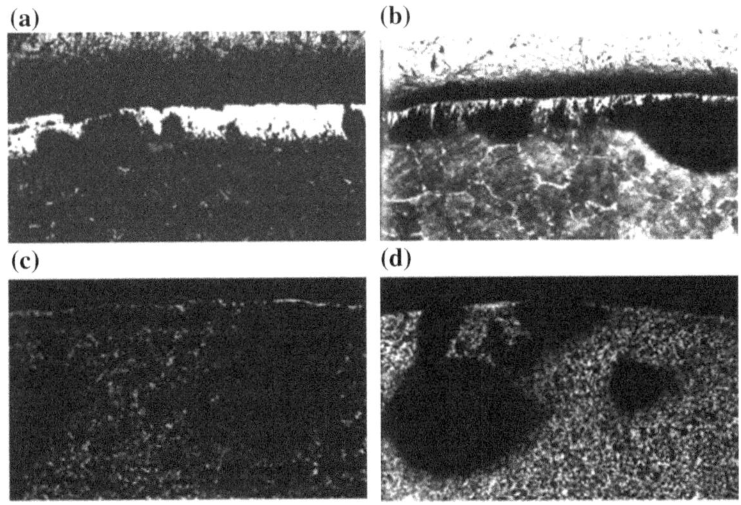

Fig. 10.16 The microstructures of the boronized one-phase (**a**, **c**) and two-phase (**b**, **d**) layers in steel 35 after the corrosion (**a**, **b**, ×200) and corrosion-mechanical (**c**, **d**, ×50) wear processes

during boriding and the formation of carbon borides inclusions [Fe_3(C, B)] in this zone shifts a stable electrode potential of the sublayer to the right in the range of stresses. Depending on the carbon and carbon boride content in the sublayer, its stable electrode potential may be equal or less electronegative than the stable potential of a one-phase boride layer. This may lead to the deceleration of the sublayer corrosion fracture or to the occurrence of the anode process in the boride layer and its resolution. In the given sample, the disturbance of the boride layer's continuity led to the anode resolution of borides on the borides-sublayer borderline which was manifested in internal ulcerations in the boride layer (Fig. 10.16a).

The corrosion fracture of the two-phase boride layers after the disturbance of the outer phase FeB continuity led to the anode undersurface resolution of Fe_2B phase; the further disturbance initiated the anode undersurface resolution of the main metal (Fig. 10.16b). The sublayer's structure in this case doesn't influence the process of corrosion fracture.

Corrosion-mechanical wear requires the analysis of different location variants in the range of interacting materials, i.e. values of stable electrode potentials of a main metal (φ_{Me}), a boride layer (φ_P) and an abrasive (φ_A) (Fig. 10.13).

The given quality polarization diagrams reflect the pattern of overpressure value changes on the interactive surfaces as a result of the differences in area and may give the information on the corrosion process for each interacting material. In some cases (Fig. 10.13) different stages of the corrosion-mechanical fracture on the surface may generate anode and cathode processes. For instance, an anode process may take place at the interaction with an abrasive and a cathode process—at the interaction with a main metal. This pattern of interaction is observed in the analyzed case with undersurface corrosion of the boronized layers in the chosen electrolyte.

Corrosion-mechanical wear of two- and one-phase boronized layers (Fig. 10.13i_j(2)) at the established order of positioning stable electrode potentials of interacting materials ($\varphi_{Fe} < \varphi_{Fe2B} < \varphi_{FeB} < \varphi_{SiC}$) goes like this. It begins with the anode resolution of the continuous layer with a higher speed than under the corrosion interaction due to the work of microgalvanic elements 'layer-abrasive' and the mechanical impact of abrasive particles on the surface. The abrasive in this case cuts crests in the contact point with the surface. The exposure of the second phase areas (Fig. 10.13j(2)) at the two-phase boronozied layer or the main metal in the one-phase boronized layer accelerates the undersurface corrosion abruptly. The process is also complicated by the fracture of left boride layers' bridges by the abrasive particles and the complete exposure of undersurface cavities [2].

It is expected that the location of interacting materials in the range of stresses would influence corrosion-mechanical fracture of two-phase layers and decrease it in comparison with one-phase layers due to the lesser difference of stable electrode potentials of the layer and abrasive. Taken that the hardness of one-phase layers is lower than two-phase layers', the corrosion-mechanical fraction of one-phase layers would accelerate even more depending on external conditions (abrasive hardness, its content in aggressive medium, circulation speed of this medium).

The studies on corrosion-mechanical fraction were carried out on the machine which guaranteed a free movement of abrasive, aggressive medium and studied the samples in a closed retort moved by the principle of "drunk barrel". The speed of spinning—480 cpm. The retort volume was half filled with the mixture of abrasive and electrolyte in proportion 3:2.

The obtained results of the experiments of two-phase and one-phase boronized layers generated in steel 35 at the saturation in salt mixture with different electrochemical reducing agents showed that the highest resistance is characteristic for the

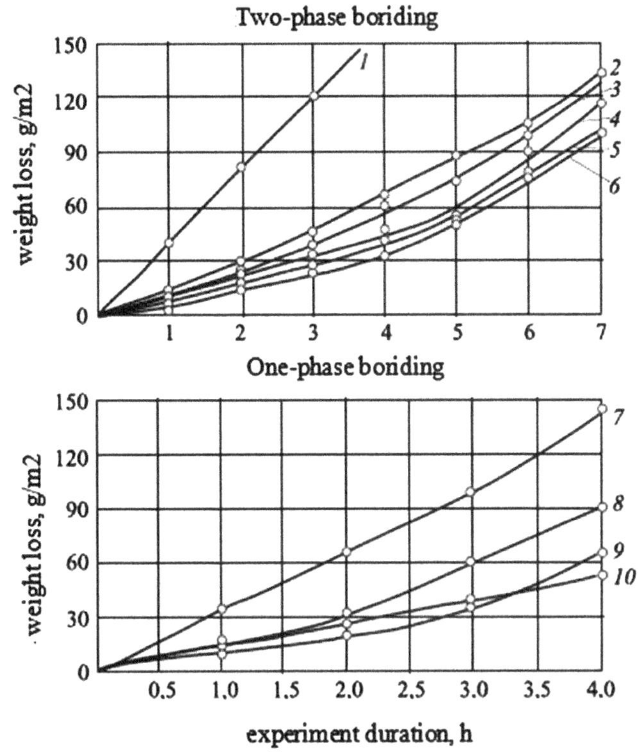

Fig. 10.17 The influence of test duration on the mass loss in boride layers obtained in salts with different electrochemical reducing agents: *1* steel 35; *2* P70; *3* KM; *4* B_4C; *5* CMM; *6* сицр (silicon-zirconium); *7* SiC (950 °C); *8* SiC (800 °C); *9* Silicon-manganese (800 °C); 10 Silicon-manganese (950 °C)

layers alloyed by rare-earth metals and zirconium. Among one-phase layers, the boride layers alloyed with manganese are the most resistant (Fig. 10.17).

The microstructural analysis of samples after wearing proved the intensive undersurface character of the corrosion-mechanical fracture in studied conditions (Fig. 10.16c, d). The total mass loss of the samples in conditions of undersurface corrosion due to the limited area of contact surface may be small, but its negative impact on the constructive hardness of the detail is obvious.

The practical importance of the obtained results is the determination of effective protection duration of the steel sublayer by different boronized layers in 10% water soliution of H_2SO_4 (Table 10.11) which for corrosion fracture reaches 60–225 h, for corrosion-mechanical—35–100 h. Exceeding these numbers would cause the development of undersurface destruction processes.

One of the types of corrosion-mechanical wear is hydro-abrasive or gas-abrasive wear. In this case the plasticity of boronized layers has a prevailing role in providing a necessary wear-resistance.

Table 10.11 The influence of alloying on the duration of effective protection in steel 35 at the corrosion and corrosion-mechanical wear [2]

Type of wear	The duration of effective protection, h, at alloying with different elements				
	Two-phase layers FeB + Fe_2B, alloyed			One-phase layers Fe_2B, alloyed	
	–	P3M	Zr	Mn	Si
Corrosion	80/80	60/60	225/120	225/225	150/160
Corrosion-mechanical	40/35	35/40	100/60	30/40	15/30

Note The numerator—time of effective protection for layers obtained at 800 °C; the denominator—at 950 °C

Thus, the best choice of the boronized layer as a protective coating from corrosion at corrosion or corrosion-mechanical wear in other mediums is to carry out an analogous analysis despite the result of high corrosion resistance properties of steel (10–100 and more times) demonstrated in many works which are based on the mass loss of studied samples.

10.7 The Mechanical Properties of Boronized Steels

The general influence of boronized layers on mechanical properties of steels and smelts is the reduction of ultimate strength by 5–20%, yield strength—by 5–10%, relative elongation—by 20–50%, relative neckling—1.5–3 times and impact strength—1.5–2 times. The important factor in this influence is the scale factor, i.e. a correlation between the layer's width and the sample's width or diameter. A small sample factor for the two-phase boride layer (<0.016) doesn't significantly effect it and a given set of estimated values fits the limits. This level of the scale factor correlates with the width of boride layers <0.05 mm for standard samples. If the scale factor is <0.016, the influence becomes significant. For the one-phase boride layer, the border number of the scale number is 0.023 [2].

The change in loadings for boronized samples causes the change in the boronized layer's behavior and its impact on the strength characteristics of steels. In particular, the brittle fracture of the material causes the transition from one-axis tension to bending and compression to be accompanied by the increase of strength characteristics; the average proportion is σ_t:σ_{bend}:σ_{comp} = 1: (2–2.5):(10–15). The difference is determined by the peculiarities of cracks generation and expansion in brittle materials. The trajectory of crack movement and loading, causing the transition of the crack from a balanced to an overcritical point, depend on the type of loading and determine the level of stresses leading to the complete destruction of the body [18].

The compression of boronized samples is accompanied by the increase of ultimate strength due to the borides' high ultimate strength (σ_{comp} = 200–

250 kgf/mm^2). The bending tests of boronized samples show that the boride layer reduces strength in stretched fibres and increases it in compressed ones. This leads to the reduction of loading causing the fracture but in the end it increases bending resistance (σ_{bend}) by increasing relative yield strength ($\sigma_{0.2}$) [19]. The influence of the scale factor in this case stays the same. The increase of loading over the elastic limit causes the fracture of boride layers and generation of cracks on the stretched side and shears on the compressed side. After the destruction of the boride layer, the plastic deformation begins to develop in the sublayer zone [1].

The increase of the core strength increases the yield strength due to the higher sublayer strength in comparison with the substrate's strength. It also provides static bending with increased resistance. Isothermal quenching of boronized steel guarantees a higher ultimate strength than at continuous cooling during quenching in one cooler.

The fatigue strength of boronized steels depends on the conditions of boriding, steel brand, its thermal processing and the scale factor. If one-phase and two-phase boride layers are narrower than 150 mkm after slow cooling, the fatigue strength increases by 20–50% in comparison with an unprocessed steel.

The fatigue strength tests in smooth samples revealed that one-phase boronized layers increase fatigue strength by 30–50%, boronized layers with separate inclusions of FeB phase on the surface increase it by 10–30%, two-phase layers with a continuous zone of FeB phase on the surface increase it by 10–20%. This pattern is observed if the width of a one-phase boride layer is less than 170 mkm and a two-phase layer doesn't exceed 130 mkm. Thicker layers and higher FeB content in the layer (for instance, up to 70% after electrolysis saturation) leads to the decrease of fatigue strength.

The tests of boronized samples with stress concentrators also causes the decrease of fatigue strength.

The accumulation of fatigue damage is closely connected to the change of the physical-mechanical properties of boronized steels in the process of cyclic loading; the whole damaging cycle is divided into four periods [1, 20] (Fig. 10.18).

The initial (incubational) period is characterized by the accumulation of elastic deformation of the main metal's crystal lattice as a result of the increase of dislocations density in the underboride zone. These processes lead to the strengthening of the sublayer and slight increase in resonant characteristics (dynamic elastic modulus and oscillation amplitude), the reduction of magnetic permeability and the increase of internal friction. The redistribution of stresses in the boride layer and transition zone are weakly developed at the initial fatigue period.

The hardening of the transition zone and the decrease of microhardness are observed in the end of the 1st and the beginning of the 2nd period. Simultaneously the maximum density of mobile dislocations is achieved and the transition zone begins to weaken. It is accompanied by the formation of submicroscopic disturbances of the sublayer continuity and a slight decrease of magnetic permeability. The level of these changes is insignificant because an abrupt fall of residual stresses in the sublayer causes counter-effects. The electrical resistance and internal friction don't change, the resonant characteristics decrease. The withdrawal of vacancies

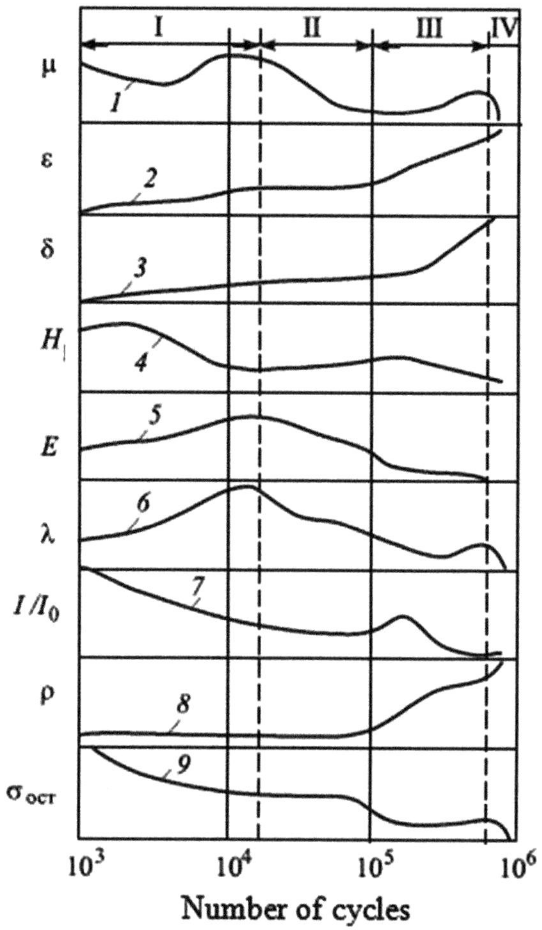

Fig. 10.18 The change of physical-mechanical properties of boronized steel 20 during fatigue tests: *1* magnetic permeability; *2* relative prolongation; *3* internal friction; *4* microhardness; *5* elastic modulus; *6* resonant amplitude; *7* integral intensity; *8* electrical resistance; *9* residual stresses

and dislocations into sub-micro-cracks and the defect annihilation reduce the defectiveness of the transition zone and residual stresses. The redistribution of residual stresses in the whole boronized layer leads to the decrease of residual stresses differential in the boride layer and is accompanied by the decrease of magnetic permeability in the end of the second period.

The third period is characterized by a further increase in the amount of sub-micro-cracks and their transformation into micro-cracks. This leads to the increase of internal friction and electrical resistance and drop in microhardness and resonant characteristics. The growth of microcracks impacts the boride layer leading to the abrupt change of all physical-mechanical properties: the decrease of magnetic permeability and resonant amplitude; the significant growth of electrical resistance and internal friction. The spontaneous expansion of cracks along the section is very fast so the third and fourth periods merge.

The metallographic studies revealed that fatigue fracture begins under the diffusion layer if the boride layer is 90–100 mkm wide. The ferrite grains begin to generate first lines of shifts at 60–90 mkm depth from boride needles to the period correlating with 10–15% durability. The growth of the shifts' number (accumulated in separate areas) and the expansion of shift-formation along the sample's length (from a fastening point to a free end) and its section (in direction to the boride layer and core) is observed later. The shift lines transform into loosening lines and microcracks. The growth of microcracks towards the boride layer takes twice more time than towards the sample's core. Upon reaching the boride zone, the expansion of cracks abruptly accelerates which leads to the complete destruction of the sample [21].

Thus, the durability of the sample is determined by the time of the crack growth towards the boride layer. The speed of crack expansion depends on the properties of the substrate material, sublayer's structure and the level of residual compression stresses. High compression stresses mean slow crack expansion. If the boride layer is thicker than an optimal norm, the possibility for the formation of critical microcracks significantly increases and the destruction center becomes closer to the boride layer—and the fatigue limit decreases.

The thickening of the initial boride layer moves the center of fatigue fracture to the surface which is explained by the pattern of residual stresses distribution along the layer's width. A certain width of the layer which depends on the type of boriding, phase composition, cooling speed and residual stresses distribution triggers the fracture to start from the sample's surface. In this case the boronized layer decreases the fatigue strength in comparison with the starting point.

Two-phase boride layers containing more than 30% of FeB phase generate cracks even after slow cooling of boronized details (20, 20X, 45); these cracks penetrate the whole width of the boride layer and are located on the surface as a net or separate cracks. They function as stress concentrators and ready centers of fatigue fracture which begins from the surface. The fatigue strength decreases [1].

It should be noted that, under other types of loading, the cracks net doesn't affect the durability of boronized details, especially when compression stresses are formed on the surface.

Further quenching and tempering of boronized steels with two-phase layers reduces the fatigue limit due to the formation of many micro- and macrocracks in the boride layer. Therefore the center of fatigue fracture moves from the transition zone to the detail's surface [22]. The tempering temperature rise increases the fatigue limit by 20% but within the limit for steels without coating.

Pseudo-eutectic structures increase the fatigue strength of steels and superalloys; it is explained by the disorientation of borides in relation to the surface, their high dispersability and minimal brittleness of the layer. The boride disorientation encourages the fatigue cracks to turn towards the surface; the plastic solid solution

Table 10.12 The characteristics of fatigue resistance in steel ЭП 718 with a pseudo-eutectic boronized layer [2, 12]

The surface type	Testing temperature, °C	Fatigue strength, σ_{-1} kgf/mm^2, 2 × 10^7 cycles
Smooth	20	34–40/52–74
	600	46/46
Notched	20	12–14/32
	600	16/20

Note Numerators contain values of fatigue strength for non-boronized samples; denominators—samples with pseudo-eutectic structure of boronized layer

decelerates their growth. Better characteristics of pseudo-eutectics structures are connected with the healing of the structure's defects which occurred as a result of previous processing during the diffusion-crystalization growth of boronized layers due to the remelting of surface layers of the detail (Table 10.12).

References

1. Voroshnin L.G., Lyakhovich L.S. Borirovaniye stali. Moscow, 1978. 240 p. [Steel boriding].
2. Krukovich M.G. Razrabotka teoreticheskikh i prikladnykh aspektov upravleniya strukturoi i svoistvami borirovannykh sloev i ikh ispolzovanie pri proizvodstve transportnoi tekhniki. Dissertatsiya doktora tekhnicheskikh nauk. Moscow, 1995. 416 p. [The development of theoretical and applied aspects of controlling the structure and properties of boronized layers and their use in producing transport machinery. Doctorate dissertation].
3. Belsky E.I., Sitkevich M.V., Ponkratin E.I., Stefanovich V.A. Khimiko-termicheskaya obrabotka instrumentalnykh materialov. Minsk, 1986. 247 p. [Chemical-thermal processing of tool materials].
4. Sizov I.G. Razrabotka nauchnykh osnov i tekhnologii elektronno-luchevogo borirovaniya zhelezouglerodistykh splavov s polucheniem na poverkhnosti boridov tugoplavkikh metallov. Dissertatsiya doktora tekhnicheskikh nauk. Moscow, 2003. 305 p [The development of scientific foundation and technology of electron-beam boriding of iron-carbon alloys aimed at obtaining high-melting point metal borides on the surface. Doctorate dissertation].
5. Lyakhovich L.S., ed. Khimiko-termicheskaya obrabotka metallov i splavov. Spravochnik. Moscow, 1981. 424 p. [Chemical-thermal processing of metals and alloys. Reference book].
6. Turov Yu. V. Issledovaniye osobennostei formirovaniya i svoistv boridnykh pokrytii na stalyakh. Minsk, 1974. 25 p [The study of peculiarities of the formation and properties of boride coatings in steels].
7. Polyakov B.Z., Babushkon B.V. Ostatochnye napryazheniya v borirovannoi stali // Khimiko-termicheskaya obrabotka metallov i splavov. Minsk, 1971. Pp. 121–125 [Residual stresses in boronized steels].
8. Birger I.A. Ostatochnye napryacheniya. Moscow, 1965. 252 p. [Residual stresses].
9. Kroukovitch M.G. Influence of Carrier // Processing Condition and Microstructure on Residual Stresses in Borid Layers. Proceedings of «Fourth European Conference on Residual Stresses (ECRS4)». June 4–6, 1996. Gluny en Bourgogne (France). P. 44–49.
10. Samsonov G.V., Neshpor I.S., Khrenova L.M. // FMM. 1969. Vol. 8. Iss. 4. Pp. 622–630.

11. Krukovich M.G. Kompleksnye tekhnologii uprochneniya iznashivaemykh detalei podvizhnogo sostava i instrumentov // Tekhnologii remonta, vosstanovleniya, uprochneniya i obnovleniya mashin, mekhanizmov, oborudovaniya i metallokonstruktsii. Saint-Petersbutg, 2004. Pp. 474–479 [The complex technologies used for hardening worn mobile details and tools].
12. Golubin A.V., Banas I.P., Krukovich M.G. et al. Poverkhnostnoye uprochneniye austenitnykh Fe-Ni-Cr splavov borirovaniyeem // MIIT university proceedings. Moscow, 1991. Iss. 843. Pp. 20–39 [The surface hardening of austenite Fe-Ni-Cr alloys with boriding].
13. Khruschov M.M., Babichev M.A. Abrazivnoye iznashivaniye. Moscow, 1970. 251 p. [Abrasive wear].
14. Vlasov V.M. Rabotosposobnost uprochnennykh truschikhsya poverkhnostei. Moscow, 1987. 304 p. [The durability of hardened surfaces under friction].
15. Golego N.N., Voroshnin L.G., Labunets V.F., Martynyuk M.N. // Voprosy povysheniya nadezhnosti i dolgovechnosti detalei i uzlov aviatsionnoi tekhnoki. Iss. 3. Kiev, 1972. Pp. 44–48.
16. Kroukovitch M.G. Surface friction and wear of refractory steels after thermo-chemical treatment // Proceedings of EUROMAT 2000. V. 1. Tours, France, 7–9 November. P. 609–614.
17. Bezruchko V.P., Korotya A.S. Vliyaniye razmera abrazivnykh chastits na iznos borirovannoi, boromednennoi i boroalitirovannoi stali 45 // MiTOM. 1977. №3. Pp. 68–69 [The influence of abrasive particles size on the wear of boronized, boron-cuprum and boron-aluminum treated steel 45].
18. Andrievsky R.A., Lanin A.G., Rymashevsky G.A. Prochnost tugoplavkikh soedinenii. Moscow, 1974. 232 p. [The durability of high melting point compounds].
19. Gurevich B.G., Balter M.A., Pirogova V.A., Dukarevich I.S. Povysheniye iznosostoikosti i sroka sluzhby detalei putem elektroliznogo borirovaniya. Leningrad, 1966. 32 p. [The improvement of wear resistance and durability of details by electrolysis boriding].
20. Babushkin B.V., Kulik A.Ya. Izmeneniye mekhanicheskikh svoistv borirovannykh stalei v protsesse ustalosti // Problemy metallovedeniya i progressivnaya tekhnologiya termicheskoi obrabotki. Minsk, 1968. Pp. 197–200. [The change of mechanical properties of boronized steels under fatigue].
21. Kogan R.L., Zemskov G.V., Dorofeev N.N., Kishinevsky E.L. Mikrostrukturnaya kartina ustalostnogo razrusheniya borirovannoi stali // Khimiko-termicheskaya obrabotka metallov i splavov. Minsk, 1971. Pp. 125–129 [Microstructural overview of fatigue fracture of boronized steels].
22. Pokhmursky V.I. Korrozionnoustalostnaya prochnost stalei i metody ee povysheniya [Corrosion-fatigue durability of steels and methods of its improvement].

Chapter 11
Modeling the Formation of Diffusive Boronized Layers and Their Wear-Resistance

Abstract The chapter is dedicated to the modeling of processes resulting in the formation, growth and wear of boronized layers. It is studied in the context of engineering construction of structures in surface layers aimed at obtaining necessary properties of detail's surface. The most important stages of structure engineering of surface layers are analyzed; the conditions defining the term "surface engineering" are outlined. The engineering structure constructing of boronized layers offers wide opportunities in modeling the properties of layers and their practical application. The appropriate choice of formation technology, phase composition and width of a particular steel brand allows boronized layers to provide the significant increase in the durability of fast-wearing machinery details, technological tools and equipment. Boriding is a multidimensional process which results are difficult to predict therefore a quite logical solution would be to model the technological process and construct a mathematical model using the combination of mathematical and object pattern modeling. The chapter studies the major stages of modeling a life cycle of boronized layers: the formation process for a diffusion boride layer, growth kinectics and layer's wear.

Chemical-thermal processing in general and boriding in particular are related to the methods of surface engineering; their main goal is to ensure the formation of the necessary structure and properties in surface layers for certain exploitation conditions by constructing the structure on the basis of their surface and core peculiarities.

The fracture of details and tools begins mainly from the surface due to the physical-mechanical and chemical reactions with the environment. The surface layers are exposed to more intensive impacts in comparison with the internal volumes of details. The surface faces all active local and general loadings which leads to the generation of maximum loadings. The presence of different constructive, technological and exploitation concentrators of stresses intensifies the process of damage accumulation and accelerates the detail's fracture. A certain role is played by the structural instability of the surface which is determined by non-compensated atom bonds on the surface of the crystal lattice.

M.G. Krukovich et al., *Plasticity of Boronized Layers*,
Springer Series in Materials Science 237,
DOI 10.1007/978-3-319-40012-9_11

Thus, boriding and the thermal treatment that follows result in the general strengthening and hardening of the primary surface; the surface generates stronger and more stable phases and structures.

In this case the distribution of hardness and strength along the detail's section correlates with the distribution of working stresses occurring during exploitation [1, 2].

The structure of the surface layers can be constructed in the process of chemical-thermal treatment and involves the following stages:

- The evaluation of exploitation conditions for details and tools and reasons for their failing;
- The introduction of specific requirements for the properties of protective coatings and their quality;
- The evaluation of the level of changes done to the sizes of details and the roughness of their surfaces as the result of the processing;
- The choice of a protective coating from the range of recommended coatings (Fig. 11.1, colour inset).
- The choice of the technological base appropriate for the chosen hardening method.

In accordance with the given data for most exploitation conditions, boronized layers of different structures are the most optimal and appropriate.

The evaluation of exploitation conditions for boronized layers is carried out with the respect to the pattern of hardness and stress state distribution along the detail's section. The most common types of hardness distribution (Fig. 11.2) are provided by the technological procedures which change their structure and phase composition (Fig. 11.3).

The main components of surface engineering are:

- The maintenance of conditions necessary for constructing multizone and multiphase composite structures along the layer's width;
- The control over the widths of these zones;
- The determination of the optimal correlation between the width and hardness levels of boronized layers and core hardness.

The results obtained in this work eloquently demonstrate the wide capacities of the engineer construction of boronized layers' structure.

The two-phase and one-phase boride layers characterized by reduced hardness on the surface (Fig. 11.2a, b) are formed as a result of the boron potential reduction in the saturation medium, the surface alloying by elements which decrease boride hardness, de-boriding and further heating for quenching and other types of thermal treatment and generation of surface porousness. Surface porousness occurs during iron cast boriding or when there is an excessive amount of oxygen in the saturation medium. Porousness may have a positive effect by keeping oil on the hardened surface.

The pseudo-eutectic structures and layers obtained during electron-beam treatment, due to the multiphase structure, have a certain hardness interval and significant thickness (Fig. 11.2c, d). The generation of α-solid solution with carbon

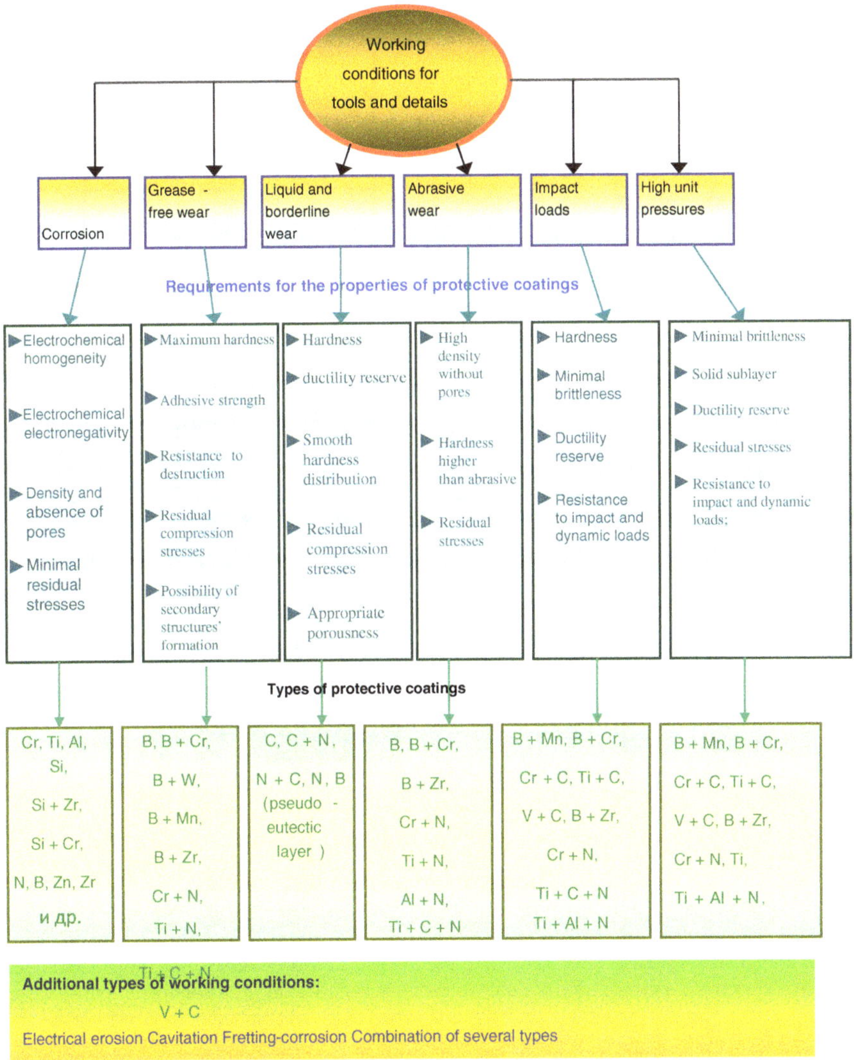

Fig. 11.1 The dependence between protective coatings and working conditions for a certain detail or tool

boride inclusions in the structure's sublayer ensures a smoother transition from the layer to the core and a favorable distribution of residual stresses (Fig. 11.2d).

The formation of one- and two-phase layers changes the sublayer's structure and ensures a smoother change in hardness from the layer to the core. These layers are obtained by creating a liquid-crystal state of the layer during saturation, the boride layer resolution during additional treatments in mediums with low boron potential and during the heating for further thermal processing (Fig. 11.2e).

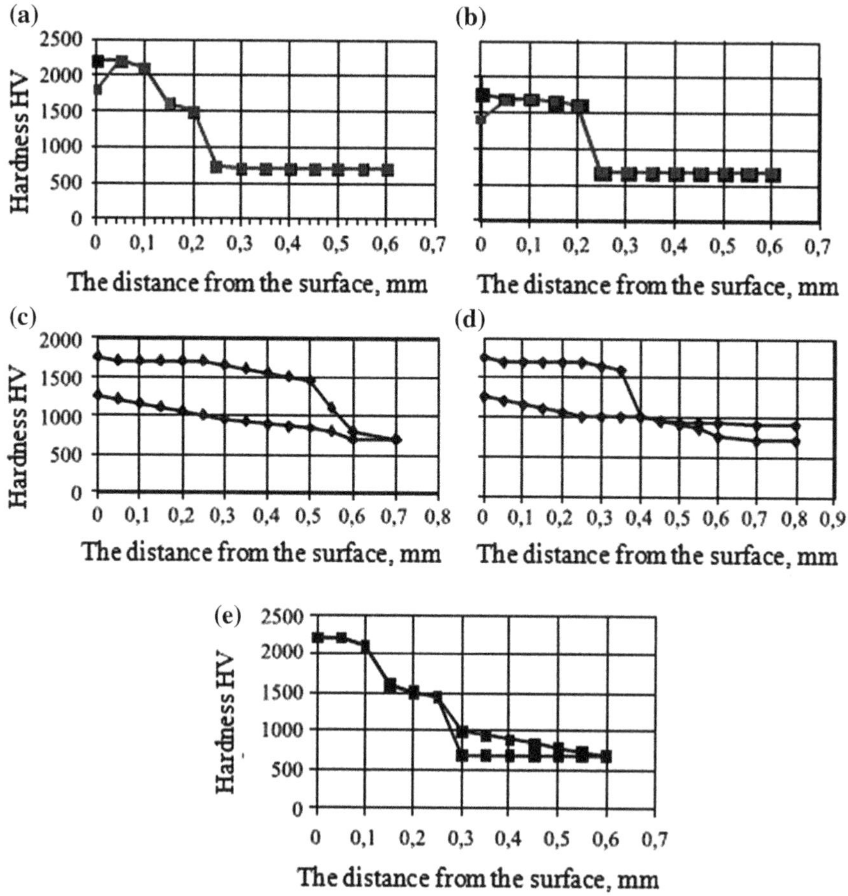

Fig. 11.2 The pattern of hardness distribution in boronized layers, **a** two-phase layer ($FeB + Fe_2B$); **b** one-phase layer (Fe_2B); **c** pseudo-eutectic layer (α-solid solution + borides); **d** pseudo-eutectic layer (α-solid solution + borides) with a sublayer (α-solid solution + carbon borides); **e** two-phase layer ($FeB + Fe_2B$) with a sublayer (α-solid solution + carbon borides)

The appropriate choice of formation technology, phase composition and layers' width for particular steel brands causes a significant increase of durability in non-durable machine details, technological equipment and tools. At the same time, due to the multiple influence factors for the results of boriding, the difficulties of process modeling for saturation mediums, the complexity of forecasting final results and many other additional technological peculiarities, the optimal choice is hard to make. One of the ways which allow to obtain the forecast results is modeling a technological process of hardening.

A mathematical model of any technological process allows to understand the mechanisms of physical influence, analyze their main patterns, quantitatively

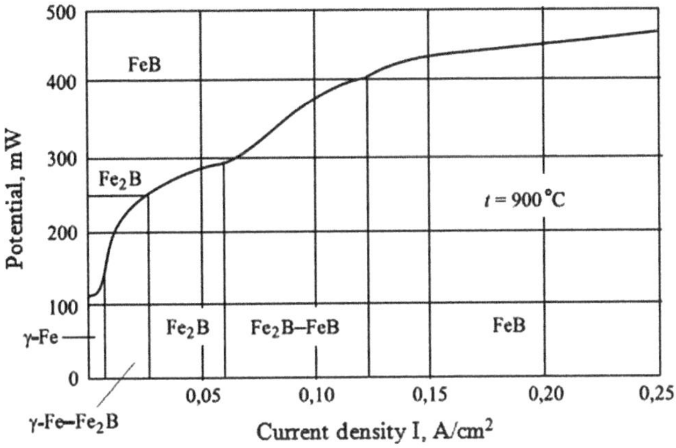

Fig. 11.3 The influence of the current density on the shift of iron potential in borax solution and phase composition of the surface

evaluate the influence of technological actors on the final results of processing. Mathematical models of diffusion processes in multicomponent systems are built by following the general methods [3].

1. The microscope method which analyzes the mechanism of the shift in diffusing elements (elementary processes) and establishes the connection between diffusion characteristics and different physical properties of solid bodies (crystal structure, melting point, element solubility, elastic constants etc.).
2. The phenomenological method based on non-equilibrium thermodynamics.

These methods result in obtaining the differential equations which approximately describe diffusion processes in multicomponent systems. They also determine the dependence between diffusion parameters and different physical properties of solid bodies.

However the use of the mentioned methods for a particular technological process often turns out impractical. The reason for that is that the mathematical model doesn't always correlate with the real technological process, requires elaborated descriptions of chemical, kinetic and technological data. Also the algorithms require complex calculations with multiple references to the experimental data. In particular, the attempt to create model "diffusion + kinetics" allowed to obtain important mathematical data but didn't result in obtaining information of clean iron boriding which would fit the experimental results [4].

Thus, the most appropriate way is to combine different models for separate stages of the process (the formation of boron subions in the saturation medium, their transportation to the processed surface and their reduction on this surface, the patterns of boronized layers' formation and their growth) and experimental data for creating the whole process model and determine the control parameters, i.e. combine mathematical and pragmatic modeling.

11.1 Modeling the Formation of Diffusive Boride Layers

The formation process of diffusion boride layers in ionic gaseous, liquid or solid mediums are determined by the following factors:

- The formation speed for ions with low valency (subions) in the saturation medium which depends on the medium composition and excessive pressure in the saturation medium;
- The transportation speed for subions in the saturation medium to the processed surface;
- The speed of redox-reaction and adsorption capacity of the processed surface;
- The reaction capacity of the redox reaction products and the saturation medium materials which determined the cycles of subion formation and continuity of their generation;
- The diffusion processes in the processed metal.

Most of the factors are connected with the formation conditions of saturation element atoms which in the end determine its concentration on the surface.

The qualitative description of these factors may include the thermodynamic theory of structure, sustainability and fluctuation. These stages of boriding in saturation mediums are studied as a system which transforms the heat energy into the energy of chemical interaction of medium components and solid body components. Similar systems in certain conditions of energy flows and substance interaction generate the processes of matter arrangement which correlate with the decrease of entropy and the formation of self-organizing dissipative compounds. Dissipative subcompounds for mass transfer self-organization under boriding are boron ions of various valencies (subions) [5].

The quantitative description of changes in these factors is quite complex as the saturation process involves the constant change of external conditions (temperature drops, changes in saturation medium compositions due to the surface elements' passivation, changes in pressure, reaction activity of the processed surface etc.). Moreover changing conditions generate new dissipative compounds of self-organization determining the process of mass transfer.

The optimal way of the qualitative description of factors is the search for integral system characteristics which would reflect the final result in real conditions as objectively as possible. Such characteristics is the boron potential of the medium which determines the concentration of boron on the processed surface.

Boron potential is an index of saturation capacity of the medium; it is a value of the stable electrode potential of the electrode, measured in relation with a neutral control electrode. Both electrodes are put into the saturation medium and the difference in potentials is measured. A carbon-oxygen electrode was used as a control electrode. Taking into account that the boron potential of the medium must be more electronegative that the potential of the processed surface and forming phase [5, 6], calibration dependencies may be built; they connect the phase composition of the surface (the boron potential) and the processing conditions.

In particular, electrolysis boriding requires a steel polarization curve as a calibration dependency (11.3), which connects the acquired value of its stable electrode potential during boriding, the layer's phase composition on the surface and the cathode current density.

The current density below 0.025 A/cm^2 guarantees the boron concentration on the surface up to 8.8% in mass and the formation of separate Fe_2B borides on the surface located in α-solid solution and the shift of electrode potential to −240 mW. The current density in the interval 0.025–0.07 A/cm^2 sets the boron concentration on the surface from 8.8 to 9.0% and ensures the formation of a continuous boride layer Fe_2B on the surface while moving the surface potential from −240 to −290 mW; the interval 0.07–0.12 A/cm^2 increases the boron concentration from 9.0 to 16.2% and the formation of FeB and Fe_2B boride mixture accompanied by the shift of the surface potential from −290 to −400 mW. The current density over 0.12 A/cm^2 triggers the formation of boride layers with a continuous FeB phase layer. The surface potential shift is more than −400 mW. The parameters for controlling the plasticity of boronized layers are salt component composition, current density, the type of processed material, the temperature and duration of boriding. Changes in these parameters lead to the change of stable electrode material potentials, i.e. the medium boron potential.

The boron potential of the saturation medium under non-electrolysis boriding is estimated by the values of stable electrode potentials of electro-chemical reducing agents and the level of shift of stable electrode potentials of the saturation medium under processing (Table 7.12). The shift of stable electrode potentials also happens under the changes in salt components and temperature (Fig. 11.4). Thus, the parameters for controlling the plasticity of boronized layers in this case are the salt component composition, the chemical composition of the electrochemical reducing agent and its amount in the bath, the type of the processed material, temperature and duration of boriding.

Boriding in gaseous mediums, obtained from the mixture of gases or obtained during contact and non-contact boriding with powder, grained or compact materials, determine the boron potential by measuring stable electrode potentials under corresponding apparatus work. Measurements take place in special high-temperature electrolysis cells where electrodes are fully isolated from the corpus details of furnaces and crucible. The parameters for controlling thc plasticity of boronized layers is the composition of gaseous medium, the circulation speed, the type of the processed material, the composition of powder and compact mixtures, the level of powder mixtures density, the level of sealing, the temperature and duration of boriding.

Thus, each particular case links the boron potential with the boron concentration on the processed surface and the phase composition of the forming boronized layer.

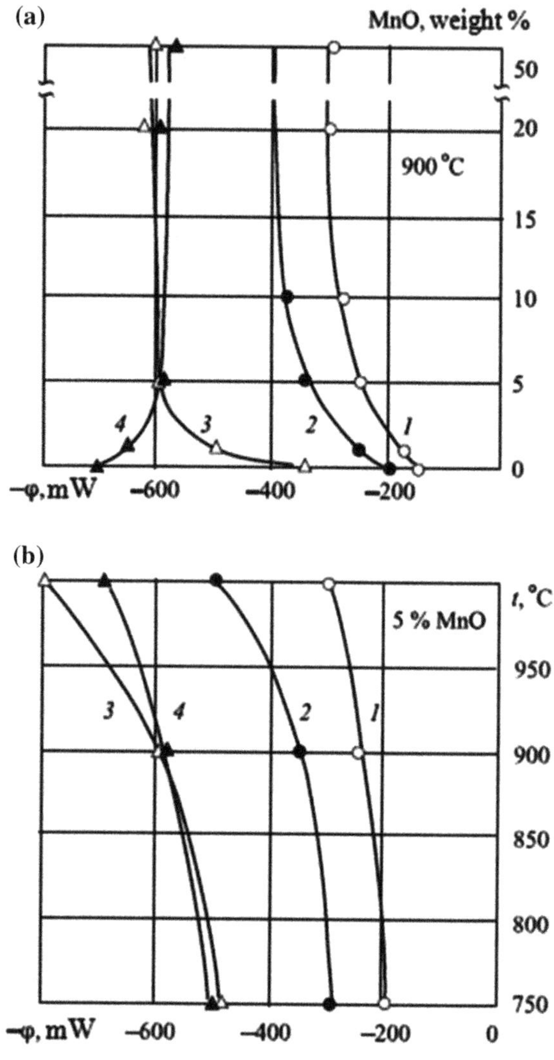

Fig. 11.4 The influence of chemical composition of smelt, **a** and temperature, **b** on the shift of stable electrode potentials: 1—Fe; 2—Fe_2B; 3—FeB; 4—Silicon-manganese 17

11.2 Modeling the Growth Kinetics of Diffusive Boride Layers

Modeling growth kinetics for boronized layers in clear iron is carried out by solving the problem of boron diffusion in iron and boride phases, using existing methods [3, 7] or experimenting with determining the layer's width in iron in given circumstances with the following use of obtained results for calculating the widths of boronized layers in construction steels. The latter way is the most appropriate as other methods still require the experimental proof of obtained calculations.

Modeling kinetics for boriding of construction steels takes into account the influence of alloy elements and carbon on the thickness of the boride layer. This influence was determined by the statistical analysis of boride layers growth in standard and specially produced medium- and low-carbon and medium-alloy steels (Fig. 11.5).

The general calculation formula is:

$$y = y_0 A e^{-ax}, \tag{11.1}$$

where y—calculated layer width in steel; y_0—calculated or experimental layer's width in technical iron (Armco-iron); A and a—empirical coefficients determined from kinetics data on the influence of one alloy elements; x—content of alloy elements, mas. %.

The presence of several alloy elements (taking into account their additive influence on the growth of the boride layer) changes the formula:

$$y = \Pi y_0 A_n \exp\left\{\sum_{k=1}^{n} a_k x_k\right\}, \tag{11.2}$$

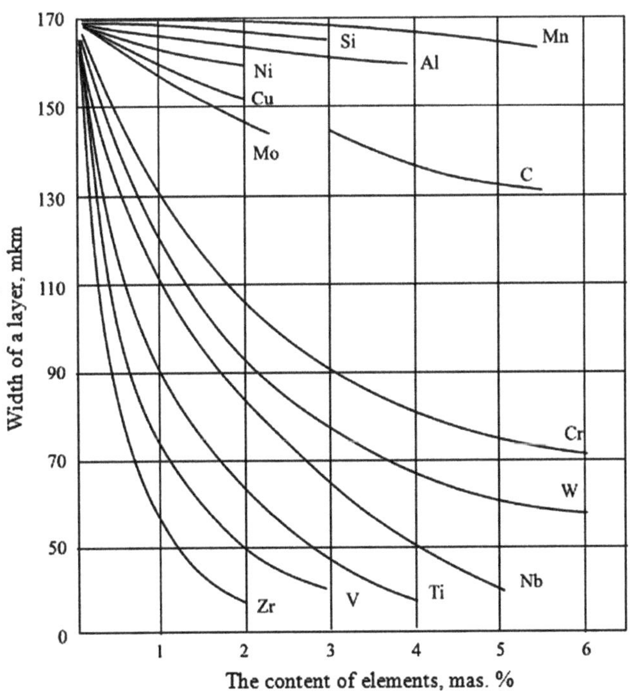

Fig. 11.5 The influence of alloy elements on the width of a boride layer (950 °C, 4 h)

or for *k*-element:

$$y_k = y_{k-1} A_k \exp(-a_k x_k) \quad (11.3)$$

where k = 1, 2, 3, … n; y_{k-1}—the width of the layer including the influence of previous alloy elements.

The assumption about the influence addictiveness doesn't intensify inaccuracies in the fluctuation of the boride layer's width; the inaccuracy is similar to the inaccuracy of temperature measurement in the saturation volume, the composition of saturation medium fluctuation, the fluctuation of chemical composition of the processed steel and other technological peculiarities.

Thus, the width of the layer, obtained in a particular moment of time on any construction low- and medium-alloy steel, can be calculated using the reliably calculated boride layer's width (incl. boron potential) or the experimental result of the boronized layer's growth speed on technical iron in particular conditions. Having the data on thermal dependence of the layer's width allows to make calculations on any temperature. Setting the layer's width in a particular steel determines the duration of necessary soaking. Having the result of boriding for any steel leads to the exclusion of alloy elements influence in a given steel and the calculation of the layer's width in another steel with another set of elements.

The influence of each element in isolation is expressed by the formulae, given in Table 11.1. The formulae were obtained by the approximation of dependencies between the changes in the layer's widths and the content of alloy elements on specially produced steels (Fig. 11.5) [5, 8].

Table 11.1 The formulae for calculating the influence of alloy elements on the total width of boride layers

Alloy element	Formula of influence
C	$y_C = y_0 \exp(-0,65 x_C)$
Cr	$y_{Cr} = y_C(1 + 0,65 x_{Cr}^3 / y_C) \exp(-0,25 x_{Cr})$
Si	$y_{Si} = y_{Cr} \exp(-0,06 x_{Si})$
Mo	$y_{Mo} = y_{Si} \exp(-0,22 x_{Mo})$
W	$y_W = y_{Mo}(1 + 2x_W^2 / y_{Mo} \exp(-0,28 x_W)$
Ti	$y_{Ti} = y_{Ti}(1 + 25 x_{Ti}^2 / y_W \exp(-0,88_{Ti})$
V	$y_V = y_{Ti}(1 + 40 x_V^2 / y_{Ti}) \exp(-1,1 x_V)$
Zr	$y_{Zr} = y_V(1 + 40 x_{Zr}^2 / y_V) \exp(-1,1 x_{Zr})$
Nb	$y_{Nb} = y_{Zr}(1 + 3 x_{Nb}^3 / y_{Zr}) \exp(-0,43 x_{Nb})$
Mn	$y_{Mn} = y_{Nb} \exp(-0,02 x_{Mn})$
Al	$y_{Al} = y_{Mn}(1 + 5 x_{Al} / y_{Mn} \exp(-0,1 x_{Al})$
Cu	$y_{Cu} = y_{Al} \exp(-0,2 x_{Cu})$
Ni	$y_{NI} = y_{Cu} \exp(-0,1 x_{NI})$

Note y—calculated width of boride layer; y_0—layer width in technical iron; x—alloy element content in steel in the range from 0 to 6 mas. %

The use of boriding in real conditions often lead to the necessity of comparing the results obtained in different saturation conditions (temperature and duration of processing) and with different technological methods. The comparison of the results should be carried out not between layer's widths in particular steels but between the values of the kinetics coefficient (*D*) combined with its thermal dependency.

Random data on the width of boride layers can be unified by processing conditions with the use of the formulae of width dependency (*y*) to the duration (*t*) and the kinetics coefficient (*D*) to the processing temperature (*T*):

$$y = D\sqrt{T}; \tag{11.4}$$

$$D = Be^{bT^K}. \tag{11.5}$$

where *y*—calculated width of a boride layer; *D*—kinetics coefficient depending on the type of the crystal lattice, the correlation of atomic radiuses of saturation of diffusing elements, the diffusion coefficient, the phase composition of the layer and processing conditions; *B, b, k*—empirical coefficients depending on the boron potential of the medium, layer's phase composition and processing method; *T*—processing temperature, °C; *t*—boriding duration, h.

The calculation of the boride layer thickness, obtained in different saturation mediums, is done according to the formula (11.4) for ARMCO-iron, using the coefficients given in Table 11.2. The coefficients are determined by the equations for the approximation of the experimental data. The validity of approximation reaches 96%.

Table 11.2 The calculated coefficients in ARMCO-iron

Saturation medium composition	Coefficients		
	B	b	k
100% $Na_2B_4O_7$ (electrolysis, $j = 0.25$ A/CM^2)	3	−17	6.2964
70% $Na_2B_4O_7$. 30% B_4C	0	−15	5.4708
70% $Na_2B_4O_7$, 30% SiC	2	−18	6.5166
75% $Na_2B_4O_7$, 25% CaC_2	1	−11	4.3036
70% $Na_2B_4O_7$, 30% KM	3	−23	8.2191
70% $Na_2B_4O_7$, 30% Silicon manganese	4	−15	5.4975
75% $Na_2B_4O_7$, 25% Silicon zirconium	3	−15	5.5531
70% $Na_2B_4O_7$, 30% P70	2	−21	7.6222
97% B_4C, 3% NH_4Cl	7	−18	6.3728
98.4 B_4C, 1.6% AlF_3	3	−20	7.2045
100% B_4C	8	−21	7.3321
70.5% Al_2O_3, 10% Calcium silicon 10% B_2O_3, 0.5% NaF	1	−20	7.3314
$B_2H_2/H_2 = 1/25$	8	−10	3.7127

The calculation of kinetics coefficients is followed by the calculation of the boride layer width in technical iron and the layer's width in a chosen steel with a corresponding set of alloy elements from Table 11.1.

The control calculation of the boride layer's widths in alloy steels involving the use of growth speeds in technical iron revealed that the fault doesn't exceed 10% in comparison with the experimental results. The reliable calculation allows to determine the saturation conditions for obtaining faultless layers in particular steel in accordance with the developed recommendation (Table 6.1). In case if particular boriding conditions for high-alloy steels, iron cases or solid smelts are necessary, other formulae should be developed on the foundation of the existing approach.

11.3 Modelling the Wear Processes of Boronized Layers

The contemporary and perspective engineering requires the maintenance of friction units and tools' working capacity at advanced levels, i.e. the working capacity in the wider range of temperatures at higher speeds and unit pressures, in complex environmental conditions. This requirement is constantly accompanied by the need for reducing production working hours and material consumption of details. The problem might be solved by creating the theoretical base for developing and choosing new combinations of tribotechnical materials according to their physical-chemical and mechanical properties.

The wear process is often characterized by its intensity which manifests itself in different parameters [9]:

- The volume of the metal removed from the friction way in a unit of time, one detail's turn or one body's oscillation;
- The volume loss per one unit of the normal force at a unit sliding distance, $mm^3/(H \cdot m)$ which is often called the wear coefficient;
- The mass loss per time unit;
- The change of a particular size per time unit;
- The relative change of size or volume in proportion to the similar changes in the control material.

Multiple wear mechanisms and their close connection prevent from dividing and classifying them into groups which would ease the process of making more reliable forecast calculations. It is worth noting that the ambivalence of friction processes has an important function in details and tools' wear that's why their types may be divided by the impact that deformation and adhesion make. The adhesion and fatigue wear are positioned in extremes because fatigue is mostly caused by deformation processes and adhesion is dominant during adhesion wear [9].

At the same time the frictional interaction always happens in particular environments which significantly influences the tribological processes connected to the chemical activation of friction surfaces, temperature rise and loosening of the undersurface layer. These processes change the deformation speed and strength of adhesion bonds in the end, but they continue to play an important role.

We don't give classification of wear processes in accordance with Russian state standards (GOST 27674-88) but it's important to understand that the tribosystems with the only type of wear don't exist. Normally several mechanisms are acting with simultaneous adhesion impact, plastic deformation and fatigue processes. Moreover real conditions generate the transition from one type of wear into another. The reasons for that are temperature fluctuations, changes in the chemical composition of the environment, the destruction of protective coating, changes in working parameters and the increase in oscillations due to the wear of the contacting details of mechanical units. Thus, a wear prognosis for a particular system based on properties of frictional couple materials and working conditions doesn't give reliable results due to the instability of structure and processes on the frictional surfaces.

One of the directions in friction surface engineering is the construction of a boronized layer's structure during diffusion boriding. This process causes the formation of diffusion layers of a different phase composition and structure in steels and smelts which is reflected in the pattern of hardness change in its width (Fig. 11.2). Stronger boronized layers in wider ranges of temperature fluctuation and mechanical impact bring this strength to the friction units' surfaces. However, if boronized layers interact with cuprum or aluminum smelts or stainless steels, significant adhesive setting is observed and these materials are carried to the boronized surface. In these cases the effect of boronized layers significantly reduces.

The classification of boronized layers according to their tribotechnical characteristics, given in Chap. 9, allows to choose layers for particular exploitation conditions more effectively.

Boronized layers are usually used for increasing the wear-resistance in details working under sliding friction with and without lubricant and under abrasive wear. The durability of a hardened detail under sliding friction is determined by the width of the boronized layer which is limited by the quality requirements as each brand steel has the optimal width of the faultless boride layer. In some cases the use of defected boronized layers is allowed. If the experimental data is missing, the width may be calculated by the given methods.

The literature provides researchers with a significant number of experimental data which process the high wear-resistance of boronized layers. However, these results are difficult to correlate with as they characterize the behavior of particular layers in particular experimental conditions. The choice of the optimal structure and the width of boronized layers requires the whole complex of tests for different layers in similar conditions. This method is expensive. The studies on the properties and usage of boronized layers allowed to develop the evaluation method for calculating their wear-resistance.

The prognosis evaluation of wear-resistance in diffusion boronized layers should include the criteria values of plasticity, friction coefficient, sliding speed and contact pressures.

The plasticity criteria for diffusion layers is their ability to resist shearing which is called shear stress (σ_{sh}) and is calculated according to the formula [5, 10]

$$\sigma_{sh} = 0.174 \cdot P/(2 \cdot l^2 + l \cdot c), \tag{11.6}$$

where *P*—loading, kgf; *l*—distance from the imprint's center to the edge, mm; *c*—diagonal length of the imprint, mm.

Shear stresses with a high level of validity correlate with hardness values of different layers (diffusion and quenched) (Fig. 9.9). This pattern allows to determine shear stress values at any distance from the surface by the hardness values which can be directly measured by the depth of a compact layer or calculated by hardness values of each phase in a compact multi-phase structure.

Hardness is a characteristics connected to the crystal structure of particular phases and the pattern of their distribution in the structure. The determination of hardness values for heterogeneous diffusion layers (with certain approximation) should involve the addictiveness principle (layer change of relative amount of each phase q_1, q_2,... q_i) along the layer's width:

$$\mathrm{HV} = q_1 \cdot \mathrm{HV}_1 + q_2 \cdot \mathrm{HV}_2 + q_3 \cdot \mathrm{HV}_3 + \ldots + q_i \cdot \mathrm{HV}_i. \tag{11.7}$$

The pattern of change in shear stresses along the width of the boronized layer is directly dependent. FeB–Fe_2B transition zone changes the shear stress from 22 to 35 kgf/mm^2, Fe_2B transition zone to the sublayer—from 35 to 100 kgf/mm^2.

Thus, the shear stress along the width (x) of boronized layers in dependence of hardness (y) can be calculated according to the following formula:
for boride layers

$$y = 4090.2 \cdot x^{-0.2832}, \quad R^2 = 0.9978; \tag{11.8}$$

for composite, eutectic and pseudo-eutectic structures

$$y = 3641.3 \cdot x^{-0.2498}, \quad R^2 = 0.9949; \tag{11.9}$$

for sublayers

$$y = 59574 \cdot x^{-0.9511}, \quad R^2 = 0.9464. \tag{11.10}$$

According to the developed method, the evaluation of wear-resistance includes the data on hardness, plastic characteristics of the layer (relative prolongation), shear stress and friction conditions; this allows to evaluate the durability of the layer or any other material under different exploitation conditions. The shear stress, hardness and relative prolongation are used with respect to their changes in the layer's width. The boronized layer is divided into parts width different hardness:

- The continuous layer of FeB boride;
- The transition zone FeB–Fe_2B;

- The continuous zone of boride Fe_2B;
- The transition zone Fe_2B-α-solid solution (sublayer);
- The sublayer.

Thus, the comparative forecast calculation can be done by the proposed empirical formula for evaluating the relative wear-resistance of diffusion layers [11]:

$$U = 10(1-\delta)\cdot\left(\sqrt{\sigma_{\text{sh}}}\right)^{-1}\text{HV}\cdot\exp\left(\sqrt{m\sigma_{\text{sh}}} - n\cdot\sigma_k - k\cdot f - e\cdot v\right), \quad (11.11)$$

where HV—hardness of a homogeneous diffusion layer or composite heterogeneous structure; δ—relative prolongation of the studied layer or material, %. In the absence of this characteristics it can be calculated by the formula:

$$\delta(\%) \approx 4\cdot 10^3\cdot(\text{HV})^{-1}; \quad (11.12)$$

Σk—unit pressure, kgf/mm^2; f—friction coefficient; v—sliding speed, m/c; m, n, k, e—empirical coefficients which may acquire following values (determined by the approximation of correlating characteristics influence on wear-resistance under sliding friction): $m = 0.01$; under liquid friction $n = 2$, $k = 2$, under grease-free friction and abrasive wear $n = 5$, $k = 4$, under borderline friction $n = 3.5$, $k = 3.0$.

The comparative evaluation of wear-resistance in diffusion boronized layers with different types of structures, done according to the proposed method, showed a good correlation with the results of the real-life tests in different friction conditions. The calculated wear-resistance is used as a comparative characteristic of durability for different layers in particular exploitation conditions.

References

1. Dubinin G.N. Strukturno-energeticheskaya gipoteza vliyaniya diffuzionnogo sloya na obyemnye svoistva splavov // Zaschitnye pokrytiya na metallah. 1976. Iss. 10. Pp. 86-90 [Structural-energetic hypothesis of the influence of diffusion layers on voluminous properties of alloys]
2. Zinchenko V.M. Inzheneriya poverkhnosti zubchatykh koles metodami khimiko-termicheskoi obrabotki. Moscow, 2001. 303 p. [Engineering of surfaces of cog wheels by chemical-thermal processing]
3. Prusakov G.M. Matematicheskiye modeli i metody v raschetakh na EVM. Moscow, 1993. 144 p. [Mathematical models and methods in computer calculations]
4. Kukharev B.S., Fisenko S.P., Shabunya S.I. Diffuzionnye modeli obrazovaniya novykh faz pri borirovanii zheleza // Preprint №2. Minsk, 1992. 28 p [Diffusion models of forming new phases while boronizing iron]
5. M.G. Krukovich, Razrabotka teoreticheskikh i prikladnykh aspektov upravleniya strukturoi i svoistvami borirovannykh sloev i ikh ispolzovanie pri proizvodstve transportnoi tekhniki. Dissertatsiya doktora tekhnicheskikh nauk (Moscow, 1995), p. 416 [The development of theoretical and applied aspects of controlling the structure and properties of boronized layers and their use in producing transport machinery. Doctorate dissertation.]

6. Krukovich M.G. Issledovaniye zhidkostnykh bezelektroliznykh protsessov khimiko-termicheskoi obrabotki. Dissertatsiya kandidata technicheskikh nauk. Minsk, 1974. 298 p. [The study of liquid non-electrolysis processes of chemical-thermal processing. Candidate of science dissertaion]
7. Voroshnin L.G., Khusid B.M. Diffuzionnyi massoperenos v mnogokomponentnykh sistemakh. Minsk, 1979. 256 p. [Diffusion mass carry in multicomponent systems]
8. Kroukovitch M.G. Surface friction and wear of refractory steels after thermo-chemical treatment // Proceedings of EUROMAT 2000. V. 1. Tours, France, 7–9 November. P. 609–614
9. Myshkin N.K., Petrokovets M.I. Treniye, smazka, iznos. Fizicheskiye osnovy i tekhnicheskiye prilozheniya tribologii. Moscow, 2007. 368 p. [Friction, lubrication, wear. Physical foundation and technical use of tribology]
10. Grigorov P.K., Katkhanov B.B. Metodika opredeleniya khrupkosti borirovannogo sloya // Inter-university Scientific Journal. Rostov-na-Donu. Iss. 16. 1972. 97 p. [The methods for determining brittleness of a boronized layer]
11. Kroukovitch M.G. Surface friction and wear of refractory steels after thermo-chemical treatment // Proceedings of EUROMAT 2000. V. 1. Tours, France, 7–9 November. P. 609–614

Chapter 12
The Prospective Boriding Technologies Guaranteeing the Improvements in the Plasticity of Layers

Abstract Boronized plastic layers are obtained with different methods. Herewith the adequate choice of boriding technology, composition of saturating mixture, substrate's material, layer's width, temperature-temporal parameters, heating and cooling conditions lead to the formation of plastic boronized layers under any type of processing. The chapter contains recommended compositions and methods for generating most perspective plastic boronized layers. The chapter also deals with the general preparation procedures and their characteristics. Thus, it is important to take into account the peculiarities of the chosen technology of boriding. The authors describe the following technological processes in order to point out the specifics of the chosen boriding technology:

- the technology of boriding in powder mediums;
- the technology of boriding in liquid mediums;
- the technology of boriding in compact materials (coatings);
- the technology of boriding in gaseous mediums.

Moreover the methods for intensifying the diffusioin saturation are proposed.

The method of chemical precipitation of borides from steam phase is analyzed (CVD-process). For instance, the precipitation of TiB_2 boride leads to obtaining 3300–4500 HV hardness on the surface.

Boronized plastic layers are obtained in different ways. The right choice of process technology, saturation medium composition, sublayer material, layer's width, thermal-temporal parameters, heating and cooling conditions ensure the formation of plastic boronized layers under any kind of treatment. It's important to note that the structures and phase compositions of plastic layers may significantly differ and can provide the different complexes of physical-mechanical and chemical properties. In all cases the determination of processing conditions and layers' purposes should be done according to the differential approach described in the previous chapters.

Boronized layers are used mainly for increasing wear-resistance; therefore the structural type of the layer would determine its use. In particular, the conditions of

M.G. Krukovich et al., *Plasticity of Boronized Layers*,
Springer Series in Materials Science 237,
DOI 10.1007/978-3-319-40012-9_12

alternating loading require the composite structure of the layer with disperse boride inclusions and the structure with disconnected borides on the surface as they are the most appropriate.

The compositions and the production methods of the most prospective plastic boronized layers are given in Table 12.1.

Boronized layers may be arranged in the following range according to their decreasing plasticity (Table 12.1):

- composite pseudo-eutectic layers;
- composite layers after combined boriding;
- composite layers after electron-beam boriding;
- composite layers after laser boriding;
- boronized layers with disconnected inclusions or colonies of borides;
- one-phase boronized layers alloyed with manganese and obtained in the medium-temperature interval;
- two-phase boronized layers obtained at low-temperature and medium-temperature boriding;
- two-phase alloy boronized layers with the high-boron phase content ≥15%;
- two-phase boronized layers with the high-boron phase content <15%;
- two-phase boronized layers with the high-boron phase content >15% obtained at high-temperature boriding.

The table contains several methods and compositions used for generating boronized layers of different plasticity. It is important to understand that most of compositions may form both plastic and brittle boronized layers (if technological recommendations are violated).

12.1 The Technological Specifics of Boriding Processes

12.1.1 The Preparation for the Surface Hardening

The process of preparing surfaces for boriding takes an important part in the whole technological process. The insufficient carefulness may lead to some irreparable fault. Preparation processes are determined by the type and conditions of hardening, the processed material and its thermal processing, the detail's configuration etc.

The process involves the following operations [1]:

1. Cleansing of the working surfaces from oxides, oil marks, dirt. Mechanical cleaning is normally used: manual, mechanized in tumbling barrels, ultrasound, chemical—degreasing in solvents: ether, benzoyl, acetone, spirit, four-chloride carbon; chemical and electrolysis etching.
2. The visual examination of the detail's surfaces in search for pests, blisters and cracks. Details with cracks are not appropriate for hardening treatment. Pests and blisters should be classified by the level of suitability.

Table 12.1 The conditions for generating boronized layers with different plasticity

No.	Structural type of layer	Technological method and composition of saturation mixture, mass%	Substrate's material and phase composition of layer	Temperature and duration
1	2	3	4	5
1	Composite pseudo-eutectic layers	Gaseous contact method in powder mixture 20% B_4C, 76% Al_20_3, 4% KBF_4	High-alloy steels and Ni-based smelts	1050 °C, 4–6 h
2	Composite layers of hieroglyphic type after combined boriding	Electron-beam processing of boride layers obtained with any method	Steels, alloys, iron casts	1100 °C, local heating, 15–50 s
3	Composite layers after electron-beam boriding	Saturation in compact materials (coatings) at electron-beam heating	steels, alloys, iron casts	1100 °C, local heating, 4–5 min
4	Composite layers after laser boriding	Saturation in compact materials (coatings) at laser heating	Steels, alloys, iron casts	1100 °C, local heating, $(1–6) \cdot 10^{-3}$ s
5	Composite-eutectic layers	Saturation in any medium at the diffusion-crystallization mechanism of layer's formation	Steels, alloys, iron casts	1100–1200 °C, volume or zone heating (high frequency current, contact), 0.1–3 h
6	Layers with disconnected boride inclusions of their colonies	Gaseous contact method in powder mixture: 60% B_4C, 10% graphite, 10% Fe powder, 20% Na_3AlF_6; saturation in granulated mixture on the base of B_4C (<65) and particle size >5 mm	Construction and tool steels, cast irons	850–1000 °C, 3–8 h
7	One-phase (Fe_2B) boronized layer alloyed with manganese and obtained at medium temperature interval	Saturation in bath (70% $Na_2B_4O_7$, 30% silicon-manganese), electrolysis saturation in bath (88% $Na_2B_4O_7$, 4% NaF, 8% MnO); saturation in powder mixture (93% B_4C, 4% silicon-manganese, 3% KBF_4)	Construction and tools steels, iron casts, smelts	750–900 °C, 3–8 h

(continued)

Table 12.1 (continued)

No.	Structural type of layer	Technological method and composition of saturation mixture, mass%	Substrate's material and phase composition of layer	Temperature and duration
8	One-phase (Fe_2B) boronized layer obtained in high-temperature interval	Gaseous contact or contact-free method: BF_3 blowing of mixture B_4C and SiC or mixture EKabor (Borudif technology). Saturation in the mixture on the base of B4C (98%), $CaCO_3$ (1%) and Na_2CO_3 (1%) in kerosene pyrolysis. Including methods and compositions from 7	Construction and tools steels, iron casts, smelts	900–1100 °C, 3–8 h
9	Two-phase boronized layers (FeB, Fe_2B) obtained in low-temperature or medium-temperature boriding	Gaseous method: BCL_3 /(75% H_2 + 25% N_2) = 1: 50. Solid method: EKabor mixture. Liquid method: electrolysis—(mixture of tetraborates, halogenides Na, K, Li and KBF_4), non-electrolysis (mixture of tetraborides, halogenides Na, K, Li, KBF_4 and B_4C)	Construction and tool steels	550–900 °C, 6–16 h
10	Two-phase layers alloyed with Si, Al, Mn, S, Cr, Cu, Ni, Zr, P3M, Sr with high-boron phase (Fe, Me)B > 15%.	Saturation in liquid, powder and compact mediums containing corresponding elements. Simultaneous and consequent saturation	Construction and tool steels, cast irons, smelts	900–1100 °C, 3–8 h
11	Two-phase boronized layers with high-boron phase (FeB) content <15%	Gaseous method: BCl_3/ (75% H_2 + 25% N_2) = 1: 25. Solid method: EKabor mixtures. Liquid method: electrolysis—the mixture of tetraborates Na, K, KBF_4. Non-electrolysis—in the mixture $Na_2B_4O_7$ (70%), B_4C (10%) and SiC (20%)	Construction and tool steels, cast irons and smelts	900–1100 °C, 3–8 h

(continued)

Table 12.1 (continued)

No.	Structural type of layer	Technological method and composition of saturation mixture, mass%	Substrate's material and phase composition of layer	Temperature and duration
12	Two-phase boronized layers with high-boron phase content >15% obtained at HTB	Gaseous method: BCl_3/(75% H_2 + 25% N_2) = 1:20. Solid method: mixture on the base of B_4C (97%) and NH4F (3%). Liquid method: electrolysis—in bath $Na_2B_4O_7$, non-electrolysis—in the mixture $Na_2B_4O_7$ (70%) and B_4C (30%)	Construction and tool steels, cast irons, smelts	800–1100 °C, 3–8 h

3. The preliminary mechanical treatment of hardened surfaces:
   - the elimination of burrs and rounding of sharp edges in order to obtain boride layers;
   - the correction of sizes of precision details in case when the hardening processing is the final stage in the detail's production;
   - the correction of working surfaces' roughness to acquire the set roughness according to the blueprint.

The minimal radius of rounding sharp edges is acquired manually or in tumbling barrels filled with a Al_2O_3 containing suspension, rubber pieces and metal objects (bent nails, nuts etc.). If sharp edges and burrs are preserved after meeting the diffusion of saturation element flow, the sharp edges and tops increase in size (Fig. 12.1, color inset); the increase is followed by the shearing of the layer in these places—either simultaneous or under an insignificant mechanic impact. The boriding of details with rounded edges leads to their sharpening within the limit.

Boronized layers ensure the average growth of a detail's size by 10% of its size on one side (Fig. 12.1, color inset). Therefore the precision details should be decreased in size by carrying out preliminary mechanic processing if the size is excessive. To reduce the possibility of the details' size change due to heating and further thermal processing, it is appropriate to have preliminary annealing or normalization or post-mechanical processing quenching before final polishing.

In general the change of roughness after boriding has following regularities:

- it is preserved if $R_a = 0.5$;
- it decreases by 1–2 classes if $R_a = 0.25$;
- it increases by 1–2 classes if $R_a = 1$.

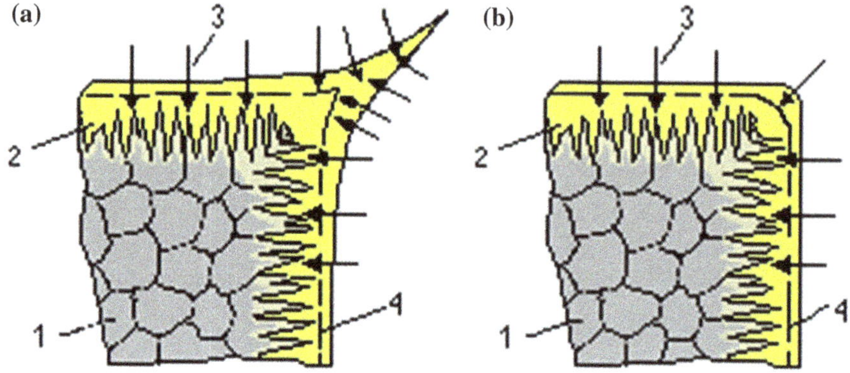

Fig. 12.1 Type of sharp edges after boriding without their rounding (**a**) and preliminary rounding (**b**): *1* the structure of processed metal; *2* boride layer; *3* the direction of saturation element diffusion flow; *4* the border of processed surface before saturation

The preparation procedures also include some types of preliminary processing of hardened details used for accelerating the growth of diffusion layers.

12.1.2 The Specifics of Boriding in Powder Mixtures

The general requirement for boriding in powder mixtures is the necessity of the careful preparation of initial components. The first stage involves the drying and chipping of all mixture components. Boron carbide powder must be thoroughly tempered at 300–400 °C for 1.5–2 h. The saturation mixture after weighing or long preservation is put into barrels-mixers and mixed for 30 min (minimum) at rotation speed ~60 c/min.

The components of the mixture, ready mixtures and reusable mixtures are sifted with electrical vibrational sieves with cells from 2 to 0.2 mm.

Boron-containing mixture components (boron amorphous, boron carbide, ferro-boron, Mg, Ca, Ba polyborides etc.) are used for boriding 10–30 times. The number of uses depends on the container's sealing level during saturation, the level of density, the processing temperature and the coefficient of the container's loading.

Metal-thermal and other mixtures should be regenerated after a triple use by introducing 10–30% of a freshly-produced mixture and further mixing in barrels-mixers. After each use, the mixture should be supported by 1–2% of the activator and mixed. This technology allows to use the mixture 1–10 times.

The container's packing involves the following stages:

- a packed bed of mixture (20–30 mm wide) is put onto the container's bottom;
- the details are positioned at the distance of 20 mm from each other and from the container's walls and between the detail's layers;

- the mixture is packed and the distance between the mixture level and upper layer of details is set (min. 50 mm);
- the operations are carried out according to the chosen scheme of sealing.

The container's sealing, i.e. the protection from the saturation mixture and detail's oxidation during heating and processing, is carried out by bedding silicon carbide powder over the saturation mixture (20–30 mm wide), producing a mixture for the fusible lock and putting it over the saturation mixture with asbestos partition, positioning it along the container's perimeter between the lid and the wall of a special pocket or using a protective atmosphere or vacuum. Boron anhydride (B_2O_3) or natrium-silicon clod (Na_2SiO_3) are used as fusible locks; they are supposed to provide the protection from air flow at temperatures >500 °C. The mixture of boron anhydride and lead oxide (PbO) is used as fusible locks at low-temperature boriding; it provides protection from 350 °C.

The prepared containers with the mixture for the fusible lock are put into the furnace heated up to the processing temperature or the temperature higher than the processing temperature by 100–150 °C. In the first case the duration of heating is determined on the principle "30 min for 100 mm of a crucible section". In the second case the heating duration is decreased by 30%. The furnace temperature regulator is set at a given mode during the container's loading into the furnace.

After finishing the process, the container is taken from the furnace and gets cooled. After cooling the fusible lock breaks; its pieces, sand layer and asbestos partition are also taken out. Then the mixture and the details are put onto a tray. Normally the separation of the details from the mixture isn't complicated especially at the container's unpacking temperature of 30–40 °C.

The use of silicon carbide eases the removal of the formed crust; it peels off together with unfused silicon carbide and asbestos partition. Silicon carbide after chipping is used multiple times.

Boriding in powder mixtures in open containers is carried out in protective atmospheres: kerosene pyrolysis products, nature gas, hydrogen, nitrogen, argon and vacuum. The number of use for boron carbide saturation mixtures increases by 50 times.

For instance, obtaining a plastic one-phase layer (Fe_2B) is guaranteed by the use of the following composition (mass%): 98% B_4C, 1.0% Na_2CO_3, 1% $CaCO_3$—in the atmosphere of kerosene pyrolysis. The container with the saturation mixture and the details is closed with leaf asbestos and put into a shaft furnace heated up to the set temperature. Kerosene supply begins from the moment of the container loading and finishes before unloading. The speed of kerosene supply in drops is determined by the volume of the furnace space. Boriding is carried out at 850–1050 °C for 4–8 h. The container is unpacked at room temperature. The width of a one-phase layer is 150–300 mkm.

The regeneration of the saturation mixture is done after the double use by adding 1.0% Na_2CO_3 and 0.5% $CaCO_3$ and further mixture agitation. The number of uses of regenerated mixtures reaches 70 [2].

The commercial powder mixtures have following compositions:

- 5% B_4C, 90% SiC, 5% KBF_4;
- 50% B_4C, 45% SiC, 5% KBF_4;
- 85% B_4C, 15% Na_2CO_3;
- 95% B_4C, 5% $Na_2B_4O_7$;
- 84% B_4C, 16% $Na_2B_4O_7$;
- amorphous boron (containing 95–97% B);
- 95% amorphous boron, 5% KBF_4.

12.1.3 The Specifics of Boriding in Liquid Mediums

The most popular boriding processes in liquid mediums are electrolysis and non-electrolysis saturation from melted salts. The main components of smelts are technical borax—$Na_2B_4O_7{\cdot}10H_2O$ and the halogenide mixture:

- the mixture KBF_4 + LiF + NaF + KF for the details boronized at 600–900 °C;
- the mixture 20KF + 30 NAF + 49LiF + (0.7–1.0)BF_2 (mol%) for boriding at 800–900 °C in the atmosphere containing 90% N_2 and 10% H_2;
- the mixture with ratio 9:1 (KF-LiF) + KBF_4 in the nitrate atmosphere;
- KBF_4 + NaCL at 650 °C;
- the mixture 90 (30LiF + 70 KF) + 10KBF_4 for boriding at 700–850 °C;
- 80% $Na_2B_4O_7$ + 20% NaCL for boriding at 800–900 °C.

The electrolysis process exploits these salt smelts as they are; the non-electrolysis process requires the addition of one of the following components in a powder state: boron, boron carbide, ferron-boron, manganese polyborides (MgB_{12}) or calcium polyborides (CaB_6) or electrochemical reducing agents (ligatures, containing Si, Mn, Al, Mg, Sr, Ba, Ti, Zr, rare earth metals) in the amounts preserving an appropriate viscosity of saturation mixtures. In some cases the mixtures of boron-containing powders and boron-free ligatures are used.

The electrolysis processes require the use of furnace-baths with internal and external heating of the crucible and their combination.

The details are included into the electrolysis chain as cathodes. Graphite or carborundum rods are used as anodes. The graphite electrodes are glazed before using (they are soaked in borax smelt for 15–20 min). The energy sources are motor-generators (АНД 250/500, 750/1500 etc.) and different rectifiers (ВКГ-100А6 ВАГГ-600/12) [2].

Borax-based compositions ($Na_2B_4O_7$) are prepared this way:

- the furnace-bath is filled with borax half the crucible's volume;
- it is heated up to 450–550 °C and soaked till the crystalized liquid completely annihilates;

- the furnace is heated up to the certain temperature (over 750 °C) which ensures melting of the borax;
- the bath mirror is evenly covered by further portions of borax without agitating it with the smelt; their complete fusing is expected;
- the latter operation is repeated multiple times until half the bath is fused;
- the electrodes are set at 300 mm from the crucible's bottom and no less than 100 mm from the walls; the current of cathode protection of the crucible is switched on (current density 0.01 A/cm^2);
- borax fusing continues until the set level;
- the stable current goes through the smelt for 10–15 h (current density 0.05–0.2 A/cm^2).

If necessary, other salt additives (well-dried and chipped) are injected in small portions.

If borax is fused in the bath with electrode heating, additional nichrome wire heating electrodes are put into the bath, half the crucible is filled with borax; the alternating heating current switches on. New portions are added when borax fuses. The additional heaters are put out of the bath after reaching the half. The further fusing is carried out according to the described technology using the heat of main electrodes.

The details for boriding are put on special devices providing them with a reliable electrical contact and preventing their mutual screening from the anode. The details are heated over the bath mirror for 10–15 min before loading. The heated details are smoothly put into the smelt without spilling the smelt. The level of sunk details should be below the smelt level by 30–50 mm. Boriding is carried out at 850–980 °C at current density 0.08–0.15 A/cm^2.

Cleansing of the details from fusing residues and final tailings is done in heated water at 80–90 °C for 2–3 h. The reduction of the cleaning period is reached by the change in the polarity of the details from cathode to anode 10–15 min before the end of boriding or by soaking the boronized details in the intermediate chlorine-barium bath for 15 min.

Fusing of borax and other salts during non-electrolysis saturation is carried out in the same way as according to the electrolysis method. An electrochemical reducing agent in powder state is added to the smelt in portions. Another portion of the reducing agent is added after the end of the rapid reaction between the reducing agent and smelt accompanied by the smelt airing and the increase of its level. As fusing continues, the level of the bath increases; to prevent the bath to run out, final portions of the reducing agent should be smaller. After adding the whole volume of the electrochemical reducing agent, the bath is used for 15 min; before loading details, it is again agitated.

The more comfortable loading and unloading of details is reached by binding the details with wire and putting them on special devices. Materials of the binding wire and devises should resemble the chemical composition of the details' material. Details, devices and binding wire shouldn't exceed the limit of the bath mirror. Boriding is carried out at 550–1050 °C for 2–16 h depending on the mixture composition.

The bath loses its saturation capacity over a prolonged working period. The maintenance of saturation capacity is reached by adding a fresh mixture instead of the mixture gone with the details. After 3 months, the bath is half refreshed or fully updated.

12.1.4 The Specifics of Boriding in Compact Materials (Coatings)

Boriding in compact materials (coatings) is used if large-scale details need processing or if boriding and heating for quenching are combined. The technological process involves the following procedures:

- the production of saturation and protective mixtures and their careful agitation;
- the introduction of bonding materials into the mixture; the amount depends on the required consistency;
- the preparation of processed surfaces for boriding;
- the application and drying of the coatings until they acquire the state of a solid compact material;
- boriding at a set processing mode;
- cooling and cleansing of the processed surfaces from the fused mixture.

The consistency of the paste added to the powder mixture of bonding materials is determined by the chosen technology of its application on the processed surface: pulverization, dipping, brushing or applying with a palette knife; this depends on the size and details' configuration. The width of the saturation coating normally reaches 2–5 mm. Thicker coatings are applied in several stages.

The use of organic bonding materials (acetone, butyl acetate, acetone-xylene, nitro-cellulose polishes, Zaponlack etc.) in coatings requires drying to be carried out at room temperature. The use of non-organic or organic-mineral coating substances requires drying at 120–200 °C.

The use of fast heating methods (the high frequency current method, electrical contact, laser, electron-beam) doesn't require any additional preparation of details for processing and protective procedures of compact materials. The use of furnace heating, i.e. prolongated heating in the furnace atmosphere, requires using protective coatings. They are applied in the same way. The most appropriate protective coating contains 50% B_2O_3 and 50% SiO_2. It is mixed on the foundation of hydrolyzed ethyl silicate.

Boriding in contact materials is carried out with the glow discharge at the corresponding temperature. The paste contains, for instance, amorphous boron (60%) and borax (40%). This process leads to a significant thickening of boronized layers.

12.1.5 The Specifics of Gaseous Medium Boriding

Gaseous medium boriding is carried out in the atmospheres containing diborane (B_2H_6), diluted by hydrogen, or boron trichloride (BCl_3), diluted by hydrogen, nitrate or ammonia. The process usually happens at 800–950 °C under pressure $23.4 \cdot 10^3$–$28.6 \cdot 10^3$ Pa for 2–6 h. The width of the obtained layer is 0.02–0.25 mm. The most popular atmospheres are:

- the mixture of diborane (B_2H_6) with H_2;
- the mixture of haloid boron with H_2 or N_2-H_2 in proportion 75:25;
- the organic compounds of boron as $(CH3)_3B$ and $(C_2H_5)_3B$.

Boriding in the mixture of diborane and hydrogen hasn't received a wide industrial use due to the high toxicity and explosiveness of diborane. Boriding in organic compounds simultaneously generates carbide and boride layers. The expensiveness and low technical efficiency of BBr_3 (due to its rapid reaction with water) and high reduction temperature of BF_3 (due to the increased stability and generation of HF steams) makes them also inappropriate. Thus the most reliable compound is BCl_3. Gaseous medium boriding at BCl_3-H_2 proportion 1:15 at 920 °C and 67 kPa (0.67 bar) pressure for 2 h may cause the formation of the boronized layers of 120–150 mkm. To obtain a minimal amount of FeB boride in the composition of the boronized layer, it is recommended to use mixture N_2 + FeB as a gas-diluter in proportion 75:25 instead of hydrogen alone. A certain influence on boriding results is done by the gas flow speed. The gas consumption is estimated as 40–50 l/h. Each setting and coefficient of working space loading requires an individual calculation.

Gaseous reaction products, generated in the working space for gaseous medium boriding, are toxic and they are withdrawn into the exhaust system or burnt or go through a range of water-filled filters. The end of the process is accompanied by the end of the boron-containing gas supply and 5–10 min gas-diluter flow. The pressure is then reduced in the working space to atmospheric, the supply of the gas-diluter stops, the machine is unsealed and the boronized details are put out. The most perfect setting for processing small and medium-size details is offered in work [2] where details are loaded after BCl_3 supply in 20–30 min through the flame generated at the moment of lid opening.

Boron gaseous medium saturation is also carried out with the plasma of glow discharge at the pressure below atmospheric (270–800 Pa). Boriding in plasma is carried out in the atmospheres containing a diborane mixture with H_2 or $BCl_3 + H_2 + Ar$. These mediums in different steels allow to obtain boronized layers at comparatively low temperatures ($\sim$600 °C). The mixture of gases $BCl_3 + H_2 + Ar$ is better controlled. The occurrence of two-phase layers on the surface is accompanied by the generation of porousness combined with the separation of amorphous boron. Porousness can be minimized by increasing BCl_3 concentration.

12.1.6 Methods for Intensifying Diffusion Saturation

The intensification of diffusion saturation presupposes the influence on particular stages of chemical-thermal processing:

- the impact on the reaction capacity of saturation medium in order to increase the number of active atoms and subions (subcompounds) by increasing circulation of gaseous and liquid mediums and acceleration of exchange reactions;
- the influence on the reaction capacity of processed surfaces;
- the influence on the metal substrate by carrying out the procedures accelerating the withdrawal of saturation element atoms from the surface deeper into the processed material.

The following methods are used in practice:

1. The chemical or electrolysis etching of working surfaces which enhances the reaction capacities of hardened metal's surfaces.
2. The preliminary application of galvanic or electrochemical coatings (Fe, Ni, Ni + P, Zn etc.) which either increase the self-diffusion of iron to the surface or generate a liquid phase during saturation; it can also provide the generation of diffusion layers at low heating temperatures for general processing (zinc coatings). In the end all of these procedures increase the amount of defects in the crystal structure of the surface and undersurface layer and the further accelerated withdrawal of boron atoms from the surface into the substrate.
3. Mechanical and thermal mechanical impacts:
    - the preliminary cold plastic deformation (hardening) of the surface which enhances the surface cleansing from scale and increases the number of crystal defects—vacancies and dislocations on the surface and undersurface layer;
    - the preliminary (combined with saturation), intermediate and further hot plastic deformation of processed surfaces with the reduction level 18–65% (chemical-mechanical-thermal processing);
    - the combination of gaseous processes with finishing mechanical processing with local friction heating of details for rotation bodies (counter body: superalloy) up to 890–940 °C with the supply of saturation atmosphere into the heating zone; cooling is done by blowing the same gas;
    - the mechanical-chemical-thermal processing which includes the creation of a rubbed surface on the processed surface with a special tool during mechanical processing and following saturation. Thick layers are obtained (>1 mm) due to the diffusion healing of ribbed surface deeps.
4. The thermal impact:
    - the preliminary quenching which causes the stabilization of structure and geometrical sizes of details after hardening, accelerates the formation of

diffusion layers by increasing defects of the crystal structure in the surface layers during the disintegration of quenched structures;
- preliminary or combined with saturation thermal circulation in the range of temperatures Ac_1 + (80 … 100)°C and Ar_1—(100 … 150)°C. The heating and cooling speeds are in the interval 1.5–3 °C/c. This 1.5× accelerates the growth of boronized layers at soakings shorter than 4 h.

5. The preliminary oxidation of working surfaces processed in 1–4% water solution NaCl at 400–950 °C for 10–60 min is better used at the saturation of gases where one of the components is hydrogen. The heating process includes the reduction of oxides by hydrogen and the generation of activated loosened (porous) surface which accelerates the speed of saturation atoms withdrawal deeper into the metal and the speed of diffusion layers growth.
6. The use of stage modes for temperature, current density, saturation medium activity (boron, nitrate, carbon potential).
7. The use of electrophysical impact methods:
   - the ionic saturation in glow discharge plasma;
   - the ionic implantation, the influence of fast ions E = 40–50 keW, irradiation doze $\sim 5\text{–}10^{17}$ ion/cm^2;
   - the preliminary electric spark processing of processed surfaces;
   - the relative electrons impact on the flow surface;
   - the stimulation of glow discharge during powder and solid composites (coatings) saturation;
   - the laser impact;
   - the use of electrical and magnetic fields;
   - the use of ultrasound;
   - the use of vibro—and fluidization of saturation mixtures with applying an electrical impact (polarized and non-polarized).

The important note is that the best effect from the use of most procedures manifests itself at a relatively short saturation duration (<2 h). Longer periods of boriding make the effect less evident due to the equilibrium structure state correlating with the boriding temperature.

12.2 Other Methods of Boriding

Non-chemical-thermal boriding methods are methods of physical (PVD) and chemical (CVD) vapour deposition. CVD methods are the most popular in the industry.

12.2.1 Chemical Vapour Deposition (CVD-Processes)

The method presupposes the application of coatings of the vapour-gaseous phase by the condensation of products of chemical reactions occurring in gases or by the chemical interaction between the gaseous phase and the substrate. The coatings based on metal borides TiB_2, ZrB_2, NbB_2 etc. are applied on steels, high temperature metals and smelts.

Some borides are well isolated in the conditions mentioned in Table 12.2. The most popular coatings are TiB_2.

CVD-processes of applying TiB_2 and ZrB_2 coatings on different substrates are carried out during the feeding of gases $TiCl_4$ (or $ZrCl_4$), BCl_3 and H_2 over a heated detail in the vacuum camera. The gaseous flow generates the subcompounds of precipitated elements $TiCl_2$, ($ZrCl_2$), BCl, BCl_2 and HCl which are reduced on the processed surface with the generation of borides TiB_2 (or ZeB_2) in certain thermal and gas pressure conditions. The gas feeding should be maintained so that the atomic correlation of elements is in the following ranges:

- B:Ti = 1:2 and H:Cl = 6–10 in case of TiB_2 separation;
- B:Zr = 1 and H:Cl = 20 in case of ZrB_2 separation.

The correlation B: (B + Cl) = 0.4 makes the layer of separated TiB_2 borides dense and oriented towards surfaces {1010} and {1120} which is often associated with the occurrence of columnar crystals and high microhardness of the surface (3300–4500 HV_{050}). In order to provide a good cohesion with the surface of steels and powder solid smelts, it is recommended to apply corrosion-resistant cobalt and TiC coating on the substrate [4, 5].

Table 12.2 The conditions for the separation of certain borides by CVD

No.	Type of boride	Gaseous medium	Temperature (°C)	Pressure (kPa)
1	HfB_2	$HfCl_4$-BCl_3-H_2	1400	0.4
2	NbB_2	$NbBr_5$-BBr_3	850–1750	0.003–0.025
3	NiB	$Ni(CO)_4$-B_2H_6-CO	150	87
4	SiB_4	SiH_4-BCl_4-H_2	800–1400	6.5–80
5	SiB_x	$SiBr_4$-BBr_3	975–1375	0.007
6	TaB_2	$TaBr_5$-BBr_3	850–1750	0.003–0.025
7	TaB_2	$TaCl_5$-B_2H_6	500–1025	100
8	TiB_2	$TiCl_4$-BCl_3-H_2	1200–1415	0.40–2.0
9	TiB_2	iCl_4-B_2H_6	600–900	100
10	TiB_2	$TiCl_4$-BCl_3-H_2	750–1050	100
11	TiB_2	$TiCl_4$-BCl_3-H_2	1200	6.5
12	ZrB_2	$ZrCl_4$-BCl_3-H_2	1400	0.4–0.8

These processes ensure a relatively high speed of precipitation of surfaces with high density, high thermal, erosion and corrosion resistance. They guarantee a significant economical effect during serial production of small details. As result, TiB_2 coatings, for instance, are widely used in the cutting of the plates made from powder solid smelts, in graphite electrodes and aluminum reducing cells and also in admission valves in carbon conversion reactors.

References

1. Krukovich M.G., Minkevich A.N., Tone E.R. Tekhnologicheskiye metody povysheniya iznosostoikosti i vosstanovleniya detalei mashin. Chast II. Povysheniye iznosostoikosti detalei mashin. Moscow, 2002. 150 p. [Technological methods of increasing wear-resistance and restoration of machinery details. Part II. The improvement of wear-resistance in machinery details]
2. Voroshnin L.G. Borirovaniye promyshlennykh stalei i chugunov. Spravochnoye posobiye. Minsk, 1981. 205 p. [Boriding of industrious steels and cast irons. Reference book]
3. Kuleshov Yu.S., Nefedov V.G., Sergienko R.I. Ustanovka dlya gazovogo borirovaniya // Zaschitnye pokrytiya na metallakh. Kiev. Iss. 15. Pp. 31–33 [A setting for gaseous boriding]
4. Chatteree-Ficher R. // Met. Prog. 1986. V. 129, N 5. P. 24, 25, 37
5. Chatteree-Ficher R. // Hart.-Tech. Mitt. 1981. Bd. 36, N 5. S. 248–254

Chapter 13
The Use of Boriding Processes in the Industrial Treatment of Details and Tools

Abstract The structural type of boronized layer of special use, saturation method and processing conditions are suggested to be chosen according to the differential approach to the analysis of several factors. The purpose is reaching the maximum effect from the hardening processes. This approach is used by the authors to study the factors which can effect the choice of boriding technology for particular types of details or tools: working conditions of a tool or a detail, reasons for malfunction, the roughness of surfaces, the limiting value of wear, steel brand and its thermal processing, the precision class, size and serial characteristics. As boriding is widely used in different industrial sectors, the chapter contains examples of different boronized details used in the following spheres: metallurgy, mining, textile industry, agriculture, food-manufacturing industry, chemical industry, mechanical engineering, porcelain industry, printing industry, railway transport.

The differential approach to the analysis of several factors is used to determine the structural type of boronized layers, the saturation method and processing conditions in order to obtain the maximal efficiency. This approach for certain types of details or tools presupposes the analysis of working conditions, reasons for a possible failure of details or tools, values of the tolerable wear, steel brands and their thermal regimes, roughness of working surfaces, precision classes, size and serial characteristics.

The presence of impact loads, high unit pressures and temperature levels are the most important factors in the area of working conditions. Impact loads require the use of composite layers or plastic one-phase boronized layers obtained in the medium-temperature interval. High unit pressures determine the necessity of the solid substrate for boride layers which is acquired by post-boriding quenching of the sublayer. Therefore, the following factors would be very important: the substrate's steel brand, the content of carbon in the sublayer and the structure of the sublayer. The recommendation for the optimal variant is to exclude silicon, aluminum or nickel α-solid solutions from the sublayer. The working temperatures of details or tools are significant for guaranteeing the necessary scale-resistance by the microalloying of borides in order to reach the maximum effect.

M.G. Krukovich et al., *Plasticity of Boronized Layers*,
Springer Series in Materials Science 237,
DOI 10.1007/978-3-319-40012-9_13

The failure of details or tools due to brittle fractures without any noticeable wear should be explained by the construction or metal's quality. The presence of stress concentrators requires additional measures apart from surface boriding. A slight improvement of durability may be reached by increasing the amount of carbon in the sublayer which allows to reduce the heating temperature for quenching and obtain a more viscous detail's core. At the same time the blunting of cutting edges in tools abruptly enhances some loadings which also cause brittle fracture. Plastic one-phase boride layers would protect sharp edges from fast blunting.

Thus, unhardened details and tools may be protected from brittle fracture by boriding combined with a certain correctly chosen hardening technology.

The maximum effect of boriding is reached when the value of tolerable wear doesn't exceed the width of the boride layer. However even at the tolerable wear level (1–2 mm), the durability of the detail significantly increases with the combined work of the boride layer and the sublayer; their width is up to 1.5 mm. The correct regime of thermal processing makes the sublayer harder than the detail's core.

The choice of steel brand and thermal treatment provides necessary constructive properties for details or tools and limiting exploitation temperatures. Therefore, boriding of stamps for hot pressure-processed metals requires that the steel brand is to be preserved. It's also necessary for cutting tools with special requirements for red hardness. In case when tools are heated during exploitation at the depth which doesn't exceed the boride layer's width, i.e. when voluminous working temperature of the tool is <200 °C, it is possible to change a high-alloy expensive brand to a cheaper and more workable one.

The optimal, more technologically functional variant of hardening is the combination of boriding with any thermal treatment used for producing boronized details or tools, i.e. the use of low-, medium- and high-temperature boriding.

The roughness of working surfaces after boriding decreases at initial values $Ra > 0.8$ mkm and $Rz > 10$ mkm. If the initial roughness is $Ra > 0.8$ mkm and $Rz < 3.0$ mkm, it increases. The optimal minimal roughness of working surfaces upon boriding is acquired by finishing the surfaces with diamond pastes.

The size of details upon boriding are increased by a layer which is 10% of the boride layer's width. Other possible changes to the detail's size, due to the heating under boriding during phase transformations in steel, should be taken into the allowance depending on the form and direction of size changes of a particular detail. In other words, details and instruments before boriding should be of a nominal size (if all the changes are within the limit) or with a plus or minus allowance taking into account the boride layer and further thermal treatment.

Low-temperature boriding doesn't influence the size of details much. However, the initial steel structure may effect it.

Boriding processes, causing the generation of plastic and composite layer's structures (eutectic, pseudo-eutectic and obtained upon combined processing) of significant width (up to 1.5 mm), allows to carry out the final mechanical treatment of boronized surfaces accompanied by the elimination of extra 0.1–0.3 mm. Thus the conditions for the first and final polishing of working surfaces are created; this allows to perfect their sizes and roughness to the values correlating with the blueprint.

The detail's size and serial properties determine the method of boriding and its technological peculiarities. For instance, the worn areas of large-scale details are better to go under boriding in compact materials (coatings, plates, saturation models etc.) with the use of local heating (high frequency current, laser, electron-beam, energy-producing coatings).

Hardening is done to the details and tools working under any possible conditions: abrasive wear, corrosion wear, high impact loads and pressures, alternative loading, aggressive environments and different combinations of various conditions.

The tests on tools after high-temperature, medium-temperature, low-temperature boriding showed the increase of strength twice and more.

The industry tests and industry exploitation reveal some effects and unpredicted possibilities. In particular, the tests on the die matrix in the cold layer punching demonstrated the effect of self-sharpening cutting edges. The effect of self-sharpening manifests itself after the boride layer is peeled off on the horizontal surface of the matrix due to the accelerated wear and further wear of the softer structure of the metal's sublayer and the base in some distance from the cutting edge.

The tests on milling cutters after low-temperature boriding revealed the possibility of repeated boriding after their regrinding. This allowed to stabilize cutting conditions. The resistance between regrindings increased at least twice.

The hot stamp dies' resistance may be increased by plastic one-phase boronized layers alloyed with manganese.

Boriding is used in many industries. The right choice of the method and boriding technology allows to cover a wide range of exploitation conditions for tools and details: dry friction in air and vacuum, low- and high-temperature exploitation, abrasive-hydroabrasive and corrosion-mechanical wear, contact with melted metals and salts etc.

We give several examples of the details hardened with boriding divided into industries. The brackets contain the increase of strength (times). Similar details and tools (by the name and working conditions) in different industries are not repeated.

Metallurgy: centrifuge sieges (1,5), axis of drag-over unit (3), supply idlers (5), tracks and wheels of a tube forming mill (5), conveyer wheels (4–6), cog wheels (3–4), rolls (2–4), dies, mandrels (cores), used for pipe dragging (6–8).

Mineral resource industry: turbodrill disks (4), slush pump rod (3–4), cog wheels of open gearing (3,5), barrels and hatch rings of sludge pumps (3,5), elements of rotary drilling chains (2), fingers of grab chains in hydraulic dredges (2–3), conveyer hubs.

Textile industry: looper (4), knives (3), presser feet (5–6), shuttle (3–5), pins (8–10), live roller (6–8).

Agriculture and food industry: details of a threshing mechanism's drum in grain combines (10–15), chains and distributors in fertilizer machines (3), barrels, cogs, knives of centrifugal beet cutters (3–5), conveyer details.

Chemistry industry: nozzles of heating elements in torsional exhaust machines for chemical fiber production (10), pump cogs for supplying a cement-bitumen mass (3), die molds (3–5), screw conveyers (3–5), draw plates for pulling wired fiber (2–2.5).

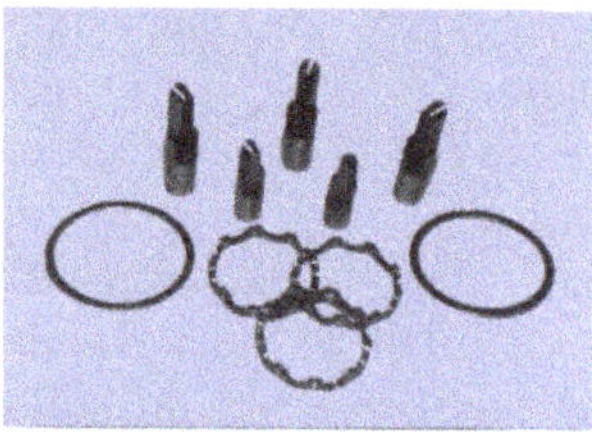

Cold bending chisels of frictionless bearing separators (steel У8). Two-phase boriding of chisels (in salt smelts).

↔ **The increase of durability–4X**

Cold heading matrix of rivet head (steel X12, changed to steel 40X). One-phase boriding (in salt mixture).

↔ **The increase of durability -10X**

Directional rings of drawing-machine for cuprum wire pulling (steel ШХ 15). Two-phase boriding (in solid powder mediums).

↔ **The increase of durability -4X**

Hard-alloy cutting plates boronizedin gaseous mediums (contact and contact-free method).

↔ **The increase of durability –1.5-2.5X**

Fig. 13.1 Types of tools

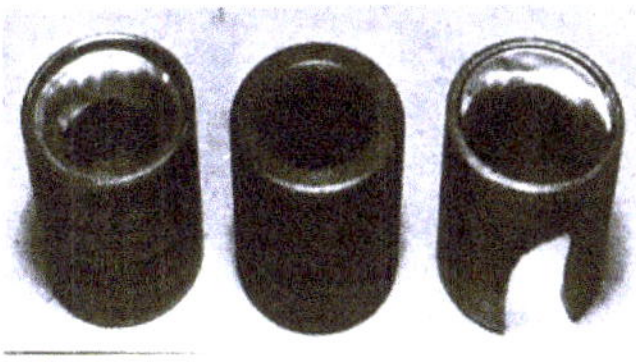

Nozzles of shot blast unit (steel 45). Two-phase boriding (in salt smelt).

↔ The increase of durability – 4-6X

Matrix and chisel for color metals hot mouding (steel 5XHB). Medium-temperature one-phase boriding (in salt smelt).

↔ The increase of durability – 3X

Eliminator of stamp for refractory metals hot moulding (steel P18). Low-temperature boriding (in salt smelt).

↔ The increase of durability – 3X

Matrixes for bolt heads hot moulting (steel 3X2B8, 45X3B3MΦC). One-phase boriding (in salt smelt).

↔ The increase of durability – 6X

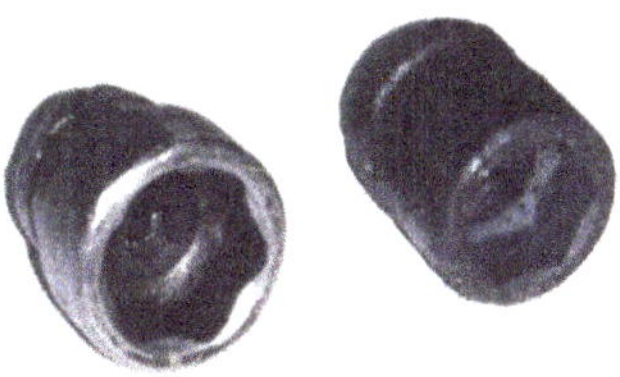

Directional rings of drawing-machine for cuprum wire pulling (steel ШХ 15). Two-phase boriding (in solid powder mediums).

↔ The increase of durability – 4X

Fig. 13.1 (continued)

Table 13.1 The examples of industrial tests for different tools

No.	Detail or tool	Steel brand	Type of boriding	Layer width (mm)	Process temperature (°C)	The increase of strength
1	2	3	4	5	6	7
High-temperature two-phase boriding						
1	Shot blast unit's nozzle	Steel 45	Non-electrolysis in salt smelt	0.15	900	4
2	Bending chisels of separators	У8	the same	0.12	950	4
3	Dress rings for cuprum wire drawing	ШХ15	In powder mixture in container	0.13	950	4
4	Chisels for opening beadling	У8	Solid composite (coating)	0.08	950	6
5	Dissectors of sand blend	У8	Powder mixture in containers	0.15	950	6
6	Form inserts for producing construction and ceramic materials	Сталь 45	Powder mixture in containers	0.18	900	6–7
High-temperature one-phase boriding						
7	Blanking chisels (d = 6–15 mm)	У8	Protector from borax smelt	0.08	950	2.5
8	Hot forming matrixes of bolt tops	3Х2В8, 4Х3В3МФС	Non-electrolysis from salt smelts	0.10	950	4–6
9	Die mold inserts for producing plastic goods	Х12	Powder mixture in containers	0.15	980	3
Medium-temperature two-phase boriding						
10	Shot blast unit's nozzle	45	Non-electrolysis in salt smelt	0.12	800	6
11	Form ejector for producing isolators	Same	Same	0.13	Same	3
12	Matrixes and chisels for forming electro-technical beads	Same	Same	0.13	Same	4
Medium-temperature one-phase boriding						
13	Taps (M16)	ПЖ2М-Х30 (19% Cr)	Non-electrolysis (salt smelt)	0.075	750	The production technology
14	Cylinder (hot gases medium 1800–2000 °C)	30ХГРА 30ХН2МФА	Same	0.10	850	3–4
15	Punching dies	ШГ15СГ	Same	0.08	850	3.2
16	Chisels and matrixes of hot forming non-ferrous metals	5ХНВ	Same	0.10	860	3
17	Upset matrix inserts for rivet tops	40Х instead Х12	Same	0.10	840	20
18	Matrixes for chipping sheet materials	Х12	Powder mixture in containers	0.08	850	2

(continued)

Table 13.1 (continued)

No.	Detail or tool	Steel brand	Type of boriding	Layer width (mm)	Process temperature (°C)	The increase of strength
1	2	3	4	5	6	7
Low-temperature boriding						
19	Glass cutting discs (Pyrex)	P18	Non-electrolysis (salt smelt)	0.02	600	5
20	Conductor for plate drilling	X12	same	0.03	680	The production technology
21	Milling cutters for grooving	P6M5	same	0.025	560	2
22	Hot stamping die ejectors for high-strength metals and smelts	P18	same	0.025	650	3
High-temperature boriding for obtaining composite structure of boronized pseudo-eutectic layers						
23	Synchronizing gear (P = 51 kgf/mm^2; T = 600 °C)	ЭП 718	Powder mixture (in containers)	0.65	1050	2.5
24	Hinge bearing	ЭП877	Powder mixture in containers	0.55	Same	3
25	Groove punching die	X12M	Combined electron-beam	0.25	1100	2.7
26	Piercing chisel	4X3B3MФC	Same	0.3	1120–1150	3.5
27	Cutting rollers for sheet material	X12M	Same	0.25	1100	2.2

Note The list contains only several boronized tools and details

Mechanical engineering: bushing for high-speed marine diesel (2,8), reinforcing plates of sandblast machine's sprayers (6), shot blast unit's blade (10), caterpillar track pins (5), machine details (3–5), plungers and liners of high pressure fuel pumps (3–10), hydraulic cylinder barrel in transmission gearbox in tractors and automobiles (3–4), control shrouds of machines (6–8), collets (5–6), cams (3–4), measuring tools (3–5), conductor barrel (2–3), axle and pins (4–8), cutting cemented carbide tool (1.5–2.5), billets and tap borers (1.5–3), cold stamping dies (2–15), hot stamping dies and inserts (1.5–4), drift pins (2.5–3), eliminators (3–4), die molds for aluminum alloys (2–15).

Whiteware production, ceramic and glass production: details and equipment for glass production (10), die molds (6–8), mixer blenders (4–6), forming chisels (8–10), brick forms (4–5).

Polygraphy: folder cogs (8), tappets (6–8), pawls (5–8), cutting elements (2–4).

Some tools and details and the examples of their use are demonstrated in Fig. 13.1 (colour inset) (Table 13.1).

Table 13.2 Examples of industry tests for different tools

No.	Factory	Tools	Steel brand	Type of boriding	Saturation temperature (°C)	Layer's width (mm)	The increase of strength
1	2	3	4	5	6	7	8
1	Lyublino foundry-mechanical plant	Dies for producing autocoupling safety devices	5XHB	Powder; open	860	0.11	1.8
		coupler key die	Same	Same	Same	0.10	2.2
		Walking beam saddle die	Same	Same	Same	0.12	2.4
		Link die СП35	Same	Same	Same	Same	2.3
2	Moscow locomotive repair plant	Matrix for clipping burrs on bolts	У8	Open method, saturation model	780	0.05	2.0 and more
		Die of harchet handle moulding	5XHB	Powder; open method	860	0.13	4.0
		Chisel and matrix of spring bending	У8	Same	840	0.15	8.0
		Chisel of hot forming bolt top, M20	5XHB	Same	860	0.12	6.5
3	Yaroslavl electric locomotive repair plant	Matrix for clipping fins of captivating bolts	У8А	Powder; container	800	0.07	2–3
		Matrix for forming captivating bolts tops (OK-10-80)	4X4M3BΦC	Same	1050	0.11	2.0
		Corner cutting matrix (OK-13-17)	У8	Powder; open method	780	0.07	2.0
		Bottom insert VENSANA	4X4M3BΦC	Same	1050	0.08	2.5
		Matrixes forming (OK-13-77)	4X4M3BΦC, 4X4M2BΦC	Same	Same	0.10	2.2

(continued)

Table 13.2 (continued)

No.	Factory	Tools	Steel brand	Type of boriding	Saturation temperature (°C)	Layer's width (mm)	The increase of strength
1	2	3	4	5	6	7	8
4	Michurinsk locomotive repair plant	Matrix of detail cutting (Rebro)	У8A	Same	780	0.08	3.0
		Dies for cutting and bending of brake block carcas	Same	Same	Same	Same	2.5 (cutting); 8.0 (bending)
		Chisel and matrix of hot forming	5XHB	Powder; container	880	0.12	2.6
		Hot forming dies of pointwork clamp	Same	Powder; open method	Same	Same	2.3
		Clasp forming dies	Same	Same	Same	Same	2.0
		Petal washer forming die	Same	Same	Same	Same	Same
		Cutters for collectors' fillister (MT 2134)	P6M5	Low-temperature; container with fused lock	560	0.02	2.2
5	Kalinin (Tver') railcar manufacturing plant	Hot forming inserts of castle nut	5XHB	Powder; open method	870	0.12	3.0

For many years boriding has been successfully used by the railroad repair service for increasing the wear-resistance of high wear tools and details (Table 13.2). In particular, open boriding was used (without sealing a saturation medium).

Chapter 14
The Development of the Theory and Practice of Boriding

Abstract The chapter studies the processes of boron mass transfer (connected with the process of boriding) and the precipitation of boron atoms on the processed surface. The classification of obtained layers is presented from the point of their growth inside the metal or from the exterior. The chapter also introduces the analysis of new developments which were stimulated by the first publication of the book. The information is systematized according to the contents of the book's chapters. The most important results are the development of fast boriding carried out by the electrolysis of salt smelts, the development of theoretical and practical calculation of eutectic temperatures and concentrations, electron-beam boriding resulting in obtaining the layers with plate morphological phase structure, the use of nano-structured saturation mixtures, modeling processes involving the interface for solution making.

Boriding is a technological process aimed at obtaining boron-containing surface layers in details and semi-finished materials. Currently there are different technologies of generating boronized layers:

- the traditional procedures at temperatures <550 °C (from gaseous, solid-powder, liquid and plasma mediums);
- the chemical precipitation from water solutions and further sintering or annealing of precipitates at temperatures >500 °C [1, 2];
- the electro-chemical precipitation from electrolytes (galvanic processes) with the further annealing of precipitates [3];
- the vacuum physical and chemical precipitation in gaseous mediums (PVD, CVD processes) in the presence of diffusion transition zones;
- the electrolysis of suspensions.

These methods differ in the technologies, used equipment, processes in saturating mediums, mass transfer mechanisms and reactions on processed surfaces which ensure the formation of different types of layers with various properties and functions. Specifically, one of these types is boronized layers with different phase compositions and structures: the precipitates of atomic boron; boron precipitates

M.G. Krukovich et al., *Plasticity of Boronized Layers*,
Springer Series in Materials Science 237,
DOI 10.1007/978-3-319-40012-9_14

combined with metals, non-metals and polymers; the precipitates of borides, carbides or nitrides of iron and alloy elements; the diffusion layers based on iron borides and borides of alloy elements or in combinations with their solid solutions. This variety causes the necessity of the exact terms referring to these layers.

The following classification is introduced:

I. the diffusion boronized layers obtained by the diffusion of boron into deep layers of the processed material. They cause a slight augmentation in the detail's size which doesn't exceed 15% of the boride layer. The processing temperature is normally below 500 °C. The major phases of the layer are iron borides and borides of alloy elements, in some cases—α-solid solution;
II. the boronized layers obtained by any method and consisting of the diffusion layer growing from the surface inward the detail's material and the layer precipitated from the outer side of the detail's surface. The augmentation is a sum of the size increase due to the diffusion zone (15% of a boride layer) and the size increase due to the outer precipitation;
III. the boronized layers obtained by any method and consisting only of the precipitated zone. The augmentation of the detail's size equals the width of the precipitated layer.

The total evaluation of the changes in the detail's size as a result of the hardening process should take into account the possibility of augmentation due to the structural transformations in metals at heating and cooling.

The given book studies mostly diffusion layers and the technology of their production; certain attention is given to the layers obtained by precipitation and further processing which ensures the presence of diffusion zone.

The layers which are precipitated from water solutions during the heating of details up to 410 °C [4] can't be defined as boronized or diffusion layers; the falsely issued C23C patent and chosen prototypes [5, 6] mess with the terminology and demonstrate the absent expertise on the processing. This process could be diffusion boriding if the temperature were above 500 °C (the diffusion of boron in iron is possible only at this temperature regime). Moreover, the studies lack the information on the layer's width, chemical analysis, which would prove the presence of boron in the layer, or phase composition. The attached illustrations might lead to the following conclusion: the colours of the surface in processed details indicate the presence of layers of unknown width which can be defined as oxide layers; the obtained colour shades correlate with the colours of tarnish. Thus, the proposed technology ("Boriding of steel surface for enameling") can't be attributed either to tranditional processes of boriding, or to the processes of chemical-thermal treatment. The obtained layers are neither boronized nor boride. These layers can be attached to the third group and can be called "boronized" only if the presence of boron is proved. However, the production method must be taken into account—for instance, boronized layers from water solution, electrochemical or electrolysis boronized layers from water solutions, precipitated boronized layers etc.

The boronized layers based on boron and alloy boride-forming elements, precipitated at 750 °C by the electrolysis of melted salt mixtures, obtained by the technology of Ali Erdemir and Osman Eryilmaz [7], are attributed to the second group as they consist of precipitated and diffusion zones. They can transform into diffusion layers due to the diffusion of precipitated boron into the surface layers of the detail; the additional conditions are prolonged soaking, turned-off circuit and a certain width of the precipitate. The obtained layer's width—100 mkm in 15 min—is the sum of widths of the diffusion zone and the zone of precipitated borides of metal and alloy elements from the outer side of the detail's surface. Boron atoms and alloy elements precipitated at 750 °C can produce reactions at the processing temperature and form a compact layer of borides due to the diffusion and chemical bonds in the precipitate. The total layer's width (200 mkm and above) is obtained in 3–4 h. In this case, it consists mainly of the diffusion boride zone.

Thus, the boom of fast boriding can be perceived as only a marketing campaign while the process of electrolysis diffusion boriding from salt smelts at the diffusion of boron and alloy elements in the processed metal is still in the traditional time frame.

The Russian edition of the present book has given a certain impulse to the improvement of boriding. Some of these new directions should be studied in accordance with the chapters of the book.

14.1 The Classification of Boriding Processes

The general classification scheme allowed to sum up the theoretical and practical results achieved over the years and present an overview of the development of boriding. The overview is based on the general interpretation of the mass transfer in saturating mediums and the formation mechanisms of saturating boron atoms. The process was studied in four directions (the formation mechanisms of saturating boron atoms; the technological properties including all known developments; the processing temperatures; the usage) which allowed to build a unified system and connect different phenomena.

The given interpretation of technological aspects is one of the most valuable assets as it allowed to work out the unified language and understanding for the development of new technologies and the usage of known technologies of boriding in particular and chemical-thermal processing in general.

A range of certain terms and definitions should be explained in more detail; this necessity is connected to the insufficient level of the understanding of boriding and the vague nature of some interpretations. Thus, boriding processes are divided into the following types:

- diffusion boriding—the process which combines boron precipitation on the processed surface and boron atoms diffusion in deeper layers of the detail;
- precipitated boriding with additional processing—boron precipitation on the exterior of the detail and further processing at temperatures (<500 °C) which ensure the formation of the diffusion zone (the use of annealing, heating for quenching, normalization);
- precipitated boriding—the process of boron precipitation on the exterior of the detail in order to obtain amorphous or other layers.

14.2 The Components and Phases of Systems 'Boron–Iron' and 'Boron–Carbon–Iron'

The components and phases in B–Fe and B–C–Fe systems stay stable in certain concentration conditions; however, their influence on the surface and volume properties differ greatly.

Although boron atoms have a small atom radius, they are too big for the formation of solid interstitial solutions in iron. In other words, this process contradicts Hagg's rule which states that solid interstitial solutions can't be formed if the radius correlation exceeds 0.59. The correlation is about 0.65–0.74 in the given cxase. Therefore, the formation of solid interstitial solutions has low probability both in α–Fe and γ–Fe. The radius of octahedral interstitial position in γ–Fe equals 0.41 of Fe radius while the radius of tetrahedral interstitial position of α–Fe is 0.29 of Fe radius.

According to the experimental data on the changes in lattice parameters at the introduction of boron in α–Fe, the parameters of boron diffusion into α–Fe and the data on the internal friction, the following hypothesis was introduced: boron forms a solid interstitial solution in α–Fe, i.e. it is soluted by the vacancies [8–10]. However, in this case the first Hume-Rosery rule on limits of continuous interstitional solubility is violated.

Value $(R_{Fe} - R_B)/R_{Fe}$ is called a size factor; for boron and iron it is 0.27 which substantially exceeds Hume-Rosery limit: $\Delta R/R_{Fe} < 0.15$. Therefore the solubility of boron in α–Fe and γ–Fe is very small as the atoms of boron create big distortions of solid solution. On the other hand, the structural peculiarities of the boron crystal lattice, its electron structure, tendency to the formation of covalent bonds also prevent the high solubility of boron in α–Fe and γ–Fe. Moreover, fast cooling of the smelt in liquid and liquid-crystal state increases the possibility of its distortion [11–13]. This state of surface layers or the sublayer may be achieved by laser, plasma, electric contact or electron beam heating.

The publications studying the influence of boron on the volume properties of roll-foundry iron and describing the nature of structural changes contain the conclusion that $Fe_3C_nB_{1-n}$ boron-carbide plays the part of hardening phase [14]. The increase of hardness in the metal foundation during high tempering (550 °C) is explained by the increase in martensite amount due to the disintegration of austenite; the increase in the hardness of hardening carbide phase takes place due to the impact of soluted boron.

The issue of FeB_2 high-boron phase existence in boronized layers is discussed. In particular, the saturation in nano-structurized mediums, obtained by self-spread high-temperature synthesis, hypothetically causes the formation of FeB_2 phase on the surface of steel 20 [15].

14.3 The Equilibrium Diagram of 'Boron–Iron' Binary System

Technological studies engage equilibrium diagrams in order to determine the phase composition of diffusion layers and the order of phase formation depending on the reaction or diffusion mechanism of boride formation on the processed surface. Simultaneously the order of phase formation and boron concentration on the surface largely depends on the mass transfer mechanism in the saturation medium.

Traditional saturating mediums have a gaseous mass transfer mechanism which involves boron subcompounds. Therefore, in these cases the diffusion mechanism of boronized layers formation is realized; boron is slowly stored on the processed surface.

A range of saturating mediums, especially nano-structured, allows simultaneous realization of diffusion mass transfer through gaseous mediums and reaction mass transfer in the contact points of mixture nano-particles and the processed surface; the condition is the presence of process activators. If the sizes of nano-particles and boride crystallites are comparable, the welding of nano-particles to the processed surface is possible due to the contact melting. This effect was seemingly observed in the work materials [15].

The demonstrated chemical reduction of boron by carbon from its oxides during boriding [16] is quite simplified as atomic boron can't move in the volume of saturating mixture. The movement and mass transfer in the volume of saturating mixture is provided by the boron compounds of low valency which are presented by gaseous compounds BO in the described conditions. Thus, the reduction is carried out gradually, saturating atoms of boron are formed on the processed surface.

14.4 Multicomponent Equilibrium Diagrams Used in Boriding Treatments of Steels and Alloys

Multicomponent equilibrium diagrams for most smelts, which could be used as technological layouts for determining concentration and temperature conditions for boriding, are still not built. At the same time the methodology of building multicomponent diagram schemes (n > 3) [17, 18] allows to achieve this by using the existing two-component diagrams. These schemes of multicomponent equilibrium diagrams are built in traditional coordinates temperature-concentration; their foundation is regular polyhedrons.

The constructed multicomponent diagrams Fe–Bi–Cr–B and Fe–Cr–W–Mo–B are still used for determining the temperature and concentration conditions for boriding of heat-resistant and high-speed steels aimed at optimizing the structural morphology of formed layers and determining the optimal conditions of their usage [19]. Furthermore, the developed divergent coordinate grid in three dimensional space with a divergence angle along the vertical axis can be also useful in crystallography for determining the coordinates of separate atoms, in construction for building structures with misplaced gravity center in the form of spiral, shifted and inclined elements, for modeling the structure of dimensional polymers etc. The transition from Cartesian to divergent coordinate system is usually carried out by simple geometrical transformation.

The proposed new direction of statistical calculation of eutectic temperatures and eutectic concentration, based on the melting points of initial components [20], has initiated the studies of these characteristics aimed at the predicting equilibrium diagrams. In particular, one of the studies [21] involves the determination of described characteristics for two-component systems based on the effects of mixing the ideal solutions of components. The heating effect and volume don't change; the change in entropy equals the change in entropy at the mixing of ideal gases. The authors have made their objective much more difficult by changing the term of eutectics into the rule of mixing at the formation of solid solutions without going into the issue of this change. The calculation of eutectic concentrations and temperature of a two-component system demands the values of internal energy of both components, Debye temperature (calculated with the use of Plank and Boltzmann's constants), changes in partial molar temperatures of solvent and solute elements and their calculations and finally effective molar parts of components in the binary eutectic point which correlate to the minimal temperature of each system.

The authors of the study [21] investigate the calculation of eutectic concentrations and temperatures of two-component systems only for pure elements though a range of studied systems form two or more eutectics between elements and chemical compounds. It means that this method can't be used for a large number of smelts or it should be improved by finding non-existing characteristics which are necessary for calculation. Moreover, the authors give the data on calculation of eutectic concentration in Al–Si system which equals 10 mol.% Si. In fact, this concentration is 11.6% in mass, 12.2 at.%, and the eutectics consists of α-solid solution and silica crystals but not the crystals of pure elements [22]. Thus, the limitations of this method are obvious as this technology ignores some basic metallurgy issues connected to the chemical compound, structure and properties of eutectic smelts.

The transition to the calculations of eutectic concentrations and temperature in multicomponent systems involves the rule of eutectic reactions and consecutive calculation method, proposed in work [17] and partially analyzed in the present book. At the same time the proposed results of calculations for eutectic systems Cr–Co–C–B, Ni–Cr–C–B, Nb–Co–C–B, Ni–Mo–Be–B and Zr–Ni–Cr–C are not acceptable as the studied eutectic systems consist not of pure elements but of carbides, borides and solid solutions; these multicomponent systems can form several eutectics. In other words, the proposed method doesn't allow to predict multicomponent diagrams of smelts with any sufficient certainty.

The oscillation theory of melting states, according to which the increase in temperature causes the increase of oscillation amplitude for atoms and the eutectic concentration of a binary system is perceived as a resonance of phase atoms oscillations, significantly complicates calculations [23]. Equilibrium diagrams can be predicted based on the calculation of eutectic temperature and the information on melting points of the eutectic's components. This method, as the calculation method in [21], offers a partial solution as it doesn't cover the possible eutectic components and doesn't calculate the concentrations of possible eutectics in two- and multi-component systems. Moreover, these methods require special databases and software which can't ensure solving the tasks of predicting equilibrium diagrams. Naturally the proposed methods [21–23] are really significant for science as they deepen the knowledge and understanding of the processes that take place in smelts at the formation of solid solutions and eutectic mixtures.

14.5 Plasticity as the Key to the Durability of Boronized Layers

Plasticity of boronized layers depends on the alloying level of the substrate and the layer itself, the phase composition of boride layers (with the minimal content of high boron phase FeB or fully consisting of Fe_2B phase), the level of compositeness of the structure consisting of boride inclusions and α-solid solution and the level of texturedness of the boride layer and its stress state.

The classifications of boronized layers by plasticity properties and exterior, given in this book, allow to choose an appropriate technology for obtaining a particular structure for the layer of certain purpose. In combination with phase qualities, this layer provides a high level of wear resistance of details and tools.

The evaluation methods applicable to boronized layers haven't changed for the last years. Properties are still evaluated by phase hardness, brittleness (determining the brittleness grade), viscosity of decomposition, stress state, tensile and bending strength, compression resistance, integral characteristics—strain stress, critical strain, critical scratch length used in scratch tests.

14.6 The Formation Conditions for Boride and Boronized Layers and Their Influence on the Layers' Plasticity

The formation of structures and properties of boride and boronized layers begins with heating when a processed surface acquires a certain reaction activity towards environment products. The environment products may be represented by atoms or compounds of oxygen, hydrogen, nitrogen, boron and other elements depending on the choice of boriding method and the composition of the saturating atmosphere. The initial stage of heating is crucial for the further development of boronized layers as it "infects" the surface in correlation with the beginning of boride crystallites formation. The further development of the layer is determined by the speed of boron precipitation, the speed of boride crystallites formation and their growth during soaking and finishes with cooling after hardening or final thermal or mechanical processing.

The mechanism of mass transfer in saturating mediums during boriding and other processes of chemical thermal processing is unified and provided by the sub-compounds (dissipative structures) containing ions of low valencies which are generated due to the misbalance in thermal systems and lead to a certain stationary state which is characterized by the minimum entropy generation [24]. Thus the mass transfer goes in self-organizing mode which can take place in any saturating medium in certain conditions. The exclusion is made for the processes of surface alloying in moulds with the application of boriding coating or boriding with the use of gasifiable patterns which are covered with boriding coating [25].

The formation of boronized layers is carried out in the conditions when the surface is in crystal, liquid-crystal or liquid state when the realization of the diffusion or diffusion-crystallization mechanism is possible. The forming structures were described earlier.

The primary structural unit of the alloyed layers obtained with the moulding by gasifiable patterns is the carbon-boride eutectics. Its dispersibility changes depending on the steel brand. The increase of carbon reduces the size of structural elements which can be explained by the appearance of multiple germinal centers of carbon-borides during crystallization (Fig. 14.1).

The formation of carbon-boride components in alloyed layers proves the surface boron cooperation with carbon and iron of moulding smelt and the products of thermal destruction of pattern's material—foam polystyrene which is used as an additional source of carbon. The decryption of the XRD pattern in steel У8 made in [25] has been done incorrectly as the formation of boron and iron carbides is highly improbable when FB_2 is the source of boron in ($B_{13}C_2$, B_4C, $B_{10}C_2$, Fe_7C_3, Fe_5C_2) compositions.

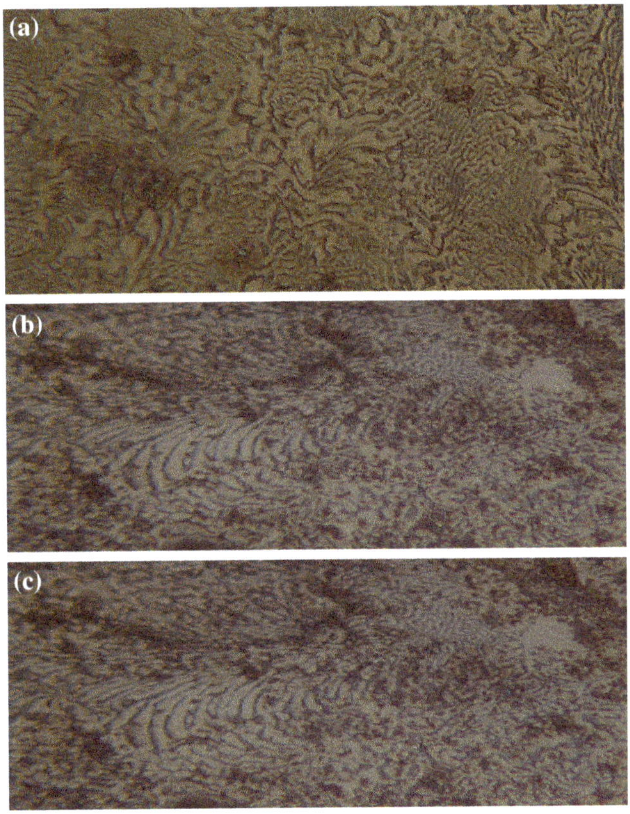

Fig. 14.1 Boride eutectics in the structure of alloy layer in steels: 10 (**a**), 45 (**b**) и У8 (**c**). ×500 [25]

14.7 Methods of Reducing the Brittleness of Boronized Layers. The Parameters of Boriding Technology Aimed at Determining the Plasticity of Boronized Layers

The methods of reducing the brittleness of boronized layers, described in Chap. 7, are currently used as the foundation for improving the exploitation wear-resistance of the details after boriding. The general directions of the control over the properties of the layers under different saturation methods are:

- obtaining one-phase structure layers on the base of Fe_2B or layers with a minimal content of brittle phase FeB;
- the micro-alloying of borides in layers of a substrate or saturating medium;
- the optimization of temperature-temporal parameters;

- the stage regime of managing process parameters: current density, the content of electro-chemical reducing agent, temperature, boron potential of a saturating medium;
- the use of concentrated energy sources under the processing or additional processing of obtained boride layers;
- simultaneous or consecutive saturation of metal surface with boron and elements partially soluted in borides (C, Si, Al, S, Cu);
- the regulation of the sealing level of a saturating medium or processing in a controlled atmosphere;
- the preliminary precipitation of metals (Fe, Ni, Ni + P, Mn etc.) on the processed surface and further boriding;
- the simultaneous or consecutive diffusion saturation with boron or alloy elements.

In particular, saturation in casting forms [25] leads to obtaining eutectic layers according to the diffusion-crystallization mechanism. The diffusion mechanism reduces the brittleness by the simultaneous precipitation of boron and alloy elements on the processed surface under salt electrolysis in the conditions of the significant growth of the layer from the exterior side of the detail [7].

The reduction of brittleness was reached by the further laser or electron-beam heating of boronized layers obtained by traditional methods [26]. The formation of fragmented layer's structure was witnessed; it ensures a higher level of plasticity (Fig. 14.2).

Additional processing is aimed at increasing the plasticity of boronized layers and expanding their application.

The major means of reaching this goal were:

- the change in the phase composition and reduction of hardness;
- the dispersion of the surface and/or sublayer's structure;
- the formation of smooth hardness transition from the surface to the core;
- the creation of unloading areas by applying the grids of laser runs on worn-out surfaces.

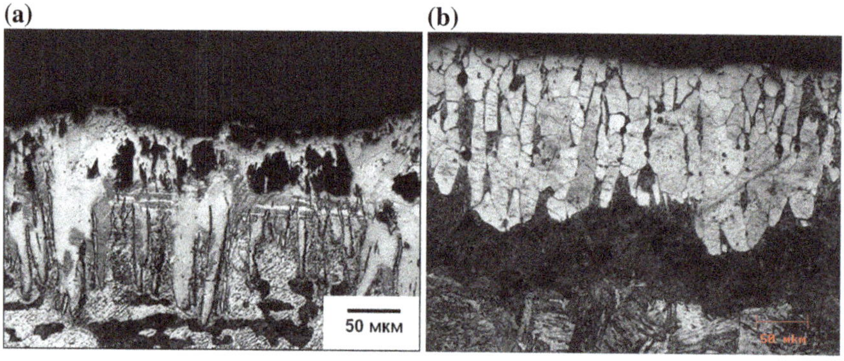

Fig. 14.2 The microstructures of boronized layers in steel 40× before (**a**) and after (**b**) laser processing

Laser processing was used for boride layers in steel 40 x, obtained in the powder saturation mixture containing 3% NH_4F, 10% Al_2O_3, 87% B_4C. Boriding was carried out in a container with a fusible lock in order to protect the models and saturating mixture from the furnace's oxidizing atmosphere. The mixture of sand (95%) and dehydrated borax (5%) was used as a fusible lock.

Laser processing was carried out in the scanner for the dimensional control of the laser beam; the beam was maintained by laser machine "Kometa-M" and several scanning devices for the high-voltage beam oscillation which ensured better opportunities for the temperature control [27]. The emission capacity changed in the range of 700 and 1300 W. The speed of beam movement—2–50 mm/s. The laser heating of the boride surface was done by the defocused or scanning beam directed inversely to its movement with 220 Hz frequency. The width of laser impact zone was 2.4–3.2 mm under the defocused beam and 5.6–6.1 mm under the scanning beam. The processing regime was chosen so that the temperature of the surface didn't exceed the melting point of Fe–B–C triple eutectics, i.e. 1100 °C.

The evaluation of surface microhardness was done by the microhardness meter M-400-H (LECO) under 100 g loading. The metallographic analysis of the surface miscrostructure was carried out by the microscope PME-3 "Olympus" equipped with a digital camera; the image analyzer IA-3001 (Leco) was also used.

The study of the microstructure of boronized layers showed that the layers consist of at least 2 zones: the large columnar crystallites of borides over 100 mkm and the grey layers positioned between and under the needles. Taking into account the boron potential of the used saturating medium, the layer of borides was determined to consist of FeB and Fe_2B phases with the microhardness of 19,500 and 16,000 MPa accordingly. High-boron phase (FeB) was positioned by separate inclusions on the surface. This could be seen in the level of etching characteristic (Fig. 14.2a). Moreover, pores of 50 mkm were witnessed in the near-surface layer. The formation of porous structure in the surface is connected to the insufficient sealing of the saturating container and the fast exhaustion of the working mixture. However, this structure of boride layers is preferable in the conditions of friction with greasing as the obtained open porousness plays the role of oil-retaining pockets.

The additional high-speed laser heating of boronized layers and the further cooling ensured by the heat dissipation into the deeper layers of the material leads to the fragmentation of boride structure and elimination of porous defects. The refilling of pores might happen due to the sharp volume increase under local heating and emergence of compressive stresses and active diffusion processes with iron and boron atoms. This structure can increase the wear-resistance under friction without greasing in the layers processed by the laser beam; the local heating changes both the structure and stress state of the boronized surface in general. The picture of the microstructure after laser heating is demonstrated in Fig. 14.2b. The average microhardness of columnar phases is 15,000 MPa, grey zones between and under boride needless have 7500 MPa microhardness. The reduction of boride microhardness is explained by the dispersion of high-boron and harder phase (FeB) and the reduction of stress state due to the formation of fragmented structure. The dulling of borides needles and the increase in their thickness are also observed. All of this provides a slight increase in the plasticity of boride layers.

The heating of the base material leads to the formation of a continuous zone of martensite in the sublayer (Fig. 14.2b). The formation of martensite layer is connected to the excessive presence of carbon in the sublayer; carbon is obtained as a result of boriding due to its edging by the growing boride layer and sufficient speed of heat dissipation. The ferrite inclusions, positioned at some distance from the boride layer, are exposed to the hammering hardening and their microhardness increases (1.5–2X) than at the initial stage. Thus, a favourable smooth redistribution of microhardness into the deeper layers of the model is created.

The defocused laser beam, applied to the surface, leads to the partial burning of boron on the surface layer. High-frequency scanning makes the topography of the surface layer more even and doesn't drastically change in comparison with the initial state. This allows the boronized surface to use the detail with a modified high-frequency scanning beam without using additional finishing or fine grinding.

Laser affected trails, due to the increased plasticity, may function as relaxation oscillators for exploitation strains in rather prolonged wearing surfaces. Their role especially increases when the trails are applied in the form of grid at 60° angle in relation to the direction of glide process.

To ensure the optimization of wear resistance, the size of grid cells for laser trails, the angle and the level of modification in boronized layers must be set individually depending on the material of the detail, foundation state, saturation conditions and the structural type of boronized layers, the value of critical wearing, the form of friction surface and exploitation conditions.

Boriding in the mixture of amorphous boron (brand A; ТУ 212-001-49534204-2003) and fluoric manganese (MgF_2) in correlation 1:3 under electron-beam heating with the smelting of steel 20 models provided the formation of deposited boronized layers [28].

The mixture, 1 mm wide, was put on the surface of the models; the density 0.2 g/cm^2 was reached. The processing was carried out while moving the table with the speed 10 mm/s under the electron beam; the beam was generated by the electron accelerator ELV-6 which ensured the electron energy 1.4 meV and the beam current 4 mA. The window, generating the beam and sending it into the atmosphere, was positioned 90 mm from the processed surface. The diameter of the electron beam near the surface was 12 mm. A part of the models after fusing was cleaned; a new layer of boron powder and stone was applied and the process of depositing was repeated. Then the same procedure was done to the models gone through two stages of depositing. The result was the analysis of the structure and the evaluation of the microhardness of surface layers after single, double and triple depositing.

The metallographic studies were carried out with the optical microscope AxioObserver A1 m. The phase composition was investigated by the X-ray diffraction meter ARL XTRA. The scanning involved the use of Cu Ka radiation in the stage regime; the accumulation time t = 3…10 s per one point and movement $\Delta 2\theta = 0.05$ и 0.02°. The microhardness of surface layers was evaluated by

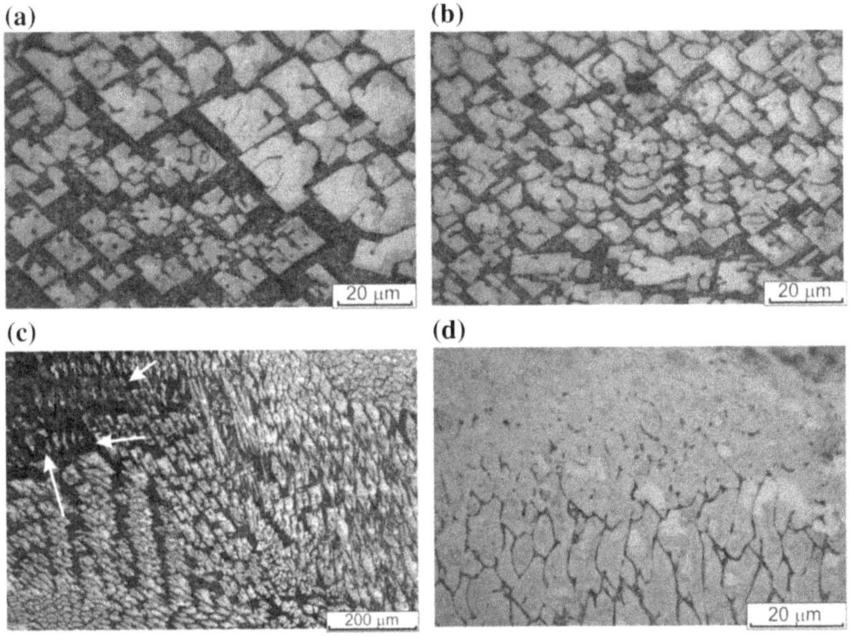

Fig. 14.3 The microstructures of boronized layers obtained by electron-beam depositing of amorphous boron (arrows indicate zones with low content of borides) [28]. **a**, **b** The eutectics of sheet-like type in the section containing FeB and Fe_2B borides. **c** The uneven content of borides in the partially melted surface layer under one-stage depositing; **d** The continuous layer of borides, positioned next to the surface of partially melted model after the second depositing

indenting the square diamond pyramid into the sections under loading 0.98 H and the use of hardometer Wolpert Group 402 MVD.

The metallographic studies of the structure of partially melted compositions allowed to conclude that the surface steel layer is melt through 0.6 mm inward. The surface layer contains eutectics (α–Fe–Fe_2B) and borides like FeB and Fe_2B (Fig. 14.3). The metallographic studies caused the determination of the structural heterogeneity of the surface layer after single depositing. Zones of increased content of borides and zones without borides are observed (Fig. 14.3c).

The microhardness of zones with increased content of borides reaches 14,000 MPa at the maximum depth of their storage—0.4 mm. The increase in the structural homogeneity of the surface layer and borides distribution is reached by repeating the process of depositing. After double and triple depositing the uneven distribution of borides is eliminated. Simultaneously a transition zone is observed; it contains a reducing number of borides and an increasing share of eutectics (α–Fe–Fe_2B). The zone of initial crystallization next to the zone of basic metal doesn't contain excessive large boride crystals. The zone of basic metal, next to the zone of the initial crystallization, has ferrite-perlite structure which is characteristic for the conditions of temporary overheating of pre-eutectoid steel. The results of metallographic analysis are proved by the data of the X-ray structure studies which were done at the front layered sections. The surface layers of coatings contain Fe_2B and boride FeB which leads to the reduction of borides. The deeper layers don't generate FeB. The presence of boron compounds with iron couldn't be found at 800 mkm by the X-ray structural analysis.

The measurement of microhardness of surface layers allows to make conclusions on the depth of the hardened layer. After the deposition of one, two and three layers of amorphous boron on steel 20, the width of the coating reaches approximately 550, 750 and 900 mkm accordingly; the level of microhardness near the surface is 1200–14,000 MPa and near the basic metal 2000–2100 MPa [28].

The use of boron carbide in the mixture with iron powder and MgF_2 process activator under electron-beam non-vacuum processing and higher values of beam current (up to 30 mA) ensures the formation of boronized structures in steel 20. They are positioned in the following order (from the surface to the deeper layers): over-eutectic, eutectic and pre-eutectic types. This structural order is connected to the fact that boron carbide takes part in the formation of deposited carbon layer in accordance with the triple equilibrium diagram of Fe–B–C.

The deposition is carried out by the following parameters: the beam current—24–30 mA, the electron beam energy—1.4 meV, the density of filling—0.2 g/cm^2, the speed of model movement in relation to the electron beam—10 mm/c.

The metallographic analysis of the structure in obtained models was carried out with the optical microscope AxioObserver Alm in 25–1000 zoom range. The study of the thin structure in formed coatings (Fig. 14.4) was done with the bit-mapped electron microscope Carl Zeiss EVO 50 XVP. The chemical etching of the models was carried out in 3% solution of nitric acid in ethanol [29].

The deposition of boron carbide powder by the electron beam directed into the air atmosphere ensures the formation of composite gradient structure of surface layers where iron borides oriented into different directions are the hardened phase.

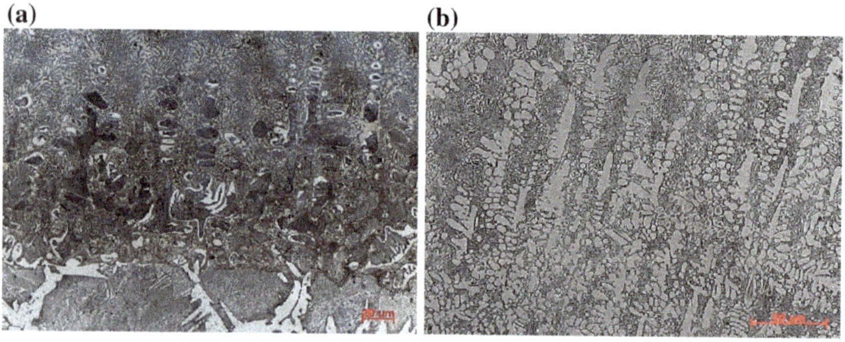

Fig. 14.4 The microstructure of deposited coating, obtained by the electron beam directed into the air atmosphere [29]. **a** The pre-eutectic zone (the border of deposited layer and substrate). **b** The over-eutectic zone (the processed surface)

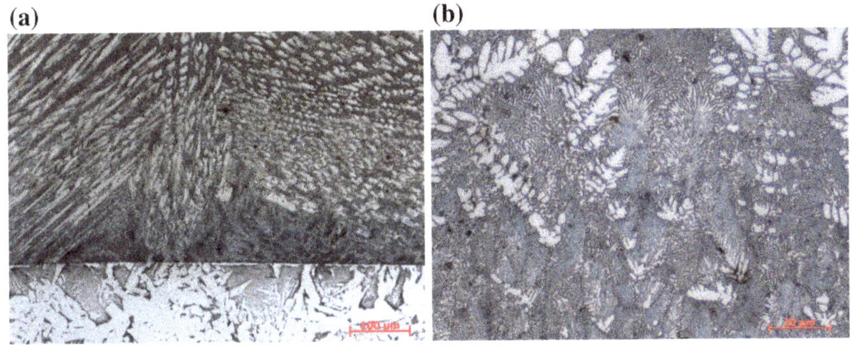

Fig. 14.5 The microstructures of boronized layers obtained by non-vaccum electron-beam depositing (**a**) and by high-frequency currents (**b**) in steel 20, processed in the mixture of boron carbide and amorphous boron [30]

The use of boron carbide and amorphous boron determines the formation of boronized layers with boride inclusions (with both sheet-like and columnar morphological structure). The processing of steel 20 was carried out in the conditions described above; the scanning of the surface was done by the electron beam (50 Hz frequency and 50 mm scale) in order to ensure the deposition of powder mixture along the total width of the model (Fig. 14.5a) [30].

High-frequency current heating generates the typical over-eutectic dendrite structure containing initial boride inclusions and eutectics of zebra-like structure (Fig. 14.5b) [30]. High-frequency current caused the formation of 0.6 mm wide layers, electron-beam heating—1.8 mm.

One of the ways to improve the plasticity of boronized layers may be the technology of obtaining multicomponent layers, for example, boron-coppered, under the use of consecutive technologies of electro-explosive alloying and electron-beam processing. At the same time there is a danger of cracks formation during the depositing of boron-copper layers [31].

The diffusion formation mechanism of boronized layers accompanied by the use of nano-structurized saturating mixtures ensured the increased plasticity of the two-phase layer [15].

The mixtures contained boron oxide and chrome oxide as basic components, aluminum powder as a reducing agent, aluminum oxide as an inert diluting agent and oxides of iron, nickel, zirconium, cuprum, tungsten and molybdenum as additional reagents. Powder mediums were obtained by the aluminum-thermal reduction of charge material mixture accompanied by the CVC process in special crucibles outside the furnace. The reduced mixture was grinded, sifted (maintaining the necessary granule-meter composition 0.3–05 mm^2); activators KBF_3 and AlF_3 were added for the further use under thermal-diffusion saturation.

The sealing of the container was done by the fusible lock based on boron anhydride. The saturation of carbon steel 20 was carried out at 950 °C for 4 h.

Table 14.1 Phase composition and microhardness of boride phases in steel 20

No.	The content of the reaction mixture, % in mass	Phase composition and the width of the layer (mkm)			Microhardness (MPa)	
		General	FeB	Fe_2B	FeB	Fe_2B
1	22% Al_2O_3 + 15% Cr_2O_3 + 26% B_2O_3 + 28% Al + 6% NiO + 3% ZrO_2 Boron sources: CrB_2, Ni_4B_3, AlB_{12}	200	60	60	12,900–18900	11,700–16,800
2	25% Al_2O_3 + 15% Cr_2O_3 + 25% B_2O_3 + 27% Al + 5% Ni +3% ZrO_2 Boron sources CrB_2, $Ni4B_3$, AlB_{12}	240	60–80	180–160	10,200–18,900	11,700–17,800
3	8% Al_2O_3 + 9% Cr_2O_3 + 23% B_2O_3 + 22% Al + 10% Ni + 25% Fe + 3% ZrO_2 Boron sources: CrB_2, Ni4B3	140–160	0	140–160	–	10,600–15,100
4	12% Al_2O_3 + 9% Cr_2O_3 + 23% B_2O_3 + 22% Al + 6% CuO + 25% Fe + 3% ZrO_2 Boron sources: FeB, CrB	140–160	0	140–160	–	11,700–18,900
5	24% Al_2O_3 + 15% Cr_2O_3 + 25% B_2O_3 + 28% Al + 5% WO_3 + 3% ZrO_2 Boron sources: CrB, AlB_{12}	240	40	200	15,100–17,800	10,200–18,900
6	24% Al_2O_3 + 15% Cr_2O_3 + 25% B_2O_3 + 28% Al + 5% Fe_2O_3 + 3% ZrO_2 Boron sources: FeB, CrB, CrB_2, AlB_{12}	320	120–160	200–160	12,900–15,100	10,000–18,200
7	22% Al_2O_3 + 15% Cr_2O_3 + 26% B_2O_3 + 6% CuO + 28% Al + 3% ZrO_2 Boron sources: CrB2, AlB12	380	220–300	160–80	12,900–20,100	13,100–20,100
8	24% Al_2O_3 + 15% Cr_2O_3 + 25% B_2O_3 + 28% Al + 5% MoO_3 + 3% ZrO_2 Boron sources: CrB_2, AlB_{12}	260–320	120	140–200	11,700–21,400	13,800–18,900

The metallographic studies of the models after chemical-thermal processing was carried out under the microscope METAM P-1 with 100–1000 zoom.

The micro X-ray-spectral analysis was carried out under the scanning electron microscope Vega II LMU (Tescan, Czech Republic) and the microanalyzer Inca Energy 350 (Oxford Instruments, UK). The studies of the phase composition in diffusion coatings, formed from the studied saturating mediums, were done by the diffraction meter DRON-3 using the method of X-ray structural analysis. The durometric analysis was carried out with the micro-hardmeter PMT-3 according to the Russian state standard 9650-76. The applied loading reached 0.98 H. The compositions of saturating mixtures and the results of the studies are demonstrated in Table 14.1.

The analysis of the obtained results shows that the suggested anomaly in the microstructure and microhardness of phases is not observed. The phases are positioned according to the distribution of boron concentrations while the microhardness of phases is in the limits of microhardness of alloyed borides. The decreased brittleness of borides is caused by their microalloying.

Fig. 14.6 The view of composite powder boronizing medium obtained by the method of non-furnace metal-thermal processing [15]

The applied technology of producing saturating mixtures allowed to transform them into nano-structural state (Fig. 14.6); this state caused the contact melting due to the small size of mixture borides in the places of contact with the processed surface. The melting led to the accelerated formation of boronized layers. Moreover, seemingly, the layer was formed also at the exterior side of the models. Thus, the total width of the layer consisted of the diffusion zone and grown boride layer from the outer side.

A certain reduction of brittleness in boronized layers is acquired as a result of the simultaneous saturation by aluminum and boron [32–34] under thermal-circular processing or as a result of combining boriding and thermal circulation [35–38].

In the first case the plasticity is increased due to the microalloying of borides and the presence of solid solution of aluminum in iron in the boronized layer. In the second case the sublayer is changed due to the phase transformations and generation of small grain structure as a result of thermal circulation. It is worth noting that the acceleration of borides growth under thermal circulation isn't observed in accordance with the data in studies [34, 38] and the existing general ideas. The acceleration of speed is observed only when the diffusion layers are formed on the basis of solid solutions which is typical for the processes of cementation, nitration, nitro-cementation etc.

The most plastic boronized layers are the layers based on the α-solid solution of boron in iron. They are obtained when the boron potential of saturating medium is quite low ($\pi_B < 4.5\%$B in mass). These layers are not wear-resistant though there are certain recommendations for their use as barrier coatings of the finishing metal [39].

The boriding of the wire 1.2 mm was carried out in the powder of ferroboron FB 17 in the special continuous acting installation under the cycle contact electrical heating. The current was passed through a wire; the current duration—0–10 s; pause duration 1–3 s. The total processing time for each microvolume—2–4 min.

14.8 The Structure Compositeness as the Foundation for the Plasticity of Boronized Layers

The composite structures of boronized layers are acquired after using concentrated energy sources from pastes and coatings [28–31], under isothermal soaking in different mediums at high temperatures ensuring the diffusion-crystallization mechanism of layer formation [25, 30] and during the formation of liquid-crystal state in the substrate due to the edging of carbon and redistribution of alloying elements.

One of the methods of concentrated energy impact on the surface of steel details is micro-arc chemical-thermal processing involving the electrical heating of the detail put into the container which is filled by coal powder [16, 40]. Coal powder has a function of electrical conducting medium in the chain of electrical supply and the source of atomic carbon under processing while ensuring a stage boron reduction from boron anhydride. Boric acid (H_3BO_3) is used as a boron-containing medium which generates boron anhydride (B_2O_3) under its thermal dissociation. The process is carried out at 260 W electrical supply and 3A electrical current for 3 min. This leads to the formation of micro-arc currents concentrated in the microvolumes of powder around the detail due to the high density of the current.

The processing of steel 20 provided the generation of composite layers 0.2 mm deep consisting of the areas of high-disperse eutectoid mixture with 4500–4700 MPa microhardness and the eutectic inclusions of carbon-borides with 15.0–17.0 GPa microhardness (light areas of boronized layers). A carbonated zone with eutectoid structure is positioned under the composite layer; this zone transits into the initial ferrite-perlite structure (Fig. 14.7).

This position of structural components determined a smoother transition of hardness from the surface to the core of the detail after quenching.

The composition structure of boronized layers is also formed in the models with coatings made from B_4C or amorphous boron under high-frequency heating up to

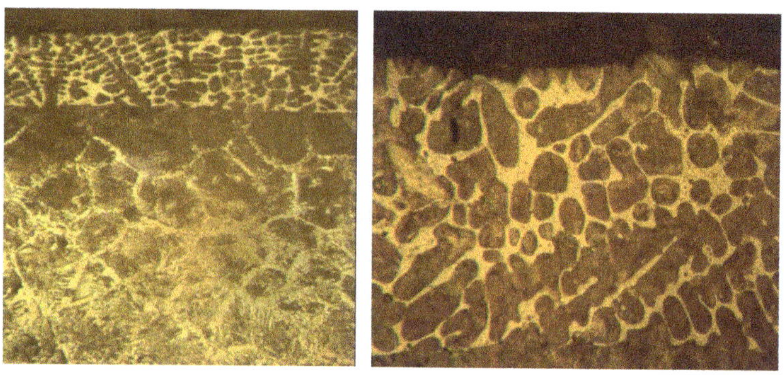

×200 × 500

Fig. 14.7 The microstructure of the boronized layer in steel 20

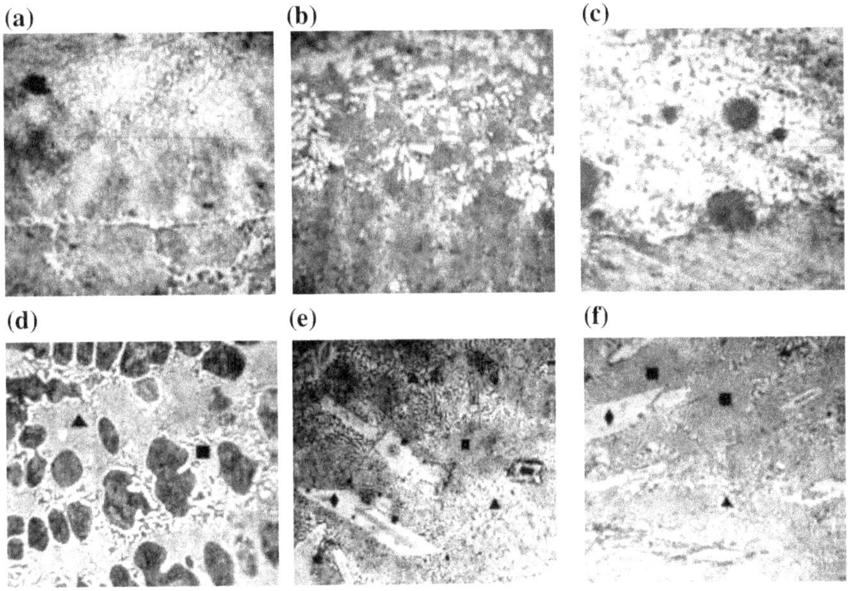

Fig. 14.8 The composite structures of boronized layers (×400) obtained in steel 65Г under the high-frequency heating up to 1200–1300 °C for 1 min (**a–c**) and 1.5 min (**d–f**) [41–43] in different coatings of saturating mixtures: **a** B_4C—84%, NH_4Cl—6%, welding flux P-0.66—10%; **b** B_4C—84%, welding flux P-0.66—16%; **c** B (amorphous)—90%, welding flux P-0.66—10%; **d** B_4C—90%, welding flux P-0.66—10%; **e** B_4C—90%, CaF_{2-5}%, welding flux P-0.66—5%; **f** B (amorphous)—90%, CaF_2–5%, welding flux P-0.66—5%

1100–1300 °C. Welding flux P-0.66, CaF_2, NH_4Cl etc. were used as activators. The major equilibrium phases in Fe–C–B system (carbon content—up to 25% in mass, boron—35%) were iron diboride FeB_2, FeB, Fe_2B, cementite Fe_3C, boronized cementite $Fe_3(C,B)$ and a special boronized carbide $Fe_{23}(C, B)_6$. The studied cases of high-speed heating created the formation conditions for meta-stable phases, phases of changing composition and solid solutions of boron and carbon in iron [41–44]. The layers had composite structures of different morphology (Fig. 14.8) and were characterized by the increased plasticity and wear-resistance.

14.9 The Connection Between the Plasticity of Boronized Layers and the Mechanical and Exploitation Properties of Boronized Steels

The major part of the book offers the materials concerning different issues: the hardness formation in one-phase and multi-phase structures, the formation of stress state, brittleness, wear-resistance, thermal fatigue resistance, scale resistance, corrosion and corrosion-mechanical fracture, the mechanical properties of boronized steels. These issues are still topical and the established regularities are proved in various studies.

The analysis of the data on hardness calculations in boronized layers, obtained by different authors, shows that the fluctuations in hardness values don't exceed the margin of error and are connected to the loading under hardness calculation and the conversion of microhardness (H100) into Vickers hardness (HV). They can also be caused by the stress state of the layer, boriding conditions and further thermal treatment etc.

The final properties of boronized details mostly depend on the stress state of the layer and detail. In particular, a favourable curve of residual stresses after boriding can be found when compression stresses are distributed evenly in the depth of the boronized layer. The maximum load of these stresses must be found in Fe_2B zone with a large specific volume; the surface should have a low level of compression stresses. This pattern of compression state is formed after quenching and tempering of boronized steels.

The study on the influence of ultrasound quenching allowed to obtain smaller blocks on the boronized surface which improved the physical and mechanical properties of boronized details [45]. The decrease in the size of blocks can be explained by the alternating field of ultrasound.

The analysis of obtained curves of macrostresses in boronized steel 45 in dependence from the amplitude of ultrasound stresses demonstrated that the best distribution of stresses takes place under the maximum amplitudes of ultrasound stresses (especially at 17.6 kgs/mm^2 amplitude and 985 °C temperature in the model). This distribution of stresses in the depth of the boronized layer under further quenching creates small resulting compression stresses in the section of the boronized layer with minimal compression stresses near the surface. This allowed to obtain a hardened surface free of crack formation tendency after boriding and quenching [45].

One of the ways to improve the exploitation characteristics of boronized layers is mechanical-chemical-thermal processing when first the surface plastic deformation takes place and then the steel goes under boriding. The result of using high-frequency heating and boriding coatings in steels is the formation of layers based on iron borides (Fe_2B) with a transition layer of needle-grain structure which ensure a smoother transition from the hardness of borides to the hardness of the core. The layers themselves are characterized by the increased plasticity [46].

14.10 Modeling the Formation of Diffusive Boronized Layers and Their Wear-Resistance

The modeling of boriding process is investigated relatively poorly although the modeling gives the opportunity to study the technological process closely and obtain the prognostic result of optimized parameters (the width of the layer, wear resistance, phase correlation, hardeness etc.) without material wastes and time costs.

The essence of prognostic calculation for the width of boride layers and further modeling of the wear-resistance is given in the book. The further studies developed this method and its practical use [47]. Thus, the mathematical planning, the statistical analysis of results, the modeling of the kinetic coefficient change allowed to determine the formation conditions for boronized layers with defined properties in construction steels within the wide interval of alloying. The method allows to calculate the layer parameters using the developed formulae alone and avoiding the need of purchasing special databases and software.

Another application of modeling is the development of a special interface for creating solutions in electron-beam processes of boriding, boron-aluminizing and hard-phase boriding [48, 49]. The solution system is a human-computer system which carries out the formation, analysis and solution making by processing large amounts of objective and subjective information.

First the initial data is put in: a brand of material, a microhardness level, a boriding method. In case electron-beam boriding is selected, the system calculates the time of processing taking into account the parameters of electron-beam heating.

The analysis on the base of experimental and calculated data results in the system giving the recommendation on the mode of electron-beam boriding (unit power, time of processing), showing the proposed structure and enumerating mechanical properties (shear stress σ_{sh} and the value of limiting deformation) (Fig. 14.9). The additional input of the given result provides its comparison with other methods of boriding (hard-phase, combined etc.). The characteristics of properties, calculation methods, references can be shown on the screen.

The advantage of the developed system is the possibility of updating and correcting the data base by a user.

The minimal requirements for using the software: Pentium 75 Hz processor; 8 Mb RAM; 30 Mb free hard disc space; Videocard 256 1024X768; Windows 9X, Me, 2000, NT, XP; installed software package Borland Database Engine.

The calculation of technological parameters for boron-alumizing can be carried out by newly-developed computer program "The calculation of optimal parameters for boron-aluminizing" [50].

The software needs the following initial values (criteria): a material, a method of boron-alumizing, a layer width, microhardness level, heat resistance, wear-resistance levels (Fig. 14.10).

Each criteria creates a variety of alternatives (A); they consist of units:

M—materials;
B—method of boron-aluminizing;
H—values of layer width;
Hy—microhardness values;
Htr—heat resistance values;
Wr—wear resistance values.

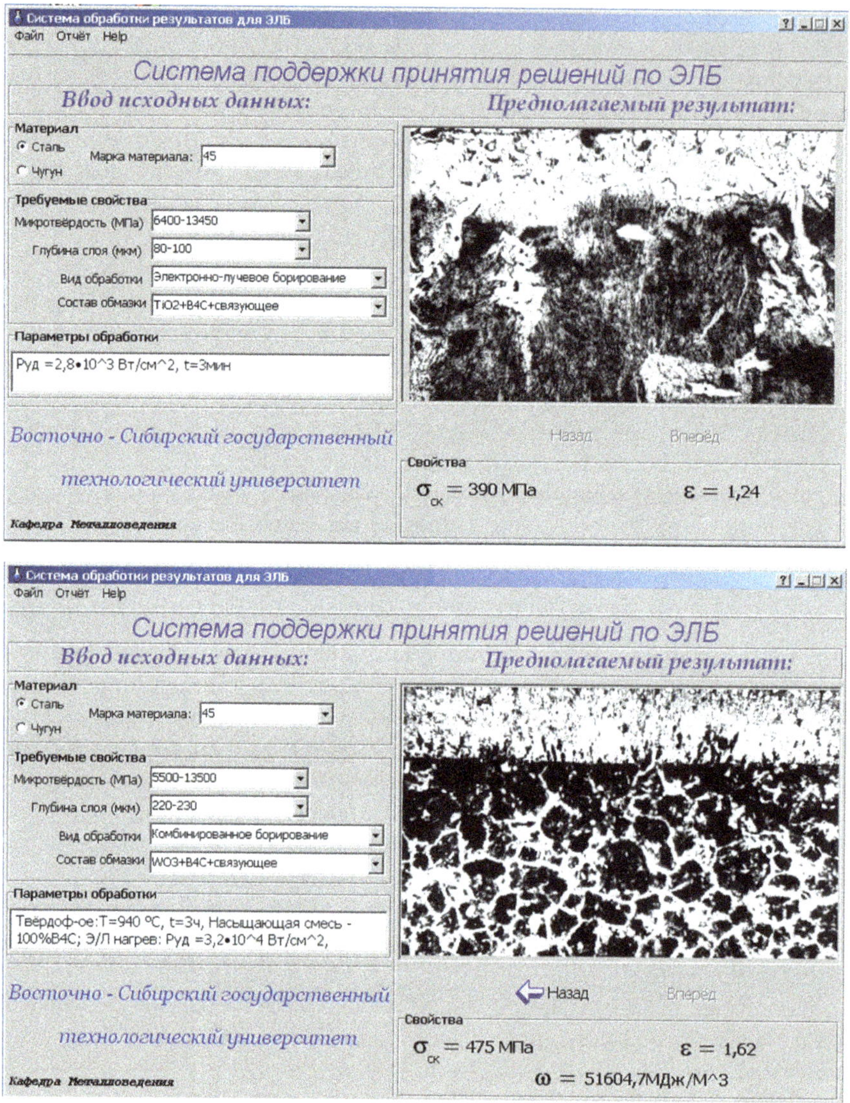

Fig. 14.9 The overview of the interface in the solution system for electron-beam and hard-phase boriding

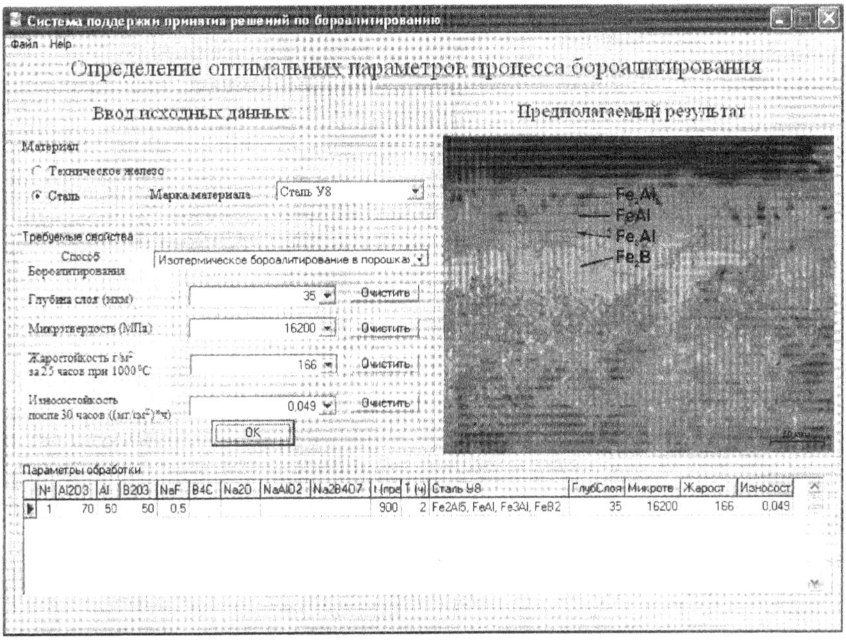

Fig. 14.10 The interface of software «The calculation of optimal parameters for boron-aluminizing»

The given abbreviations form the corpus of elements—(M, B, H, Hv, Htr, Wr). A variety of alternatives is built according to the largest priority:

$$\mathrm{A} = (\mathrm{M}, \mathrm{B}, \mathrm{H}, \mathrm{Hv}, \mathrm{Htr}, \mathrm{Wr}).$$

The priority of alternatives is set in the variety of weights which is calculated by the formula:

$$W = \frac{\lambda_i}{\sum_{i=1}^{n} \lambda_i},$$

where W—the weight of criteria; λ—the priority unit; n—the number of criteria; i = 1,…, n.

$$\lambda_i = \sqrt[n]{\sum_{j=1}^{n} C_{ij}},$$

where C—the matrix of criteria priority; n—the number of priorities; i = 1,…, n, j = 1,…, n/

The experimental and calculated data and their analysis allows the software to offer the recommendations on processing parameters: mixture composition, saturation conditions (temperature and soaking), phase composition, microstructure.

The data is kept in SQL Server 2000. All functional principles are realized in Delphi. The database is open to additions.

The method of boriding modeling offered in study [51] is based on the well-known and effective approach of the mathematical planning of experiments which allows to obtain adequate regression equations of optimized parameters and reduce the time of data acquiring and processing. However, it immensely complicates the modeling procedure. The modeling was carried out in S-Plus 8.0 software. The study doesn't analyze the influence of alloy elements in steel on the formation of boronized layers and doesn't take into the account the existing investigations in this area. A Simple computer correlation of subjective results of boriding is used for obtaining the model of impact of technological variation factors on the optimization parameters (a layer's width, hardness, wear resistance, impact viscosity levels). The wear-resistance is selected as the major indicator of surface priorities but the conditions of resistance were ignored. Moreover, the model doesn't include the phase compositions of boronized layers, their morphology, stress state (though these are the determining factors for wear resistance).

Different conditions of wear resistance require different optimal boronized layers with different structure morphology (as shown in study [52]). The abrasive wear requires two-phase ($FeB + Fe_2B$) boronized layers; the abrasive wear under impact or alternating loading requires one-phase (Fe_2B) boronized layers; the wear in conditions of bending or twisting require composite layers. Thus, the proposed method [51] turned out to be unproductive and scientifically incorrect.

14.11 The Prospective Boriding Technologies Guaranteeing the Improvements in the Plasticity of Layers

Different exploitation conditions require different morphologies of boronized layers; thus the formation technologies need a wide variety. The technologies studied in the book are still up-to-date. One of the most promising newly developed technological processes is the technology of "high-speed electrolysis boriding" [7], which allows to carry out the process in a wide temperature interval and obtain boronized layers with a various set of properties. This can be done by choosing the salt compound of the bath, the anode material and the powder components of the saturating bath. According to this technology, boronized layers consist of the diffusion part which grows inside the metal and the part obtained from the exterior of the processed surface. The formation speed of the diffusion part is positioned in the traditional timeframe.

The processing of melted salts creates some complications connected to the bath preparation, its refreshing and refilling during the use, cleaning the details from smelts, waste disposal. They increase the technological cycle of processing and costs. However, these disadvantages can be overcome by the possible benefits.

The new technologies of powder boriding include a very promising saturating process from nano-structured mixtures [15]. The self-organizing mass carry itself increases the saturating ability of the mixture; its combination with concentrated energy sources or electrophysical impact creates the conditions for the generation of composite boronized layers with a wide specter of properties and relatively short processing time.

14.12 The Use of Boriding Processes in the Industrial Treatment of Details and Tools

Boriding finds its use in the area of improving the exploitation resistance of details and tools which wear out relatively quickly working in various conditions. The interest in this sphere is explained by the practical necessity.

One of the factors for the effective use of boriding is the control over the size change due to the formation of a layer. This issue can become quite problematic when the layer is formed by hard borides of small size (0.08–0.15 mm). This issue is less complicated with composite boronized layers of a significant width (0.5–1.0 mm) with eutectic and pseudo-eutectic structures which have lower hardness and sufficient allowance for finishing. Normally the incremental diffusion boride layer in most construction and tool steels can't exceed 15% of the initial layer before tempering after boriding. This boride layer consisting of the diffusion zone and the precipitated zone outside the initial surface makes the detail's sizes depend on the processing mode and must be selected individually for a particular situation.

Other cases of size change are determined by the initial state of the substrate material, the cooling speed after boriding, the type of further thermal treatment. Therefore, the precision details of complex form require the experimental determination of size change pattern which should be taken into account while producing boronized goods.

The given recommendations allowed to increase the exploitation properties of details and tools and expand their nomenclature [53, 54].

One of the new uses of boriding is boriding of details with gaseous-thermal coatings [55]. The aim of this treatment is the increase of wear-resistance, improvement of the bond between coatings and substrates and the increase in fatigue resistance. The latter is provided by creating compression residual stresses on the surface of the detail. In the end this approach allows to double increase the fatigue resistance.

Other new directions are boriding of powder materials in order to obtain self-fluxing welding mixtures and boriding of metallic fillings of metal- or polymer-based composite materials.

References

1. Koskov V.D. Povysheniye dolgovechnosti detalei burovoi tekhniki putyom formirovaniya na poverkhnosti kompleksno-legirovannogo boridnogo sloya. Avtoreferat dissertacii kandidata nauk. Sverdlovsk. 1986. 20 p. [The improvement of wear-resistance of the surface of complex-alloyed boride layer in drilling equipment. Candidate of science dissertation thesis].
2. Dalisov V.V., Mardarevich R.S., Brodyak Ya. P. Formirovaniye mnogosloinogo nikel-bor-khromovogo pokrytiya // Zaschitnye pokrytiya na metallakh. Kiev. 1988. Iss. 22. pp. 58–61 [The formation of multi-layered nickel-boron-chrome coating].
3. Shmatko I.O. Strukturoobrazovaniye galvanoforeticheskikh pokrytii na osnove tugoplavkikh soedinanii pri zhidkostnom spekanii i razrabotka tekhnologii polucheniya iznosostoikikh pokrytii. Avtoreferat dissertacii kandidata nauk. Kiev. 1991. 26 p. [The structure formation of galvanic-phoresis coatings on the foundation of refractory metals under liquid sintering and the technology development for obtaining wear-resistant surfaces].
4. Veropakha N.V., Zubekhin A.P. Patent RF No. 2135631, C23C8/42, C23D3/00. Publ. 27.08.1999.
5. Voroshnin L.G., Lyakhovich L.S. Borirovaniye stali. Moscow. 1978. 239 p. [Steel boriding].
6. Borisenok G.V., Vasilyev L.A., Voroshnin L.G., Gorbunov N.S., Dubinin G.N., Zhunkovskiy G.L., Krukovich M.G. et al. Khimiko-termicheskaya obrabotka metallov i splavov. Moscow. 1981. 424 p. [Chemical-thermal treatment of metals and smelts].
7. Erdemir A., Eryilmaz O.L. The Proceedings of Argonne National Laboratory. 2012.
8. Grinberg E.M. Metallovedeniye borsoderzhaschikh konstruktcionnykh stalei. Moscow. 1997. 198 p. [The metal technology of boron-containing construction steels].
9. Krishtal M.A. Diffuzionnye processy v zheleznykh splavakh. Moscow. 1963. 278 p. [The diffusion processes in iron smelts].
10. Krishtal M.A. Mekhanizm diffuzii v zheleznykh splavakh. Moscow. 1972. 400 p. [The diffusion mechanism in iron smelts].
11. K., Fujimori H., Khakimoto K. Amorphous materials. Moscow. 1987. 328 p.
12. Kovneristyi Yu. K., Osipov E.K., Trofimova E.A. Fiziko-khimicheskiye osnovy sozdaniya amorfnykh metallicheskikh splavov. Moscow. 1983. 145 p. [The physical and chemical foundations in producing amorphous metal smelts].
13. Gaskell P.H. Modeli struktury amorfnykh metallov // Metallicheskiye stekla. Iss. 2. Atomnaya struktura i dinamika, elektronnaya struktura, magnitnye svoistva. Moscow. 1968. pp. 12–63. [Models for the structure of amorphous metals].
14. Vdovin K.N., Zaitceva A.A. Vliyaniye termicheskoi obrabotki na valkovyi chugun, modifitcirovanii borom // MSTU Bulletin. 2011. No. 4. pp. 13–15. [The impact of thermal treatment on boron-modified roll-foundry iron].
15. Kukhareva N.G., Galynskaya N.A., Petrovich S.N. Diffuzionnye pokrytiya, polychennye iz kompozitsionnykh sred // Nauka i tekhnika. No. 5. 2013. pp. 21–26 [Diffusion surfaces obtained from composite mediums].
16. Dombrovsky Yu. M., Stepanov M.S. Sozdaniye kompozitnykh diffuzionnykh boridnykh pokrytii pri mikrodugovom uprochnenii v poroshkovykh sredakh // VolSTU Bulletin. No. 3. 2014. pp. 41–42 [The production of composite diffusioin boride boatings under micro-arc hardening in powder mediums].

17. Krukovich M.G. Principy postroeniya mnogokomponentnykh diagram sostoyaniya v koordinatakh concentratciya-temperatura. Moscow. 1994. 24 p. [The principles of building multicomponent equilibrium diagrams in "concentration-temperature" coordinates].
18. Krukovich M.G. Razrabotka teoreticheskikh i prikladnykh aspektov upravleniya strukturoi i svoistvami borirovannykh sloev i ikh ispolzovanie pri proizvodstve transportnoi tekhniki. Dis. dokt. tekh. nauk. Moscow, 1995. 416 p. [The development of theoretical and applied aspects of controlling the structure and properties of boronized layers and their use in producing transport machinery. Doctorate dissertation].
19. Krukovich M.G., Baderko E.A. Raschyot evtekticheskikh temperature i koncentratsii mnogokomponentnykh system i postroeniye skhem diagram sostoyaniya // Aktualnye napravleniya fundamentalnykh i prikladnykh issledovanii. North Charleston. 2014. Vol. 1. pp. 123–130. [The calculation of eutectic temperature and concentration of multicomponent systems and building equilibrium diagram schemes].
20. Krukovich M.G. Raschyot evtekticheskikh koncentratcii i temperatury v dvukh- i mnogokomponentnykh sistemakh // MiTOM. 2005. No. 10. pp. 9–17. [The calculation for eutectic concentrations and temperature in two- and multicomponent systems].
21. Ganeev A.A., Khalikov A.R., Kabirov R.R. Razrabotka metodiki rascheta evtekticheskikh kontsentratsii i temperatur diagram sostoyaniya // USATU Bulletin. 2008. Vol. 11. No. 2 (29). pp. 116–122 [The development of the calculation method for eutectic concentrations and temperatures of equilibrium diagrams].
22. Lakhtin Yu. M., Leontyeva V.P. Materialovedeniye. Moscow. 1990. 528 p. [Material Studies].
23. Khalikov A.R. Modelirovaniye evtekticheskikh kontsentratsii mnogokomponentnykh diagram sostoyaniya // USATU Bulletin. 2010. Vol. 14. No. 2 (37). pp. 188–194 [Modeling eutectic concentrations in multicomponent equilibrium diagrams].
24. Krukovich M.G. Mekhanizm formirovanya diffuzionnykh sloev // Novye materialy i teknologii v mashinostroenii. Bryansk. 2012. Iss. 15. pp. 69–76 [The mechanism of formation for diffusion layers].
25. Dementyev V.B., Ovcharenko P.G., Leschev A.Yu., Makhneva T.M. Borirovaniye uglerodistykh stalei metodom litya po gazifitsiruemym modelyam // Khimicheskaya fizika i mezoskopiya. 2013. Vol. 15. No. 3. pp. 393–396 [Boriding of carbon steels by moulding according to gasified models].
26. Krukovich M.G., Klochkov N.P., Savelyeva A.S., Biryukov V.P. Strukturnaya modifikatsiya boridnykh sloev // Proceedings of the XIXth Online Conference "Novye materialy i teknologii v mashinostroenii". Bryansk. 2014. pp. 73–77 [Structural modification of boride layers].
27. Biryukov V.P. Promyshlennoe primeneniye CO2 – lazerov dlya naplavki, uprochneniya detalei i tochnogo raskroya listovykh materialov // Tyazheloe mashinostroeniye. 2006. No. 4. pp. 25–28 [The industrial use of CO_2-lasers for welding, hardening of details and precise cutting of sheet materials].
28. Teplykh A.M. Ispolzovaniye elektronno-luchevogo nagreva dlya formirovaniya na stalyakh iznosostoikikh poverkhnostnykh sloev // Materialovedeniye. Obrabotka metallov. 2011. No. 3 (52). pp. 91–94 [The use of electron-beam heating for the purpose of forming wear-resistant surface layers in steels].
29. Krivezhenko D.S., Drobyaz E.A., Zimoglyadova T.A. Osobennosti strukturnoobrazovaniya borsoderzhaschikh pokrytii, poluchennykh v protsesse vysoskoskorostnoi obrabotki // Proceedings of the 1st International Scintific Conference "Aktualnye problemy v mashinostroenii". Novosibirsk. 2014. pp. 489–492 [The specifics of structure formation in boron-containing surfaces obtained after high-speed processing].
30. Zimoglyadova T.A., Krivezhenko D.S., Komarov P.N. Formirovaniye funktsionalnykh boridnykh pokrytii na stalyakh pri vozdeistvii vysokokontsentrirovannykh istochnikov energii // Proceedings of the XXth International Scientific Conference "Sovremennye tekhnika i tekhnologii". Section on Material Studies. Tomsk. 2014. pp. 37–38 [The formation of functional boride coatings in steels under the influence of high-concentrated energy sources].

31. Vaschuk E.S., Budovskikh E.A., Teplykh A.M., Gromov V.E., Ivanov Yu. F. Osobennosti vliyaniya elektronno-puchkovoi obrabotki na poverkhnost stali 45 posle elektrovzryvnogo boromedneniya // Obrabotka metallov. 2011. No. 4 (53). pp. 67–73 [The specifics of impact of electron-beam processing on the surface of steel 45 after electroexplosive boron-copper treatment].
32. Mishigdorzhiin U.L., Sizov I.G., Leintz K., Vetter B., Furman T. Vliyaniye soderzhaniya ugleroda v stali na strukturu i svoistva boroalitirovannogo sloya // ESSUTM Bulletin. 2011. No. 2 (33). p. 11 [The impact of carbon content in steel on the structure and properties of boron-aluminized layer].
33. I.G. Sizov, U.L. Mishigdorzhijn, D.M. Makharov, A study of thermocycling boroaluminizing of carbon steels. Metal science and heat treatment. **53**(11–12), 592–597 (2012).
34. I.G. Sizov, U.L. Mishigdorzhijn, C. Leyns, B. Vetter, T. Fuhrmann, Influence of thermocycle boroaluminizing on strength of steel C30. SURFACE ENGINEERING. **30**(2), 129–133 (2014).
35. Lygdenov B.D., Butukhanov V.A., Mei Shunchi, Tsyretorov B. Sh. Uprochneniye poverkhnosti shtampovoi osnastki iz doevtektoidnykh stalei // Sovremennye naukoemkie tekhnologii. 2014. No. 4. pp. 76–79 [The surface hardening of stamp fittings of pre-eutectoid steels].
36. Lygdenov B.D., Greshilov A.D., Kharaev Yu. P. et al. Issledovaniye fazovogo sostava i defektnogo sostoyaniya gradientnykh struktur borirovannoi stali 45 // Fundamentalnye problemy sovremennogo materialovedeniya. 2010. Vol. 7. No. 1. pp. 79–83 [The study of phase composition and defective state of gradient structures of boronized steel 45].
37. Lygdenov B.D., Mosorov V.I., Mizhitov A. Ts. Issledovaniye fazovogo sostava i defektnogo sostoyaniya gradientnykh struktur borirovannykh stalei 45 // ESSUTM Bulletin. 2011. No. 1 (32). pp. 25–31 [The study of phase composition and defect state of gradient structures in boronized steel 45].
38. Guryev A.M., Lygdenov B.D., Malkova N. Yu., Shametkina O.V., Mosorov V.I., Radnaev A. R. Vysokoeffektivnyi sposob khimiko-termicheskoi obrabotki instrumentalnykh stalei // Polzunovskii almanakh. No. 4. 2004. pp. 91–93 [The effective method of chemical-thermal processing of instrumental steels].
39. Semenchenko M.B. Borirovaniye stalnoi provoloki s primeneniem tsiklicheskogo elektrokontaktnogo nagreva. Minsk. pp. 113–115 [Boriding of steel wire using the circular electric contact heating].
40. Dombrovsky Yu. M., Stepanov M.S. Mikrodugovaya tsementatsiya stalnykh izdelii v poroshkovykh sredakh // Uprochnyayuschie tekhnologii i pokrytiya. 2013. No. 12. pp. 25–29 [Micro-arch cementation of steel details in powder mediums].
41. Ishkov A.V., Krivochurov N.T., Mishustiv N.M., Ivanaiskii V.V., Maksimov A.A. Iznosostoikiye boridnye pokrytiya dlya pochvoobrabatyvayuschikh organoc selkhoztekhniki // Altai State Agricultural University Bulletin. . 2010. Vol. 71. No. 9. pp. 71–75 [Wear-resistant boride coatings for agricultural soil equipment].
42. Ishkov A.V., Krivochurov N.T., Mishustin N.M., Ivanaiskii V.V., Maksimov V.V. Vliyaniye tekhnologicheskikh faktorov na iznos poverkhnostno-uprochnennykh strelchatykh lap // Altai State Agricultural University Bulletin. 2010. Vol. 72. No. 10. pp. 92–96 [The influence of technological factors on wearing of surface-hardened A-blades].
43. Aulov V.F., Krivochurov N.T., Ivanaiskii V.V., Sokolov A.V., Ishkov A.V. Novaya konstruktsiya strelchatykh lap // Selskii mekhanizator. 2013. No. 10. pp. 34–35 [New constructions of A-blades].
44. Schukin V.G., Marusin V.V. Induktsionnoy borirovaniye stalei // Materialovedeniye. 2015. No. 4. pp. 27–32 [The inductive boriding of steels].
45. Nechaev L.M., Fomicheva N.B., Kanunnikova I. Yu., Markova E.V. Vliyaniye ultrazvukovoi obraborki na fiziko-mekhanicheskiye svoistva borirovannogo sloya // Sovremennye naukoemkie tekhnologii. No. 9. 2008. pp. 125–128 [The influence of ultrasound treatment on the physical and mechanical properties of boronized layers].

46. Zavolokin O.A. Razrabotka mekhaniko-khimiko-termicheskoi obrabotki detalei pri uprochnenii borirovaniem v obmazkah TVCH-nagrevom. Dissertatsiya kandidata teknicheskikh nauk. Kursk. 2007. 194 p. [The development of mechanical, chemical and thermal treatments for details under boriding processing in coatings by high-frequency current heating. Candidate of science dissertation].
47. Krukovich M.G., Baderko E.A. Modelirovaniye iznosostoikosti borirovannykh sloev // Novye materialy i tekhnologii v mashinostroenii. Bryansk. 2013. Iss. 18. pp. 66–69 [Modeling wear-resistance of boronized layers].
48. Sizov I.G. Razrabotka nauchnykh osnov i tekhnologii elektronno-luchevogo borirovaniya zhelezouglerodistykh splavov s polucheniem na poverkhnosti boridov tugoplavkikh metallov. Dissertatsiya doktora tekhnicheskikh nauk. Moscow, 2003. 305 p [The development of scientific foundation and technology of electron-beam boriding of iron-carbon alloys aimed at obtaining high-melting point metal borides on the surface. Doctorate dissertation].
49. Mishigdorzhiin U.L. Razrabotka tekhnologii termotsiklicheskogo boroalitirovaniya uglerodistykh stalei. Dissertatsiya kandidata tekhnicheskikh nauk. Ulan-Ude. 2012. 171 p. [The development of the technology of thermal-circular boron-aluminizing of carbon steels].
50. Mishigdorzhiin U.L., Dyshenov B.A., Sizov I.G. The determination of optimal parameters for boron-aluminizing. Certificate of software state registration 2010615252 RF. Publ. 13.08.10.
51. Galaa O., Lygdenov B.D., Butukhanov V.A. Matematicheskoye modelirovaniye rezhimov borirovaniya stalei // Problemy mekhaniki sovremennykh mashin. 2015. Vol. 2. pp. 181–187 [Mathematical models for the modes of steel boriding].
52. Krukovich M.G., Baderko E.A. Modelirovaniye iznosostoikosti borirovannykh sloev // Novye materialy i tekhnologii v mashinostroenii. Bryansk. 2013. Iss. 18. pp. 66–69 [Modeling of wear resistance of boronized layers].
53. Krukovich M.G. Borirovaniye iz kompaktnykh materialov // Novye materialy i tekhnologii v mashinostroenii. Bryansk. 2013. Iss. 18. pp. 40–45 [Boriding from compact materials].
54. Krukovich M.G., Savelyeva A.S. Tekhnologicheskiye processy borirovaniya dlya polucheniya zaschitnykh sloev razlichnogo tselevogo naznacheniya // Novye materialy i tekhnologii v mashinostroenii. Bryansk. 2013. Iss. 18. pp. 73–77 [Technological processes of boriding for obtaining multi-purpose protectice layers].

Appendix A
The Calculation of Eutectic Temperatures and Concentrations in Two- and Multi-phase Systems

A.1 The Purpose of Calculation Methods

A range of theoretical, technological and exploitation problems of metallurgy, foundry, thermal and chemical-thermal processing, powder metallurgy, welding and fusing, chemistry production etc. and their solutions are connected with the fusion of metals, oxide and salt systems and require the knowledge of equilibrium states or temperatures of eutectic reactions for these systems. Simultaneously many systems, especially multi-component, still lack experimental equilibrium diagrams. The construction of quasi-binary profiles of real multicomponent systems requires advanced software and data banks which are still incomplete for most systems. The thermal-dynamic criteria calculations for two-component diagrams of alloys are characterized by their poor accuracy.

The choice of eutectics-forming components can be done according to the statistical analysis of double equilibrium diagrams [1, 9]. Depending on the type of the chemical bond between components and the similarity of their physical-chemical properties, eutectics is generated if only one inequality is not respected:

$$n_S \leq 1.10 \text{ или } n_T > 0.55\, n_v^2, \tag{A.1}$$

where n_S, n_T and n_V—entropy, temperature and volume factors correspondingly; $n_S = S_A/S_B$ is the correlation of melting entropies (where $S_A > S_B$);

$n_T = 1 - T_A/T_b$, where T_A and T_B are melting points for A and B components (where $T_A < T_B$); if T_A and T_B are equal, $n_T = 0$, $n_V = 0$, i.e. the system can't be defined thermodynamically as eutectic.

$n_V = (d_A/d_B)^3 + V_A/V_b - 2 + b$, where d_A and d_B—atomic diameters of the components, V_A and V_B are volumes of the components;
b—valency difference correction.

M.G. Krukovich et al., *Plasticity of Boronized Layers*,
Springer Series in Materials Science 237,
DOI 10.1007/978-3-319-40012-9

The mentioned conditions are only qualitative and sometimes are not fully respected. However, the use of these results allows to evaluate the possibility of eutectic reactions in two-component systems of clean elements when the equilibrium diagram is missing.

Eutectic temperatures in two-component systems may be determined statistically. The existing recommendations and formulae (Kordes, Gulyaev, Mednitcky, Vasilyev, Vozdvizhensky) [1–8] help to evaluate some two-component systems with a certain accuracy. Despite the established connection between general characteristics of some two-component systems (T_{eut}; T_1; T_2; C_1; C_2), these approaches don't cause the prediction of basic characteristics of systems by certain parameters (e.g., the melting points of eutectic components). Thus, these approaches are not appropriate enough for most two-component systems and especially multi-component systems.

The statistical analysis of more than 100 double and triple systems with eutectic reactions revealed the connection of the eutectic temperature with the melting points of the eutectic's components and provided investigators with formulae for calculating eutectic temperature coefficients and eutectic temperatures in two- and multicomponent systems (involving geometric and algebraic transformations).

The increase of calculation accuracy was reached by classifying the elements by their physical-chemical properties and electron structures in accordance with Mendeleev periodic table. This allowed to construct calibration dependencies of eutectic temperature coefficients and scale temperature parameters for different groups of elements and their combinations and discover the mathematical expressions of these dependencies. The scale temperature parameter is developed in order to distribute eutectic temperature coefficient values more evenly along the calibration diagrams. Modeling eutectic forming patterns in multicomponent systems involved the use of the similitude rule; according to the rule, double eutectics are considered as "components" when melting points and concentrations of a new system's eutectics are calculated. The results of these calculations have a 10% accuracy increase in comparison with the experimental data.

The test on the method was carried out during the boriding process with high-speed steels and smelts, based on Ni; they showed good correlations.

A.2 The Calculation of Eutectic Temperatures in Two- and Multicomponent Metal Systems

The basic data for the calculation are melting points of eutectic components. They can be pure elements, chemical compounds or solid solutions of limiting solubility. The liquidus temperature is taken for the solid solutions of limiting solubility.

According to the existing theories [11, 12], the formation of eutectics is a complicated physical-chemical process which leads to the generation of special columnar structure in the system's crystals accompanied by their simultaneous crystallization at a certain temperature from the liquid solution of a certain chemical composition. The special trait of eutectic crystallization is a rapid hashing function;

therefore, the oversaturation of liquid solution and generation of crystals of only one phase is not observed. The eutectic colonies are grown through each other and look as strongly branched system's phase crystals. The occurrence and growth of these colonies is controlled by one eutectics phase, called "leading phase", which determines the type of eutectics. The leading phase is characterized by a large number of directed and stronger bonds between atoms in crystals. Second and further phases are generated on the surface of the leading phase. Thus, each eutectic colony of the multicomponent system is a multi-phase poly-crystal of mutually intertwined crystals of the system.

Multi-phase poly-crystals are formed in the systems with a non-variable eutectic equilibrium, in systems with congruently and incongruently melting double and triple compounds which often generate several eutectic surfaces on multicomponent system diagrams. In the latter case the triangulation of the system should be carried out in accordance with the recommendations [1, 9, 10].

The statistical analysis of several equilibrium diagrams of two-, three- and multicomponent systems allowed to formalize the connection between the eutectic temperature and eutectic components' melting points (in K). This connection is called the temperature rule of eutectics and has the following formula.

- For two-component systems

$$T_{eut} = K_{eut}(T_1 + T_2), \text{ K}, \tag{A.2}$$

where K_{eut}—the coefficient of eutectic temperature in the range 0.25–0.494, i.e. tends to 0.5 value. This connection correlates with the results in [1]: T_1 and T_2 are melting points for eutectic components, in K. Equal temperatures break thermal conditions for eutectic reaction in the given system in accordance with the formula (A.1). It's important that this equality is relative and connected with the accuracy of melting points' determination. There are no absolute equal melting points for different elements.

- Modeling multicomponent systems involves the mechanistic similarity principle in order to facilitate the process. Melting points for double eutectics are considered as melting points for separate "components" forming new eutectic mixtures with other components or eutectics. The correctness of this method is explained by the structure of triple equilibrium diagrams with a non-variable eutectic equilibrium, when triple eutectics is formed in combination with double eutectics of the system. The recommended general approach to the determination of eutectic temperatures allows to calculate it efficiently without taking into account the peculiarities of interactions between components and eutectics types.

In particular, the direct calculation results in:

- for a three-component system:

$$T_{eut} = K_{eut}[K_{eutT}(T_1 + T_2) + T_3] = K^2_{eut}(T_1 + T_2) + K_{eut}T_3; \tag{A.3}$$

- for a four-component system:

$$T_{eut} = K_{eut}[K_{eut}(T_1 + T_2) + K_{eut}(T_3 + T_4)] = K^2_{eut}(T_1 + T_2 + T_3 + T_4); \tag{A.4}$$

- for a five-component system:

$$\begin{aligned} T_{eut} &= K_{eut}\{K_{eut}[K_{eut}(T_1 + T_2) + K_{eut}(T_3 + T_4) + T_5] \\ &= K^3_{eut}(T_1 + T_2 + T_3 + T_4) + K_{eut}T_5; \end{aligned} \tag{A.5}$$

These peculiarities can be generally noted down (involving the direct calculation of eutectic temperature, i.e. equal values of K_{eut}):

- for an even number of eutectics components *(2n)*

$$T_{eut} = K_{eut}^{\frac{n}{2}} \sum_{i=1}^{n} T_1 \tag{A.6}$$

- for uneven number of eutectics components *(2n+1)*

$$T_{eut} = K_{eut}^{\frac{n+1}{2}} \sum_{i=1}^{n-1} T_i + K_{eut} * T_n \tag{A.7}$$

This method facilitates the calculation but at the same time it is characterized by its poor accuracy. It can be used for the initial evaluation of T_{eut}.

The consequent (stage) determination of eutectic temperature for a multicomponent system should involve the amplification of eutectics temperature coefficients (K_{eut}) at each stage. The first stage is the calculation of eutectic temperature of components divided into pairs in a decreasing order of melting points, with the eutectics coefficient calculated by the formula (A.2). The next stage involves the calculation of eutectic temperature between eutectics or with consideration of left unpaired components; it is done by taking into account the amplified values of the coefficient. This method may be used for any systems (metal, salt, oxide) and gives minimum calculation errors.

The relative equality of melting points in a multicomponent system is balanced by the combination of components with different melting points. If melting points of eutectics, used in calculations, are equal, it is recommended to make new combinations of initial components and calculate their melting points. The similitude rule defines them as melting points for "components" in calculations of the new eutectics reaction temperature.

A.3 The Calculation of Eutectic Temperature Coefficients

The description of multicomponent eutectic systems requires a search for some objective characteristics which can be calculated on the basis of minimal initial parameters. The statistical analysis of different equilibrium diagrams for metallic systems with one or several eutectic reactions allowed to reveal one temperature characteristic for multicomponent systems which combines impacts of all eutectic elements. It is the sum of modules for melting temperatures' gradient of the system's components comprising eutectics. In particular, a four-component system with a non-variant diagram type has the following sum formula:

$$R = |T1 - T2| + |T1 - T3| + |T1 - T4| + |T2 - T3| + |T2 - T4| + |T3 - T4|. \tag{A.8}$$

Each particular system has a stable value of this sum and the sum of melting points of system components, therefore it characterizes the system. It allows to calculate a scale temperature parameter (X), which is further used for the eutectic temperature coefficient calculation (K_{eut}):

$$X = \frac{\sum\limits_{1 \prec i \prec j \prec n}^{n} |T_i - T_j|}{\left(\sum\limits_{i=1}^{n} T_i\right)^{0.74}}. \tag{A.9}$$

According to the given formula, a scale temperature parameter doesn't have size and can be used for constructing calibration dependencies with a minimal scattered field of real values K_{eut}, calculated according to particular diagrams [13, 14]. The list of studied double and multicomponent systems is given in tables A1, A2. The choice of diagrams was random.

The maximum amplitude of K_{eut} values for all systems at certain X is ±0.035. The amplitude is caused by the presence of unaccounted factors. The reduction of this gradient is reached by excluding calculated equilibrium diagrams and dividing interacting elements into particular metal groups (elementary, transition, intermediate) and non-metals in accordance with Mendeleev periodic table which reflects electron structures of chemical elements and their activities.

Non-metals are elements from B to F, from Si to Cl, Se, Br, Te, J, Po, At. Such elements as Si, Ge, As, Se, Te are considered to be transition elements between metals and non-metals [15]. The separation of this group is caused by smaller values of their ionization potentials which establishes their specific interaction with other elements.

Metals are normally divided into elementary (having fully filled *d*- and *f*-levels, or metals with small atomic numbers with free *d*-levels (Li, Be, Na, Mg, Al, K, Ca, Rb, Sr, Cs, Ba, Ra, Cu, Zn, Ga, Ag, Cd, Jn, Sn, Sb, Au, Hg, Tl, Pb, Bi) and transition metals with incomplete *d*- and *f*-levels (from Sc to Ni, from Y to Pd, from La to Pt, from Ac to Ku).

The constructed calibration dependencies connect the interval of eutectic temperature coefficient changes (K_{eut}) and scale temperature parameters (X) (Fig. A.1). The example demonstrates a calibration dependency diagram for a combination of elementary metals with elementary metals and their compounds. Other systems have similar diagrams and are close by values.

The calculation of K_{eut} may be carried out using methods and formulae offered in [16, 17] and taking into account the recommendations based on the statistical probability:

- pure eutectics-forming elements (metals and non-metals) require the choice of K_{eut} by average or minimal values (K^{mid} and K^{min});
- chemical compounds and combinations "pure element (metal and non-metal)—chemical compound" requires average or maximum values (K^{mid} and K^{max}).

The final value of eutectic temperature is determined by the average value of eutectic temperature coefficient between two others that are mentioned above. The respect for these peculiarities allows to decrease values spread up to ± 0.015.

The computer approximation of calibration dependencies (Fig. A1) allowed to obtain formulae for calculating K_{eut} with precision $R^2 < 0.9$:

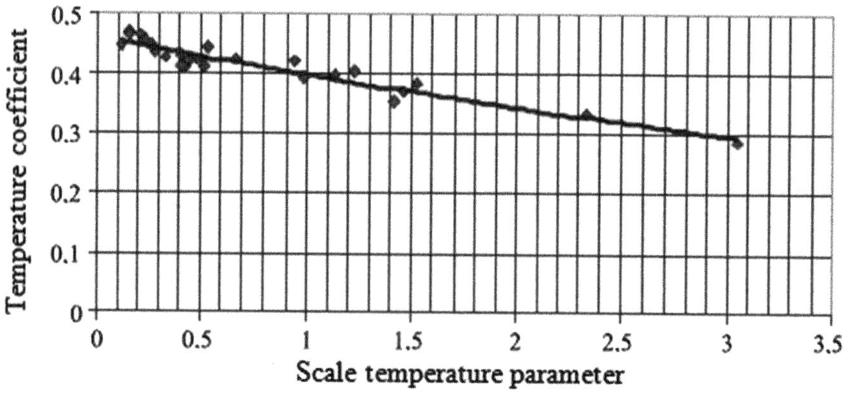

$$y = 0.4608e^{-0.1473x}, \qquad R^2 = 0.9088$$

Fig. A.1 Calibration dependencies in the determining eutectic temperature coefficients (K_{eut}) at thecombinations of elementary metals with elementary metals and their compounds forming eutectic mixtures

- for combining elementary metals with elementary metals and their compounds (I, Table A.1)

$$K_{eut} = 0.4608 \exp(-0.1473X), \ R^2 = 0.90; \quad (A.10)$$

- for combining elementary metals with transition metals and their compounds (II, Table A.1)

$$K_{eut} = 0.4847 \exp(-0.1782X), \ R^2 = 0.9229; \quad (A.11)$$

Table A.1 The systems studied during statistical processing of double equilibrium diagrams of alloys

No.	System				
	I	II	III	IV	V
1	2	3	5	6	7
1	Zn(β)–Al(α)	Th–Tl	Nd–Ni	Ga–Zn	Pd–Si
2	Al–$CuAl_2$	Ni–Zr	Ni–Ta	Al–Si	Rh–Si
3	Hg–Hg_5Tl_2(γ)	Lli–In	Ni–Ta	Sn–Te	Fe–S
4	Bi–Pb(β)	La–Tl	Ni–W	Ag_2Se–$GeSe_2$	Mg–Mg_2Ga
5	KNa_2–K	Bi–Yb	Ni–Zr	Ag_2Se–As	Nd–Ge
6	Cu–Bi	Cu– La	Fe–Zr	$AgAsSe_2$–As	V–Ge
7	$CuAl_2$–Al(α)	Cu–Zr	Pt–Hf	Ag–Si	Ni–Si
8	Pb(α)–Sb(β)	Al–Ta	Hf–Rh	GaSe–Ga_2Se_3	VSi_2–Si
9	Sn–Zn	Ag–AgCe	Co–Sm	Bi_2Te_3–$AuTe_2$	Fe(δ)–TiB_2
10	Pb–Sr	Sm–Bi	Cr–Am	Mn(δ)–Mn_2B	Fe(δ)–ZrB_2
11	In–Zn	Ag–La	Cr–Lu	GaSe–Ga_2Se_3	Pd_4S–FeS
12	Pb(α)–Sn(β)	Ag–Nd	Eu–Pd	Au–Bi_2Te_3	B–Hf
13	$AuPb_2$–Pb	Ag–Pr	La–In	SiS_2(α)–Cd_4SiS_6	C–Ni
14	Ag(α)–Cu(β)	CoAl(α)–Ca(β)	La–La_4Ir	SiS_2(α)–Cd_4SiSe_6	C–Nb
15	Cu(η)–Sn	Fe_3Al–Al	ZrIr–$ZrIr_2$	Ag–Ge	Ce–Ge
16	Cu–Pb	Mo_3Al–Mo_3Al_8	$ZrIr_3$–Ir(α)	Cu–As	Si–Ce
17	Cd(α)–Zn(β)	Sm–Cd	Am–Pt	Au–Ge	Gd–Ge
18	In(α)–Sn(γ)	Hf–Cu	Fe–Nb		Gd–Si
19	In(α)–In_2Bi	Al–La	Fe–Nd		Ge–Nd
20	Sn(α)–Bi	Al–Nb	Fe–Th		Ge–Pr
21	Cd–In(α)	Al–Pr	Ni–Pr		Sm–Ge
22	Cd–Pb(α)	Al–Sc	Ni–Yb		Ge–Ti
23	Sr–Hg	Al–Sm	Th–Co		Ge–Tu
24	Al–Ca	Al–Ce			Nd–Si
25	Au–Sn	Al–Mo			Tb–Se
26	Ca–Ga	Al–Yb			Sm–Si
27	Ga–Mg	Be–Ni			Si–Ta
28	Ga–Sb	Gd–In			
29	Li–Sr	Gd–TI			

- for combining transition metals with transition metals and their compounds (III, Table A.1)

$$K_{eut} = 0.0022 \cdot X^3 + 0.0214 \cdot X^2 + 0.0944 \cdot X + 0.4804, R^2 = 0.9357; \quad (A.12)$$

- for combining elementary metals with non-metals or intermediary elements (IV, Table A.1)

$$K_{eut} = 0.4792 \exp(-0.1818X), \; R^2 = 0.9418; \quad (A.13)$$

- for combining transition metals with non-metals or intermediary elements (V, Table A.1)

$$K_{eut} = 0.497 \exp(-0.2657X), \; R^2 = 0.9183. \quad (A.14)$$

The accuracy of calculated eutectic temperatures for multicomponent alloys, determined directly, is quite poor and errors reach 10% in the interval 0-373 K. In particular, system Pb, Sn, Cd, Bi at $K^{mid} = 0.4$, $T_{eut} = 358$ K; at $K^{min} = 0.373$, $T_{eut} = 310$ K; the average value is 334 K. The eutectic temperature established during the experiments is 341 K (error 2.1%, in °C—10%) (Table A.2).

A.4 The Essence of the Eutectic Concentration Calculation Method

The graphic interpretation of formulae for calculating eutectic concentrations in double systems [1, 5] presupposes the proportionality of eutectic line areas (characterizing eutectic weigt concentrations) and temperature differences (system elements' melting points and eutectic temperature). The development of this method led to obtaining the formulae for the approximate evaluation of eutectic concentrations. The type of liquidus lines doesn't matter. If necessary, known methods are used for constructing these lines.

The calculation may be done in different ways. The data from [1] and the equation of direct dependency in formulae (A.10), (A.11) lead to:

$$C_B = [(T_A - T_B)/(T_A - T_{eut})] \cdot 100\%, \quad (A.15)$$

$$C_A = (100 - C_B)\%, \quad (A.16)$$

Table A.2 The interaction of triple eutectics

System type	T_{melt} of eutectics components (K)			T_{eut}(calc)/T_{eut}(calc)^2 (K)	K_{eut}	X
As–Pb–Sn	e1	e2	e2	E4		
	584.5	564	581	563/538/514	0.4768	0.0281
Bl–Au–Te	e(Skl)	el	e(Sk2)			
	749	801	690	675/644.6/656.7	0.4463	0.3915
Tl–S–Ga	P2	e2	e3	El		
	1195	1093	1068	1005/881.5/921.6	0.4277	0.6248
Te–Tl–Sb	e2	P2	P3	El		
	498	495	545	485/446/474	0.4504	–0.4119
	P4	e6	e5			
	715	783	685	680/660/636.4	0.4618	0.2029

Notes A minus symbol in calculated values of the scale temperature parameter means that a smaller temperature value was deducted from a larger temperature value. Further calculations or diagram constructions require the use of modulus of this value. T_{eut}(calc)/T_{eut}(calc)$_2$—calculated temperature values of eutectic reaction at different combinations of components

at $T_A > T_B$ and $C_A < C_B$; or according to the formulae proposed in [16] at known melting points of components (T_1 and T_2) and calculated values of eutectic temperature (mass%):

$$C_{eut}^{T1} = [(T_2 - T_{eut})/(T_1 + T_2 - 2T_{eut})]\ 100\%, \tag{A.17}$$

$$C_{eut}^{T2} = [(T_1 - T_{\ni BT})/(T_1 + T_2 - 2T_{\ni BT})]\ 100\%. \tag{A.18}$$

The approximate evaluation of eutectic concentrations in multicomponent systems (for instance, in a five-component non-variant system of elements E_1, E_2, E_3, E_4, E_5 with correlating melting points T_1, T_2, T_3, T_4, T_5) is carried out in a stage mode. The first stage involves the determination of eutectic concentrations in combinations of elements divided into pairs in a decreasing order of melting points at calculated eutectic temperatures. The next step is to determine eutectic temperatures and concentrations during interaction of obtained eutectics between each other with a new eutectic temperature coefficient. These eutectics are considered as "components" of the new eutectic mixture (Fig. 2). For instance, the final element concentration E_3 in five-component eutectics (E_1-E_2-E_3-E_4-E_5) is calculated by the formula (Fig. A.2)

$$C_{E3} = C^{IV} \cdot C^{III} \cdot\ C_{E3}^{II} \cdot 100\%. \tag{A.19}$$

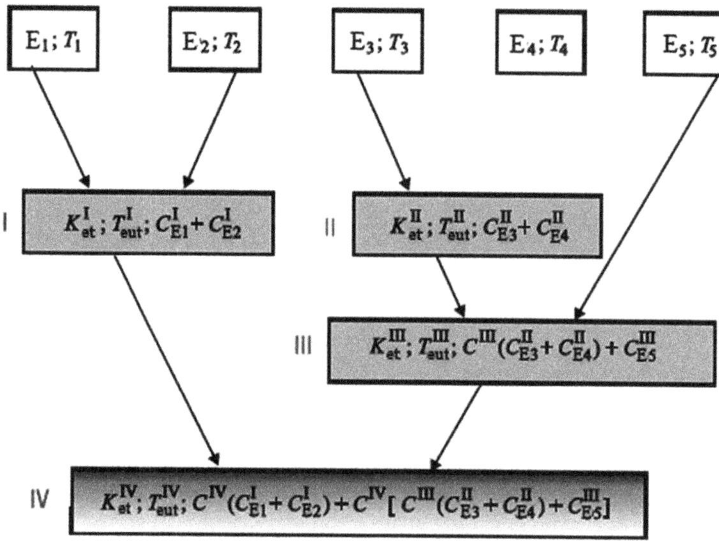

Fig. A.2 The scheme of consequent calculation of eutectic temperature coefficients: eutectic temperature (T_{eut}) and concentrations (C^{II}) in the five-component system

A.5 The Sample Calculations of Eutectic Temperatures and Concentrations in Multicomponent Systems

Different technological processes often require the determination of eutectic temperatures and concentrations in multicomponent systems which still lack equilibrium diagrams. These cases may be connected to the following technological processes: defining heating temperatures for thermal and chemical-thermal processing, defining fusing temperatures for powder materials in a solid state or in the presence of a liquid phase, defining temperature intervals for using fluxes in metallurgy and welding, setting limiting temperatures for using different materials, determining necessary energy consumption for welding, fusing and surfacing, using salt smelts in technological processes etc.

In particular, the developed method was used to determine the optimal boriding temperature in order to obtain the pseudo-eutectic structure of layers which is formed by the liquid-crystal mechanism, on Ni-based alloys and high-speed steels. The use of obtained values of eutectic temperatures allowed to construct schemes of four- and five-component equilibrium diagrams in concentration-temperature coordinates [16].

A.5.1 System Fe-Ni-Cr-B

Double and triple equilibrium diagrams of Fe, Ni, Cr, B in different combinations don't provide scientists with sufficient recommendations on boriding conditions and plausible explanations for general rules in the formation of diffusion layers. The patterns of constructing these diagrams with congruently and incongruently melting compounds and values of several experimental points allow to make an assumption on the existence of three eutectic isothermal surfaces in the studied system.

Areas of low boron concentrations (3.5–5 mass%) in all double diagrams generate eutectics based on low borides and solid solutions of boron in each metal. Consequently, a 4-component diagram would also generate eutectics based on the mentioned eutectics (Table A.3). The eutectic temperature, calculated according to the temperature rule of eutectic reactions, reached 1022–1035 °C for all combinations of components (double eutectics). The eutectic area in the 4-component diagram, according to the calculation, lies in the following intervals (mass%): 21–25% Fe; 50–60% Ni; 4–11% Cr; 4.1–4.8% B.

The increase in boron concentration to 20–50% in mass shifts the solidus surface into the area of higher temperatures. This area contains the second temperature minimum as the result of interaction between eutectics FeB-FeB_x ($x > 1$), CrB-Cr_3B_4 and a vast area of incongruent melting of NiB boride. The calculated eutectic temperature was 1240–1290 °C. The eutectic concentration in this level supposedly looks like this (mass%): 30% Fe, 30% Ni, 19% Cr, 21% B.

The transition into the area of high boron concentrations is accompanied by the formation of the third eutectic mixture as the result of interaction between eutectics CrB_4-B(Cr); hypothetically FeB_x-B(Fe) and NiB_4-B(Ni). Thus, the formed eutectics includes NiB_4, FeB_x, CrB_4 borides and solid solution of system metals in boron in the following calculated proportion (mass%): 28–30% Cr, 6–8% Ni, 3–5% Fe and 58–61% B. The melting point for the eutectics lies in the range 1690–1755 °C.

Table A.3 The calculation of eutectic temperature in Fe-Ni-Cr-B in the area of low boron concentrations

No.	The system type	T_{melt} for eutectics or components (°C)	$\sum T_{melt}$, K	$\lvert R \rvert$ T_{melt}, K	X	K_{eut}	T_{eut} (°C) (calculated)
1	Cr-Ni	1300	3446	300	0.72	0.42	1174
	α-solid solution Cr(B)-Cr_2(B)	1600					
2	α-solid solution Ni(B)-Ni_2B	1100	2846	100	0.276	0.46	1036
	α-solid solution Fe(B)-Fe_2B	1200					
3	Eutectics 1	1174	2756	138	0.39	0.47	1022
	Eutectics 2	1036					

Notes K_{eut}—the coefficient of eutectic temperature; X—the scale temperature parameter. The calculation for Fe-Ni-Cr-B system was carried out at K_{eut}^{mid}

The existence of FeB_x boride (where X > 1; i.e. FeB_2, FeB_4 etc.) was mentioned in [18–20]. However, only the Mössbauer spectroscopy allowed authors of [20] to evidently prove the existence of phase FeB_{1+x} (where x > 0.4) in the boronized layer in the studied processing conditions. Other coefficient values haven't been proved yet.

Thus, a pseudo-eutectic structure can be generated in a liquid state of the nickel alloy surface at T_{sat} = 1050 °C and 4.5–5% surface boron concentration.

A.5.2 System Fe-Cr-W-Mo-B

The eutectic temperature in this system was calculated in order to build a five-component equilibrium diagram in concentration-temperature coordinates and determine the boriding conditions to prevent the deformation of processed surfaces and create the environment for generation of pseudo-eutectic structures [16]. The diagram area of boron concentration below 10% is the most important from the point of surface saturation. Therefore this work offers a calculation only for this area which is necessary for predicting a phase composition and determining formation rules of pseudo-eutectic structures in boronized layers of high-speed steels (Table A.4).

Table A.4 The calculation of eutectic temperatures in Fe-Cr-W-Mo-B system

No.	The system type	T_{melt} for eutectics or components (°C)	$\sum T_{melt}$, K	|R| T_{melt}, K	X	K_{eut}	T_{eut} (°C) (calculated)
Variant 1							
1	α-solid solution Fe(B)-Fe_2B	1200	3346	400	0.98	0.498	1165
	α-solid solution Cr(B)-Cr_2B	1600					
2	α-solid solution Mo(B)-Mo_2B	2180	5426	520	0.89	0.43	2060
	α-solid solution W(B)-W_2B	2700					
3	Eutectics 1	1174	2756	138	0.39	0.47	1022
	Eutectics 2	1036					
Variant 2							
4	α-solid solution Cr(B)–Cr_2B	1600 2180	4326	580	1.75	0.42	1543
	α-solid solution Mo(B)–Mo_2(B)						
5	α-solid solution Fe(B)–Fe_2B	1200 2700	4446	1500	2.97	0.315	1127
6	Eutectics 4	1543 1127	3216	416	1.05	0.426	1097
	Eutectics 5						
Variant 3							
7	Eutectics 4	1543 1200	3289	343	0.85	0.435	1157
	α-solid solution Fe(B)–Fe_2B (w/out taking into account W)						
Variant 4							
8	Eutectics 7	1157 2700	4403	1543	3.01	0.31	1091
	α-solid solution W(B)-W_2B						

Carbon in high-speed steels is found in a bound state as carbides which are preserved in the structure even during long heating for quenching. This fact led to a hypothesis that its influence on boride formation is insignificant. Therefore it wasn't included into the diagram and calculation of critical points.

The next reason for the calculation is the fact that tugsten being the most high melting point metal of the system and a strong carbide generator is found in annealed steels in the form of carbide. The solid solution of steel P6M5 has at least 0.3% in mass [21]. Therefore the calculation of the eutectic temperature defining saturation conditions was done in two ways: taking into account the influence of W in boride formation and without it. In the first cast T_{eut} was 1091–1097 °C; in the second—1157 °C.

The experiments showed that a sharp acceleration of the formation of boronized layers is seen at 1160 °C with the appearance of a liquid phase in the structure of a layer. Thus, the second method of calculation (Table A.4, variant 3, no. 7) turned out to be more correct. The temperature of eutectic structure formation lies in the interval 1157–1200 °C at the surface boron concentration 4–7% in mass.

The saturation process in the powder mixture at 1180 °C and correlating concentration conditions ensured the formation of pseudo-eutectic heterogeneous structures which contain an alloyed Fe_2B-based boride and a solid solution.

The presence of non-melting carbides in the layer's structure and foundation increases the amount of solid phase and reduces the danger of deformation of the surface. However it works only during short-term soakings. Long-term soakings over 3 h increase the amount of liquid phase and makes the dilution of carbides more visible. The amount of liquid phase is also influenced by the dispersibility of the initial structure of high-speed steels and temperature, concentration and temporal saturation conditions.

References

1. Vozdvizhensky V.M. Prognoz dvoinykh diagram sostoyaniya. Moscow, 1975. 224 p. [The prediction of double equilibrium diagrams].
2. Kordes.E. // L.Anorg. und allzgem. Chem. 1927. H 167. P. 97.
3. Gulyaev B.B. // DAS USSR. 1965. Vol. 164, №1. Pp. 103–105.
4. Meditskiy L.B // ZhFKh. 1967. Vol. 41, №11. Pp. 2769–2771.
5. Vailycv M.V. // Univ. bulletin "Nonferrous metallurgy". 1969. №6. Pp. 90-95.
6. Vozdvizhensky V.M. // ZhFKh. 1963. Vol. 37, №11. Pp. 2455–2461.
7. Vozdvizhensky V.M. // Univ. bulletin "Nonferrous metallurgy". 1972. №6. Pp. 105–108.
8. Vozdvizhensky V.M. // ZhFKh. 1964. Vol. 38, №12. Pp. 2848–2855.
9. Zakharov A.M. Diagrammy sostoyaniya dvoinykh i troinykh sistem. Moscow, 1990. 240 p. [The equilibrium diagrams of double and triple systems].
10. Petrov D.A. Dvoinye i troinye sistemy. Moscow, 1986. 256 p. [Double and triple systems].
11. Taran Yu. T., Mazur V.I. Struktura evtekticheskikh splavov. Moscow, 1978. 455 p. [The structure of eutectic alloys].
12. Novikov I.I. Metallografiya. Moscow, 1984. 125 p. [Metallography].

13. Eroshenkova I.G., Olenicheva V.G., Petrova V.A. Diagramy sostoyaniya metallicheskikh sistem, opublikovannye v 1974 godu. Vipusk XX. Moscow, 1976. 360 p. [The equilibrium diagrams of metallic systems published in 1974. Issue XX].
14. Zakharov A.M., Olenicheva V.G., Petrova L.A. Diagramy sostoyaniya metallicheskikh sistem, opublikovannye v 1985 godu. Vipusk XXX, chasti I i II. Moscow, 1986. 742 p. [The equilibrium diagrams of metallic systems published in 1985. Issue XXX, parts I and II].
15. Lakhtin Yu. M. Metallovedeniye i termicheskaya obrabotka metallov. Mosocw, 1977. 407 p. [Physical metallurgy and thermal processing of metals].
16. Krukovich M.G. Razrabotka teoreticheskikh i prikladnykh aspektov upravleniya strukturoi i svoistvami borirovannykh sloev i ikh ispolzovanie pri proizvodstve transportnoi tekhniki. Dissertatsiya doktora tekhnicheskikh nauk. Moscow, 1995. 416 p. [The development of theoretical and applied aspects of controlling the structure and properties of boronized layers and their use in producing transport machinery. Doctorate dissertation].
17. Krukovich M.G. Raschet evtekticheskoi temperatury i kontsentratsii v mnogokomponentnykh sistemakh. Moscow, 1993. 10 p. [The calculation of eutectic temperature and concentration in multi-component systems].
18. Voroshnin L.G., Lyakhovich L.S. Borirovaniye stali. Moscow, 1978. 240 p. [Steel boriding].
19. Pomelnikova A.S. Razrabotka teorii i tekhnologii nizkoenergeticheskikh i drugikh poverkhnostnykh uprochnyayuschikh obrabotok stalei i splavov. Dissertatsiya doktora tekhnicheskikh nauk. Novokuznetsk, 1998. 387 p. [The development of theory and technology of low-energy and other surface hardening treatments of steels and alloys. Doctorate dissertation.]
20. Novakova A.A., Sizov I.G., Gvozdover R.S. // Poverkhnost'. Rentgenovskiye, sinkhronotronnye i neitronnye issledovaniya. 2003. №3. Pp. 99-103.
21. Gulyaev A.P. Metallovedeniye. Moscow, 1986. 544 p. [Physical metallurgy].

Appendix B
Terms Used for Describing Boriding Processes in Accordance with the Classification Scheme

B.1 General Terms

Non-deformation boriding—boriding at temperatures below phase transformations in the processed steel.

Hot boiling—creating a fluidized medium by feeding a hot gaseous medium into the retort through a powder mixture.

Saturating reusable model—detachable reusable technological equipment covered with a layer of saturatimg mixture. It repeats the relief of a part or a whole processed surface.

Saturating cooling medium—a saturating medium for low-temperature boriding is used as a quenching cooling medium, e.g. for quenching of high-speed steels.

Lined equipment—containers, retorts and muffels whose inner walls are covered with a saturating boronizing mixture (lining)

Cool boiling—creating a fluidized medium by feeding a cold gaseous medium into the retort through a powder mixture; heating of details is reached with any known method (radiation, high frequency current, electrical contact, electrical spark)

B.1.1 The Conditions Determining the State and Types of Saturating Mediums

The types of saturating mediums

Non-electrolysis self-organizing medium—a liquid medium where saturating atoms are formed according to electrochemical laws in a self-organizing mode without electrical currents.

Water solutions of salts and suspensions—liquid mediums containing boron salts in a melted state and/or boron compounds or colloid boron or precipitated boron.

M.G. Krukovich et al., *Plasticity of Boronized Layers*,
Springer Series in Materials Science 237,
DOI 10.1007/978-3-319-40012-9

Granulated mixture—a solid multicomponent mixture in a form of grains of certain sizes (normally 0.5–5.0 mm).

Compact mixture—a composite mixture acquiured from coatings, slips and suspensions after they are applied on a processed surface and dried.

Powder mixture—a solid multicomponent mixture in a form of powder.

Fluidized medium (fluidized layer, boiling layer)—a state of grainy friable material (or its layer) which intensifies the movement of solid material particles (in relation to one another) in case if a gas flow or liquid (liquefying agents) goes through it. In this state, the medium (layer) is similar to a boiling liquid and acquires some of its properties; its behavior goes according to the hydrostatics laws. A tense contact between a granulated material and a liquefying agent is reached in fluidized mediums which makes it very effective for chemical-thermal processing (incl. boriding).

Metal smelts—metal smelts which contain boron in a melted state and ensure the formation of a layer on the processed surface due to the concentration gradient alone.

Self-organizing closed medium—a gaseous medium which is concentrated in one closed space where the formation of saturating atoms and ions and saturation itself are carried out in a self-organizing mode.

Suspension—a disperse system which includes particles of a solid body (disperse phase) distributed in a liquid disperse medium. It is obtained by dispersing solid bodies in liquid mediums, mixing dry powders with liquids, enlarging colloid particles as a result of coagulation or condensation growth.

Electrical field—a particular form of electromagnetic field which determines the force action on the electrical discharge; the force doesn't depend on its movement. It accelerates boriding.

Electrolysis medium—a liquid medium where saturating atoms are formed when an electrical current acts according to electrical-chemical rules.

Electromagnetic field—a special form of matter which ensures the interaction between electrically charged particles. External electromagnetic field is applied on the reaction volume with details in order to activate the formation of saturating boron atoms and increase their diffusion mobility in metal. It accelerates boriding.

Electrostatic field—an electric field of immobile electric charges which provides the interaction between them. A electrostatic field in a gaseous medium on the borderline between the processed surface and saturating medium (gaseous, liquid, solid and their combinations) forms glow, corona and electrical spark discharges; their plasma medium on the processed surface generates saturating atoms. It accelerates boriding.

Boriding methods

Non-contact—a method of saturating a metallic surface with boron in the absence of a direct contact between a processed surface and saturatigg medium, e.g. while boriding in a container: a saturating mixture is divided from details and put into a special gas-producing container (gas-generator).

Non-electrolysis—a method of saturating a metallic surface with boron:

1. in metal smelts where boron is in a melted or precipitated state;
2. in smelts of boron-contatining salts with additives of powder electrochemical reducing agents which are needed to form extra boron atoms in a self-organizing mode during the work of short-circuit galvanic elements;
3. in smelts of boron-containing salts with additives of powder boron-containing electrochemical reducing agents which are needed to provide a maximum formation speed of boron subions in a self-organizing mode during the work of short-circuit galvanic elements;
4. in smelts of neutral salts with additives of powder boron-containing electrochemical reducing agents where the formation of boron atoms is carried out spontaneously due to the electrochemical mass carry during boron subions formation.

Boriding in a vibration-fluidized medium—a method for saturating a metallic surface with boron in a powder mixture in a fluidized which was acquired as a result of vibration of a detail or a container (furnace, retort, container etc.) with details and mixture.

Boriding in a fluidized medium—a method for saturating a metallic surface with boron in a powder mixture in a fluidized state which was acquired by feeding a gaseous medium through it.

Boriding in a closed medium as a self-organizing process—a method for saturating a gaseous medium in a closed container, muffle or retort; their inner walls are lined with a saturation mixture which serves as a source of boron in saturation gaseous medium.

Ionic implantation (ionic alloying)—a surface of a metal, half-conductor or di-electric is bombarded with ions of metals and non-metals which leads to changes in composition, structure and properties of the surface; ions penetrate the target stronger due to increases of ions' energy; ions with energy $\sim$10–100 keW penetrate the surface by 0.1–1 mkm. Ionic implantation with a flow of high-speed ions-implants (with energy of several mega electron volts) allows to implant any element (incl. boron) intro the surface of any solid body.

Contact boriding—a method for saturating a metallic surface with boron by a direct contact of the processed surface and saturating medium (solid or liquid) in the presence of boron concentration gradient.

Direct-flow consequent—a saturation method of permanent blow of the reaction volume in a certain consequence by different gases or with some changes in blow modes.

Direct flow—a saturation method of permanent blow of the reaction volume by a gas-carrier and a boron-containing gas.

Divided in a solid medium—a method for saturating a surface with boron in mixtures while dividing the process of metal-thermal reduction of boron from its oxides and the process of further exploitation of these powder mixtures for boronizing details.

Combined in a solid medium—a method for saturating a surface with boron in metal thermal mixtures while combining the process of boron reduction from its oxides and the process of boriding.

Circulatory—a saturation method used in the process of continuous circulation of a gaseous saturating medium in a closed volume with a consequent transportation through a zone with boron ions source.

Electrolysis—a method for saturating metals with boron in smelts of boron-containing salts under electrolysis or in smelts of boron-containing salts with addition of powder electrical-chemical reducing agents necessary for additional formation of boron subions which are easily reduced on the processed surface under electrolysis.

Electrolysis-plasma—a boronizing method in water solutions of boron-containing salts (and/or containing boron in colloid and/or atomic state and/or in the form of compounds) in conditions of generation of electric spark plasma in a steam coat on the border metal-solution under electric current.

B.1.2 The Technological Specifics and Temperature-Temporal Parameters of Boriding

Boron sources

Alkide boron compounds—organic compounds of boron $\{(CH_3)_3B, (C_2H_5)_3B\}$ etc.

Boron halogenides—boron compounds with halogens (BCl_3, BF_3, BJ_3, BBr_3).

Boron hydrides (boron-hydrogen, boranes)—boron compounds with hydrogen (B_2H_2, B_2H_6).

Heating methods

Corona discharge—a type of glowing discharge: electrical corona. It appears when an electrical field around one or several electrodes with a strong curvature of the surface (sharp, thin wires) is characterized by a strong heterogeneity. The electrons are surrounded by a typical glowing (corona). A current in corona discharge (small value) is proportionally mobile in relation to gas ions formed in the discharge. It accelerates boriding.

Laser treatment—heating of a processed surface covered with a boronizing solid medium (coating) or preliminary boronized; boriding might be carried out with any method or laser emission in order to obtain a boronized layer or charge its structure.

Glowing discharge—one of the types of stable autonomous electrical discharge in gases. It is witnessed at low temperature of a cathode. The gas is ionized with electrons in intra-electrode gaps. It activates the formation of saturating atoms on the processed surface and accelerates the process of boriding.

Electrical spark non-polarized heating—heating of details with spark discharges on the border between a powder mixture and a processed surface under the influence of an alternating current.

Electrical spark polarized heating—heating of details with spark discharges on the border between a powder mixture and a processed surface under the influence of a stable current.

Electrical spark discharge—a spark, one of the forms of electrical discharge in gases. It emerges if the capacity of the energy source is not sufficient for maintaining a stable arc discharge or glowing discharge. An electrical spark discharge is a beam of bright spark channels. These channels are filled with plasma which contain not only ions of initial gas but ions of electrode substance which evaporates intensively under the impact of the discharge. It is effective in chemical-thermal treatment processes, incl. boriding.

Electron-beam processing—heating of a processed surface covered with a boronizing solid medium (coating) or preliminary saturated with boron or electron emission in order to obtain a boronized layer or to change its structure.

Boriding temperature

High-temperature boriding (HTB)—boriding which is carried out at temperatures <900 °C;

Low-temperature boriding (LTB)—boriding which is carried out in the temperature interval 550—Ac_1, °C.

Medium-temperature boriding (MTB)—boriding which is carried out in the temperature interval Ac_1—900 °C.

The types of thermal treatment (combined with the operation of thermal processing)

High-temperature tempering—steel tempering at temperatures >450 °C;

Quenching (of steel)—thermal processing which involves austenitization and further cooling in the conditions ensuring a full or partial transformation of austenite into martensite or, possibly, bainite.

Normalization—thermal processing of iron smelts which involves austenitization and cooling in normal air.

Annealing—thermal processing which involves heating of a metal (detail) till a certain temperature, soaking and further slow cooling; the aim is to obtain a structure close to the equilibrium.

The types of protection (sealing) of reaction volumes

Unprotected—saturation is carried out in an open container where details are partially protected from oxidation by oxidizing the surface of a mixture and generating a protective peel as result of its melting and fusing.

Pastes—multicomponent mixtures or individual products which are characterized by the properties of viscous-plastic or elastic-viscous-plastic bodies. They are usually obtained as a result of dispersion of solid bodies in surrounding liquids.

Often surface-active substances are added. Boriding involves using pastes as protective coatings for sealing reaction volumes.

Powder filling—additional powder medium put above a saturating mixture which is aimed at preventing air from getting into reaction volumes with details due to its similarity to oxygen and other parts of air.

Fusible lock—melted medium in a container above a saturating mixture aimed at preventing air from getting into reaction volumes with details.

B.1.3 The Structure, Phase Composition and Hardness of Boronized Layers

The structure and phase composition of layers

α-solid solution—a specific layer structure which includes a solid solution of carbon and alloy elements in α-iron. It has a distinct border with a substrate.

Pseudo-eutectic structure—disperse structure of iron borides and alloyed α-solid solution without a distinct border with a substrate due to the high resemblance of alloy elements and boron. This structure is formed as a result of the diffusion-crystallization mechanism of generating boronized layers.

Eutectic structure—specific eutectic structure which consists of iron borides and α-solid solution; it is characterized by a low level of dispersability and a wavy borderline.

The types of borderlines, compactness of layers

Compact structure with needle (tooth-like) borderline—a specific structure of boronized layers in iron and steels based on phases $FeB+Fe_2B$ or Fe_2B alone.

Non-compact structure of a layer with needle (tooth-like) borderline—a structure of a layer which looks like separate needles or tooth-like zones; spaces between them are filled with α-solid solution.

Smoothed needle (tooth-like) borderline—smoothed borderline between borides and a metallic substrate which can be found due to the decelerating effect of alloy elements and carbon on the growth of borides.

Hardness

Hardness—a material's property to resist the plastic deformation without destruction under the invasion of a more solid body into it.

B.1.4 The Use of Boriding

The exploitation conditions of boronized details

Abrasive medium—a medium in liquid and solid (compact or powder) states; its elements are substances of high hardness aimed at mechanical processing (grinding, honing, polishing, regrinding, sharpening etc.) of metals and their smelts, ceramic materials, rocks, minerals, glass, wood, leather, rubber etc. Abrasive materials being parts of these mediums can be arranged in the following order in dependence with their abrasive potential: diamond, cubic boron nitride, silicon carbide, monocorundum, electrocorundum, emery, flint.

Corrosion medium—external medium where the destruction of metals (corrosion) happens due to the chemical or electro-chemical interaction.

Thermal-circular tensions—thermal tensions which appear in solid bodies under multiple (mostly intentional) changes of their temperature. These tensions emerge in connection with the change in thermal states of the bodies under multiple heating and cooling and long exposures to increased or decreased temperatures.

Boronized materials

Ferrous metals—metals and their smelts which are characterized by a black-gray colour, high density, high melting points, relatively high hardness and mostly polymorphic structure. Two groups of metals are usually exposed to boriding: iron (typical representative of the group), cobalt and nickel and high melting point metals and their smelts (metals which melting points are higher than iron's): tungsten, molybdenum, titanium, niobium, tantalum, zirconium, hafnium, vanadium, chrome.

Ceramic materials—materials obtained by fusing clays and their mixtures, oxides, carbides and other non-organic compounds with mineral additives. Boriding is applied to several types of technical ceramics (beryllium, zirconium etc.).

Index

M.G. Krukovich et al., *Plasticity of Boronized Layers*,
Springer Series in Materials Science 237,
DOI 10.1007/978-3-319-40012-9